T0327620

Risk Assessment

Risk Assessment

Tools, Techniques, and Their Applications

Second Edition

Lee T. Ostrom
Idaho Falls, Idaho, USA

Cheryl A. Wilhelmsen
St George, Utah, USA
Idaho Falls, Idaho, USA

Registered Office
John Wiley & Sons, Inc., 111 River Street, Hoboken, NJ 07030, USA

Editorial Office
111 River Street, Hoboken, NJ 07030, USA

For details of our global editorial offices, customer services, and more information about Wiley products visit us at www.wiley.com.

Wiley also publishes its books in a variety of electronic formats and by print-on-demand. Some content that appears in standard print versions of this book may not be available in other formats.

Library of Congress Cataloging-in-Publication data applied for

ISBN: 9781119483465

Cover Design: Wiley
Cover Image: Courtesy of Ryan Haworth

Set in 11/13pt and TimesLTStd by SPi Global, Chennai, India

Contents

Acknowledgments

We wish to acknowledge Jeff Heath for providing Chapter 31, Laura Ostrom and John Ostrom for providing Chapter 20, Ryan Haworth for providing the idea and graphics for the cover art, and Julie Moon for editing several chapters.

About the Companion Website

This book is accompanied by a companion website:

www.wiley.com/go/Ostrom/RiskAssessment_2e

The companion website contains solutions manual for instructors. The solutions manual contains a sampling of solutions for problems at the end of the chapters in the book. In some cases the solutions manual contains suggestions as to how an instructor might use a case study as an effective teaching tool.

CHAPTER 1

Introduction to Risk Assessment

On any given day, in every corner of the world, people are actively working, going to school, driving or taking mass transit to work, relaxing at home or on vacation, or even working at home. Some people are even finding the time to sleep. Those who are working perform jobs that range from cleaning animal kennels to serving as the head of state of a country. Every job, in fact every activity a human performs, has a hazard associated with it. The common hazards we all are exposed to include:

- Slips, trips, and falls.
- Illness and disease.
- Food-borne illness.
- Transportation: car accidents, pedestrian accidents, and bicycle accidents.
- Sports: organized sports (football, basketball, soccer) accidents and individual sports accidents (skiing, water sports, skate boarding).
- Electrical-related accidents.
- Fires.
- Weather-related accidents.
- Identity theft.
- Internet intrusion.

On top of these more common hazards are specific/major hazards. For example, cleaning animal cages include:

- Being attacked by the animal.
- The bacteria, viruses, and parasites that might be in the animal waste.

Risk Assessment: Tools, Techniques, and Their Applications, Second Edition. Lee T. Ostrom and Cheryl A. Wilhelmsen.
© 2019 John Wiley & Sons, Inc. Published 2019 by John Wiley & Sons, Inc.
Companion website: www.wiley.com/go/Ostrom/RiskAssessment_2e

- The design of the cage might pose problems: size, shape, material of construction, and sharp edges.
- The maintenance of the cage might pose problems: cleanliness, jagged metal or wood, and faulty locks/latches/gates/door.
- The condition of the floor.
- The electrical and/or HVAC system in the building.
- The building environmental conditions.

The major hazards associated with being a head of state include:

- Stress from decision making.
- Stress from the potential for war.
- Stress from political rivals.
- Potential for assassination.
- Potential for transportation accidents: airplane crashes (i.e. the President of Poland died in an airplane crash in Russia in 2010 (1).).

Hazardous occupations, for instance, firefighting, have numerous hazards associated with day-to-day activities. Risk assessment tools and techniques can be used to analyze individual jobs for risks. It is obvious that every activity the president does is analyzed for hazards. Jobs or tasks like firefighter, chemical plant worker, electrician, and even office workers are usually analyzed using tools such as job hazard analysis (2).

The focus of this book is analyzing complex systems, tasks, and combinations of tasks for hazards and the associated risks. Most of the major accidents that occur each year result from a series of events that come together in an accident chain or sequence and result in numerous deaths, environmental consequences, and property destruction. These accidents can occur anytime in the system's life cycle. One of the events from history that demonstrates this is the sinking of the Swedish ship Wasa (pronounced Vasa) on 10 August 1628 (3). The ship was fabricated between 1626 and 1628. In those days engineering of the ships was performed by the shipwright and he used his experience to determine factors such as the center of mass and the amount of ballast the ship should have. Because of various events, pressure was put on the shipbuilders to complete the ship ahead of the planned delivery time. The ship was completed and ready for sail on 10 August 1628. The ship was very ornately decorated and was heavily laden with armament. As the ship left port on its maiden voyage on that calm morning, a gust of wind hit the ship, filling her sails. The ship heeled to port and the sailors cut the sheets. The ship righted itself, but then another gust of wind hit the ship and it tipped to port far enough that water entered the gun ports. This was the event that led to the loss of the ship and approximately 30–50 lives. However, the loss of the ship was probably due to one of two design flaws: first the ship was probably too narrow for its height and, second, the ship did not carry enough ballast for the weight of its guns on the upper decks. A contributing factor was the height above sea level of the gun ports that allowed water to enter the ports when the ship listed to port. Since, as stated above, engineering of ships was more seat of the pants than a systematic design process, the real reason(s) for the disaster could only be speculated. The ship was raised from her watery grave in 1959 and has since been moved to a beautiful museum facility in Stockholm. The ship itself can be studied, but other factors such as whether the guns were properly secured, how many provisions were on the ship, and so forth will remain a mystery. Accidents can occur in any phase of a system's life cycle. For the Wasa accident, it occurred in the ship's initial phases.

A much more recent accident occurred on 23 February 2018, in Dallas, Texas, in which Atmos Energy, the country's largest natural gas distributor, caused an explosion and fire via a natural gas leak, killing one 12-year-old girl and injuring others (4). Atmos Energy also operates in Colorado, Kansas, Kentucky, Tennessee, Virginia, Louisiana, and Mississippi. The incident involved piping that leaked due to pressure on the piping. The accident investigators found that heavy rain caused underground pressure that pushed rock formations upward, which in turn caused pressure on the system, which caused the leakage. Aging pipes have since been replaced with a more flexible, high-grade plastic. Weather and aging pipes were the primary drivers in the event.

Risk assessment tools and techniques, if applied systematically and appropriately, can point out these types of vulnerabilities in a system. The key term here is "systematic." A risk assessment must be systematic in nature to be most effective and should begin early in the life cycle of complex systems. Preliminary hazard analysis (PHA) is an example of a tool that can be applied at the earliest phase of system development. As the design of a system progresses, other tools can be applied, such as failure mode and effects analysis (FMEA) and fault tree analysis (FTA). Probabilistic risk assessment (PRA) and human reliability analysis (HRA) are techniques used to analyze very complex systems. These tools usually require a well-developed design, an operating philosophy, and at least working copies of procedures to provide enough material to perform analyses. However, even mature systems benefit from risk assessments. The analyses performed on the Space Shuttle program after the Columbia accident are a good example (5). These assessments pointed out vulnerabilities of the spacecraft that were previously unidentified or viewed as being not as important.

Using the Six Sigma/total quality management philosophy of continuous improvement, risk assessment techniques applied throughout the design life of a system can provide insights into safety that might arise at various points of the system

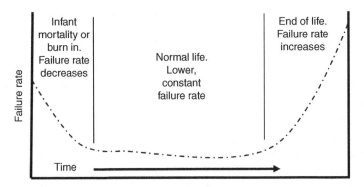

FIGURE 1.1 Bathtub curve.

life cycle (6). Reliability engineers use the bathtub curve to illustrate the classic life cycle of a system (Figure 1.1) (7). In the first part of a system's life, there is a higher potential for early failure. The failure rate then decreases to steady state until some point in the future when systems wear out or old age failure occurs.

Manufacturers usually warranty a system (a car, for instance) for the period of time from birth till just before system wears out. This way they maximize their public image while minimizing their risks or obligations.

Risk analysts are also interested in such curves, but from a safety perspective. Accidents commonly occur early in a system's life cycle because of several reasons including:

- Mismatch of materials.
- Hardware/software incompatibilities.
- Lack of system understanding.
- Operator inexperience or lack of training.

The system then enters a long phase of steady-state operation that as long as no changes perturbate the system, it remains safe. In later system life, accidents occur for the same reason as why systems wear out – components wear out. However, in terms of accident risk when a component fails in old age, it might lead to a catastrophic failure of the system, for instance, the Aloha Flight 243 accident (8). In this case the aircraft structure had become fatigued with age and failed during takeoff. In addition, latent conditions can lay dormant for many years in a system (9). These conditions could be a piece of bad computer code or a piece of substandard pipe that when challenged leads to failure. Performing risk assessments on systems throughout their life cycle can help elucidate these vulnerabilities. Once these vulnerabilities are found, measures can be taken to eliminate them and/or measures can be taken to mitigate the consequences of failures. This is the most important step of any risk assessment, that is, eliminating the vulnerabilities and reducing the risk of a system.

1.1 TERMINOLOGY

Risk assessment terminology will be presented throughout the book. However, at this time several key terms will be defined.

Risk has been defined many ways and for the purposes of this book risk is defined as "the probability of an unwanted event that results in negative consequences." Kaplan and Garrick use a set of three questions to define risk (10):

1. What can go wrong?
2. How likely is it?
3. What are the consequences?

Chapter 2 of this book defines risk in depth.

Probability: A measure of how likely it is that some event will occur (11).

Hazard: A source of potential damage, harm or adverse health effects on something or someone under certain conditions at work (12).

Severity: The degree of something undesirable (11).

Consequence: The effect, result, or outcome of something occurring earlier (11).

Vulnerability: A weakness in a system or human that is susceptible to harm (11).

Threat: Source of danger (11). Threat and hazard are considered analogous.

1.2 PERFORMING RISK ASSESSMENTS

There is no absolute rule as to how a risk assessment should be performed and to what depth it should be performed. The NASA PRA Guide (13) provides some recommendations, and the Nuclear Regulatory Commission (NRC) provides numerous guideline documents on the topic (14). The Occupational Safety and Health Administration's (OSHA) "Process Safety Management of Highly Hazardous Chemicals" regulation (29CFR1910.119) (15) requires that hazard analyses be performed for certain types of chemical operations, and the Department of Energy (DOE) specifies risk assessments for certain types of facilities (16). However, it is still up to the organization to decide how in depth the analysis should be. This book discusses tools that are effective for performing risk assessments, but the decision as to when to use the tools is up to the risk analyst. Table 1.1 provides a list of the risk assessment tools discussed in this book and at what point in an analysis they are traditionally used. In addition, this book provides other techniques that can be used to enhance a risk assessment, such as task analysis for determining human actions in a process, the Delphi process for eliciting human error probabilities and the critical incident technique for developing risk scenarios.

TABLE 1.1

Risk Assessment Tools

Tool	Traditional use	Book chapter
Preliminary hazard analysis (PHA)	This tool is used in the very beginning of a risk assessment and/or on a conceptual design of a new system, process, or operation. It is used to determine the potential hazards associated with or the potential threats poised to a system, process, or operation. This tool is also useful for organizations to evaluate processes that have been performed for years to determine the hazards associated with them	
Failure mode and effects analysis (FMEA)	This tool is used in system, process, or operations development to determine potential failure modes within the system and provides a means to classify the failures by their severity and likelihood. It is usually performed after a PHA and before more detailed analyses	
Failure mode, effects, and criticality analysis (FMECA)	FMECA extends FMEA by including a criticality analysis that is used to chart the probability of failure modes against the severity of their consequences. FMECA can be used instead of an FMEA, in conjunction with an FMEA, or after an FMEA has been performed	
Event trees	Event trees are very useful tools to begin to analyze the sequence of events in potential accident sequences. They also have utility in analyzing accidents themselves. Many variations of event trees have been developed. This book presents some of the more common ones	
Fault tree analysis (FTA)	FTA is a risk analysis tool that uses Boolean logic to combine events. The lower-level events are called basic events, and they are combined with Boolean logic gates into a tree structure, with the undesired event of interest at the top. This event is called the top event. Though this analysis tool is used to quantitatively determine the overall probability of an undesired event, it is also useful from a qualitative perspective to graphically show how these events combine to lead to the undesired event of interest. FTA has a wide range of use from determining how one's checking account was over drawn to determining why a space shuttle crashed	

TABLE 1.1
(*Continued*)

Tool	Traditional use	Book chapter
Human reliability analysis (HRA)	HRA is related to the field of human factors engineering and ergonomics and refers to the reliability of humans in complex operating environments such as aviation, transportation, the military, or medicine. HRA is used to determine the human operators' contribution to risk in a system	
Probabilistic risk assessment (PRA)	PRA is a systematic and comprehensive methodology to evaluate risks associated with complex engineered systems, processes, or operations such as spacecraft, airplanes, or nuclear power plant. PRA uses combinations of all the other risk assessment tools and techniques to build an integrated risk model of a system. A fully integrated PRA of a nuclear power plant, for instance, can take years to perform and can cost millions of dollars. It is reserved for the most complex of systems	

1.3 RISK ASSESSMENT TEAM

Risk assessment is a systematic, step-by-step approach for evaluating risk. It is the process for determining the probability of a risk occurring and the consequence of that risk. It is a fundamental component of an effective risk management program, which is a basic management tool consisting of risk Assessment and risk control. Risk assessment is the data gathering component, while risk control is the application of the risk assessment evaluation.

1.3.1 Team Approach

Individuals will respond to a risk or perception of a hazard based on their influences, environment, and biases. What one person may perceive as a relatively low risk, another may consider highly dangerous no matter what controls are in place. If one or two individuals are asked to perform an assessment, some relevant factors may be missed or ignored. When determining the best course of action for performing a risk assessment, it is important to remember that people will use their own perceptions. Even different experts will perceive different risks and, from those perceptions, conclude different results or controls. In general, bringing together a group of people who work in the environment to work together as a team is ideal.

1.3.2 Team Representatives

Before any assessment of the risks can be started, who should be involved or who makes up the risk assessment team needs to be determined. The team should consist of the right group of people with the right mix of experience. The team will be composed of 5 people, up to 10–15 if necessary, and will include people with different jobs and experiences.

Risk assessment is never a one-man show; it should be conducted by a multidisciplinary team, who has a thorough knowledge of the work to be undertaken. Team members should include management, process or facility engineers, technical personnel, supervisors, production operators, maintenance staff, and safety personnel, if available.

The team members will vary from assessment to assessment, company to company, and industry to industry, but the following elements are common:

1. *Management* should be involved to give practical application in the decision-making process of risk reduction controls and accepting residual risk level.
2. *Engineers* should be involved in the risk assessment process as they are involved in the development of the design decisions that will impact the overall risk and risk controls.
3. *Workers/operators/supervisors* should be involved and included in the team, as they are the most familiar with the tasks and uses for which the assessment will directly affect. These are the individuals best suited to identify the possible hazards associated with the end use. They can provide valuable insights on the possible controls and the practical application of those controls.
4. *Health and safety professionals*, if available, can offer valuable insights into what control measures might be available. They can identify possible hazards and propose risk reduction methods and their application.
5. *Maintenance* is another component that needs to be represented to understand the ramifications of implementing controls on the system or process being assessed.

The goal of the risk assessment team is to reduce risks to tolerable or acceptable levels. This assessment is completed by:

1. Identifying hazards and/or potential hazards.
2. Identifying users and/or tasks.
3. Determining the level of risk:
   a. Low: acceptable
   b. Medium: moderately acceptable
   c. High: not acceptable

4. Evaluating potential controls – elimination, substitution, engineering controls, administration controls, and/or use of personal protective equipment.

5. Developing a report.

6. Implementation and review.

The resulting risk assessment report must be evaluated, approved and endorsed by senior management.

Self-Check Questions

1. How do most of the major accidents occur?

2. What are the following acronyms?

   PHA

   FMEA

   PRA

   HRA

   When should they be applied?

3. What is the goal of a risk assessment team?

4. Who should be involved in the risk assessment team?

REFERENCES

1. CBS (n.d.). Poland's President Killed in Plane Crash. *CBS News*. https://www.cbsnews.com/news/polands-president-killed-in-plane-crash (accessed 4 June 2018).
2. OSHA (n.d.). Job hazard analysis guide. https://www.osha.gov/pls/publications/publication.html (accessed 4 June 2018).
3. Cederlaund, C.O. (2006). *Vasa I, The Archaeology of a Swedish Warship of 1628*. Stockholm: Swedish State Maritime Museum.
4. David, W. (n.d.). Gas service stops for thousands of Dallas homes due to leaks. *The Seattle Times*. https://www.seattletimes.com/nation-world/gas-service-stops-for-thousands-of-dallas-homes-due-to-leaks (accessed 18 June 2018).
5. CAIB (2003). Columbia Accident Investivgation Board, Final Report. NASA.
6. IIARF (n.d.). A global summary of the common body of knowledge 2006. https://na.theiia.org/iiarf/Public%20Documents/2006-CBOK-Summary.pdf (accessed 4 June 2018).
7. Ostrom, L.T. (2018). Bath tub curve diagram.
8. NTSB (n.d.). Aloha Airlines, Flight 243 Boeing 737-200, N73711, Near Maui, Hawaii. https://www.ntsb.gov/investigations/AccidentReports/Reports/AAR8903.pdf (accessed 4 June 2018).
9. Reason, J. (1990). *Human Error*. New York: Cambridge University Press.
10. Kaplan, S. and Garrick, B. (1981). On the quantitive definition of risk. *Risk Analysis* 1 (1): 11–37. https://www.nrc.gov/docs/ML1216/ML12167A133.pdf (accessed 4 March 2019).
11. Farlex (n.d.). The Free Dictionary by Farlex. https://www.the freedictionary.com/severity (19 February 2019).

12. CCOHS (n.d.). Hazards. http://www.ccohs.ca/http://www.ccohs.ca/topics/hazards (accessed 18 June 2018).

13. SMA (n.d.). NASA policy directives. https://sma.nasa.gov/policies/all-policies (accessed 18 June 2018).

14. US NRC (n.d.). NRC regulatory guides: power reactors (Division 1). http://www.nrc.gov/reading-rm/doc-collections/reg-guides/power-reactors/rg (accessed 18 June 2018).

15. OSHA (n.d.). 1910.119: Process safety management of highly hazardous chemicals. https://www.osha.gov/laws-regs/regulations/standardnumber/1910/1910.119 (accessed 19 February 2019).

16. DOE (n.d.). Risk Assessment in Support of DOE Nuclear Safety, Risk Information Notice, June 2010. https://www.energy.gov/ea/downloads/risk-assessment-support-doe-nuclear-safety-risk-information-notice-june-2010 (accessed 19 February 2019).

CHAPTER 2

Risk Perception

2.1 RISK

Simply defined, the word "risk" denotes some measure of uncertainty. In casual use, risk implies negative consequence, while opportunity implies positive consequence (1).

Perception is the process of interpreting sensory stimuli by filtering it through one's experiences and knowledge base. Note that perception is not the same as sensation, as the latter term is physiological and the former is learned (2).

Taken together, risk perception is an individual or group assessment of the potential for negative consequence. Within emergency management (EM) circles, understanding the public's general perception of risk (for instance, the isolated opinion contains peaks and valleys) is useful in establishing the necessary level of pre-incident emergency training, public relations and instructions/recommendations during the incident, and post-incident continuing communications. Risk perception plays in the choices made as to what information is to be provided and what format – both inside the affected organization and outside. From the company's management structure to its blue-collar workers, to its colocated workers, to the neighboring suburbs and beyond, risk perception is as close to "the facts" as each person gets until their vision is altered by some later greater truth.

Like the proverbial two-edged sword, risk perception both serves and hinders EM organizations and, subsequently, those protected and supported by the EM function. At one tapered end of the spectrum lies a band of Chicken Littles pointing at the blue sky and warning of dire consequences; at the opposite, there's a huddle of frumpy white-coated scientists swaddled in disdain. Most us lay somewhere between these two, our placement in demographics split into hundreds, even thousands, of layered and skewed bell curves based on age, income, experience, education, and innumerable

Risk Assessment: Tools, Techniques, and Their Applications, Second Edition. Lee T. Ostrom and Cheryl A. Wilhelmsen.
© 2019 John Wiley & Sons, Inc. Published 2019 by John Wiley & Sons, Inc.
Companion website: www.wiley.com/go/Ostrom/RiskAssessment_2e

other facets. It's the endless variability of the public that makes risk perception such a difficult management issue. But it can be managed.

While it is safe to say that there's a near-endless variation in perception, cataloguing allows one to build boxes around like perceptions until they're within a defined set. To build these boxes, you need to know what factors affect the audience's perception. Is the subject matter highly technical and beyond the bulk of the area's laymen? Does the vicinity you're considering have decades of experience with your industry? Has that experience resulted in bad blood?

To understand what knowledge is relevant to your evaluation of the audience's risk perception, you have to consider the business you're in. Examples of heightened sensitivity (and negative perception) are those where the technology involved borders on black magic to the layman. Fission and fusion and government weapons facilities come immediately to mind, followed by sprawling laboratories where the workers say little about what goes on inside their white walls. The less he knows, the more Joe Q. The public thinks, wonders and, ultimately, worries. As a matter of comparison, let's consider two hazards commonly found in high population areas – liquid petroleum gas and lawn care chemicals.

Liquid petroleum gas (LPG), aka propane, is a common fuel in the city and the country, in industrial areas and rural areas, at homes, and in businesses. Small containers, such as those used for barbeques, propane-powered vehicle fuel tanks, and camp trailers, can be found literally everywhere. Larger containers, such as 30 000-gal bulk tanks for distributors, can be found in most cities across the country. At the pier side, where massive ships unload, tanks may carry as much as 50 000 tons of propane.

Decades of use and familiarity have made the modern man comfortable with propane. The 250- and 500-gal tanks available at many gas stations are no cause for concern (to the point that few of those operating them even wear protective gloves or a face shield during propane handling). The racks of refillable barbeque propane bottles stored outside most grocery stores aren't either. Nor are the multiple tractor-trailers carrying thousands of gallons of propane across town and on every freeway.

A quick calculation shows that the 6000 gal of propane on any given tanker truck contains approximately 549 million British thermal units (BTUs) – equal to 138.4 tons of TNT or a small nuclear device (3). Still, it's just propane, right?

Suburban America believes in chemistry when it comes to their pretty yards, colorful flowers, and heavily laden vegetable gardens. For just the care and feeding of their lawns, they rely on sprayed fertilizers, pesticides, selective herbicides, fogs for trees, and powders for annoying ants. If you watch Joe Q. publicly working with the likes of these, often as not you'll find just a bit of care included. Service companies warn you to keep your animals away from sprayed areas until dry. Folks generally wear gloves and avoid spreading chemicals on windy days. Some go as far as wearing dust masks. The point here is that they think about it. Why? What's the difference between the hazards of LPG and lawn care chemicals? They both are hazardous, but in different ways and, generally speaking, both are accepted as acceptable risks by the public.

2.2 KNOWLEDGE LEVEL

In this day and age, we're sensitive to the use, misuse, and abuse of chemicals. Situation-stained landscapes like Love Canal and Bhopal are at the near edge of our memories and, if not, there's always a chemical release or Resource Conservation and Recovery Act (RCRA) violation or factory fire spewing noxious fumes somewhere on the news. With these constant reminders, we realize that while chemicals are necessary, they're also dangerous. And since there hasn't been a recent prominent example of a propane truck leveling a city, it's just assumed to not be as dangerous. After all, it's just propane, right?

To lay this issue solely at the feet of education is oversimplifying it. First off, this approach assumes that a well-informed public will perceive the risks as the experts do, that a low probability of an incident's occurrence is enough to quantify the risk as insignificant. This approach, while mathematically valid, doesn't consider the emotional component the general public includes when they weigh in. Teach them about a hazardous process, derive its accidents and initiators, describe potential mitigations, and a definite chance exists that you'll just end up with hyper-aware protestors. If for no other reason than they now can quantify the results, regardless of the fact that that particular set of results is considered highly unlikely in the view of engineers and scientists. That's the power of emotion.

Then there's the cognoscente – those same scientists and engineers mentioned earlier. This "trust me" crowd can undermine all efforts an EM plan puts together with lackadaisical commitment and faint praise. Instead of objectively quantifying the issues and assessing the necessary response, these types enable the denial and Pollyannaish attitude often adopted in the burgeoning moments of an incident. In comparison to all possible reactions, scientific apathy is probably the most dangerous. Those smitten with it are often the last to realize that the water's risen over their collective chins and they don't know how to swim.

If data cannot adequately focus on one's perception, then those factors that can are of some importance. They're at least a subset of the knowledge necessary to engage in successful EM (assuming successful EM requires a targeted risk perception). Subjectively speaking, the greatest influence on a layman's risk perception is probably experience. It's common for mankind to live by the ideals, ideas, and rules of thumb that are either established or rooted in some past learning obtained solely through the act of living or listening to those that lived before you. Unscientific? Certainly. But, over time, the impact of experience has more history than even science (witness the power of lore throughout time), so it's worth listening to.

Ideals (or principles, if you'd rather) are beliefs stemming from some personal philosophy. They represent good vs. bad and right vs. wrong – binaries and shades of gray that we choose to use to guide our lives. One's ideals are founded from so many avenues that their mapping would look like a spider web. There's parental influence, religious posturing, peer pressure, social paradigm, economic stratification, and a thousand other variables to contend with. Luckily for the analysts, it's also a social paradigm that most folks establish their ideals early in life and stick to them. Additionally, birds of the same feather flock together – or like-thinking folks revolve within defined cells that are easier to measure and speak to than the individuals making them up. For example, after confirming the target ideal, it is possible to predict the group or groups who weave this ideal into the foundation of their belief system. For example, groups generally assigned the "green" label can be expected to place harmony with the environment above corporate profit. Likewise, the technically minded will usually couch their positions in concrete facts when queried for a position on any contentious new theorem. Given a group's anticipated ideals, one has a leg up on pegging their expectations and, ultimately, their perceptions.

Ideas are a bit of a quandary because, by definition, they don't fit into the accepted mold. Still, using a bit of risk analysis brainstorming, a talented group of individuals can tag a good percentage of the possible directions new thought may take a given emergency situation and therefore consider most of the new ideas. Of all the unknowns, ideas are easily the easiest to account for. Even if the risk analysis doesn't consider every possibility (an impossible feat itself), it can create envelopes where all the possibilities fall into with a reasonable degree of confidence. This answer here is mostly left to science and easy to control.

Rules of thumb (also termed as "cognitive heuristics") are unconsciously applied by most folks every day. Sailors predict the weather with poetic prognostications formed on cloud structure and moon auras. Investors buy and sell based on seemingly irrelevant political and emotional trends that predicate the bullish or bearish market. The military constructs offensive and defensive plans based on the adversary's holiday calendar and home country's politics. The impact of cognitive heuristics is the hardest to gather up and quantify, as these rules vary from person to person based on their teacher's experience base. What was law to the teacher when the teacher was the student may now be revised or rescinded simply by fresher experiences. Rules of thumb are much like common sense in the sense that they're not common – they change and adapt with use and over time (4).

With all this said, the above list of risk perception influences is by no means conclusive; so many other factors come to bear that the subject is worthy of tomes. Those mentioned here are only the tips of the iceberg.

So, armed only with a few thoughts and the knowledge that risk perception varies to an infinite degree, how can one account for its impact when developing an EM process? To start, plan on providing the right information to the right ears. Select the facts each group needs to hear and deliver the appropriate details for each audience. This includes, among others, the general public, the elected officials, and those within the EM system. The public wants to hear what the hazards are and know how and why there's not a threat to little Johnny when he's running about in the backyard. Elected officials need about the same information, but they need an entirely different delivery. Pepper your discourse with solid and supportable statistics. Provide sound bites and concise thoughts. The EM folks are the ones you need to save your details for. And, while it is possible to give them too much detail, it's better to err in this direction than the opposite way.

Communications aren't what they used to be; there's much more to work with beyond radio spots, television commercials, and door-to-door fliers. With a modest investment, those interested in maximizing their safety posture's exposure can put websites up describing the processes in question, the protections provided, and accident/release potentials framed in real-world terms. Professional organizations allow for presentations at conventions and training classes. Local centers for higher education can be involved in preparing single day, multi-day, and semester-long coursework that supports the professional growth of those in the EM system and those employed within the industry concerned about the hazard in question (5).

Establishing EM tours and fire and rescue plans ahead of time goes a long way toward giving your emergency response organizations a deeper level of confidence in your, and their, ability to respond to a problem. If you have a large or complex facility, establishing a realistic emergency drill program that involves emergency services will take that confidence even further.

Facilities with a genuine ability to impact the habitability of surrounding areas, such as large chemical plants, petroleum processing facilities, and nuclear facilities, should establish a training and response program for the general public within their affected areas (sometimes termed as "emergency planning zones" or EPZs). EPZs describe to the public what facility warning alarms sound like and what actions should be taken when the alarm occurs and provide them with materials to keep their knowledge level fresh.

All of the above efforts play to both sides of the information feeding into the public's perception – the knowledge side and the experience side.

With the broad spectrum of perspectives available, it's important that one choose the desired audience, as it's doubtful that you'll find an approach that pleases everyone. Statistically speaking, an organization's best return for their dollar will come from targeting the center of the risk perception bell curve and finding a way to include as much of the standard deviation as responsible spending will allow. The downside to this approach is that someone's left out, and that someone will undoubtedly be courted to the edge by the Chicken Little or White-Coated Doubter. To add injury to insult, this third sigma demographic is likely to beat the loudest and be the most inclined to play whatever cards they have up their sleeve to make their point. What should be done about them?

Short of something illegal, there are not many effective options available. Optimally, the dissenters would be ignored, and they'd go away, because dealing with their proselytizing somehow imbues it with significance. When that's not possible, for whatever reason, one has to once again consider what the right answer is from a risk perspective standpoint.

Remember that perspective is learned. The audience that is considering the discordant rumblings of the dissenter believes they're learning something valuable. In order to combat that impression, you have two choices: you'll undermine the dissenter's teachings or you'll provide more powerful teachings of your own.

There is an inherent failure in taking the negative approach (i.e. casting doubts on your dissenter), because interested parties paying attention may see that as an attack on the person and not the idea. That leaves option 2 – providing more powerful teachings.

Assuming that one's already splayed the relevant facts for all the audience to see, the best "high road" option here may be speaking about the criticisms of your detractors in a positive and fact-filled manner. Be grateful that they are providing a new and exciting perspective to the discussion. Embrace their thoughts, and then discuss them thoroughly while sifting through your bag full of data. Examine both and graciously consider the difference between your perspective and theirs. In the end the audience, those whose perception you wish to influence, will make their minds up based on ideals, ideas, and rules of thumb anyway. And if you've given them the education they require, that's the best you can hope for.

Self-Check Questions

1. What makes risk perception such a different management issue?

2. What is the greatest influence in a layman's risk perception?

3. What are cognitive heuristics?

REFERENCES

1. Berger, B. (1994). *Beating Murphy's Law: The Amazing Science of Risk*. New York: Dell Publishing.
2. Fielding, R. (n.d.). Is it worth the risk? Risk perception. http://www.pitt.edu/~super1/lecture/lec4011/001.htm (accessed 9 July 2018).
3. CBS News (2011). Burning rail car a "small thermal nuclear bomb". https://www.cbsnews.com/news/burning-rail-car-a-small-thermal-nuclear-bomb (accessed 19 February 2019).
4. Taylor, I. (2008). Scientific management. https://www.bartleby.com/essay/Scientific-Management-FKC4RN7K8RVA (accessed 9 July 2018).
5. Johnson, B.B. (1992). Advancing understanding of knowledge's role in lay risk perception. https://scholars.unh.edu/cgi/viewcontent.cgi?referer=https://www.bing.com/&httpsredir=1&article=1135&context=risk (accessed 9 July 2018).

CHAPTER 3

Risks and Consequences

3.1 INTRODUCTION

In March 2011, a terrible 9.0 earthquake occurred off the coast of Japan, and this generated an equally devastating tsunami (1). Subsequent to these two natural events, several nuclear power plants had catastrophic failures (see Tables 3.1 and 3.2) (2). The natural events could not be predicted. In fact, earthquake prediction is much less understood than is forecasting the weather (3). However, the risk of a nuclear plant having a catastrophic failure, even a meltdown after an earthquake, has been analyzed (4). At some point the benefit of having nuclear electrical power was found to be greater than the risk of a nuclear reactor meltdown. In this chapter the concept of risk and consequence will be discussed.

3.2 RISK AND CONSEQUENCE

Newton's first law states that every action has an equal or greater reaction (5). Relating this to risk and consequence, it can be stated that every human action has a consequence. Human actions, of course, range from drinking a glass of pure water that has very positive consequences to launching a nuclear war that has very detrimental consequences. With every action humans undertake, they are making a calculation that the results of that action will have a positive, neutral or a minimally negative outcome. The actions are the risks and the outcomes are the consequences. In the course of a day, humans make hundreds of decisions, and some of these decisions have major implications on the rest of a person's life.

Driving is always a good model of risk. Harold Blackman once stated that if every human activity were as safe (unsafe) as driving, risk analysts would not be needed.[1]

---

[1]Private communication with Harold Blackman, director of the Center for Advanced Energy Studies.

Risk Assessment: Tools, Techniques, and Their Applications, Second Edition. Lee T. Ostrom and Cheryl A. Wilhelmsen.
© 2019 John Wiley & Sons, Inc. Published 2019 by John Wiley & Sons, Inc.
Companion website: www.wiley.com/go/Ostrom/RiskAssessment_2e

TABLE 3.1
Nuclear Power Station Accidents and Incidents

			Nuclear power station accidents and incidents	
Year	Incident	INES level	Country	IAEA description
2011	Fukushima	5	Japan	Reactor shutdown after 2011 Sendai earthquake and tsunami; failure of emergency cooling caused an explosion
2011	Onagawa		Japan	Reactor shutdown after the 2011 Sendai earthquake and tsunami caused a fire
2006	Fleurus	4	Belgium	Severe health effects for a worker at a commercial irradiation facility as a result of high doses of radiation
2006	Forsmark	2	Sweden	Degraded safety functions for common cause failure in the emergency power supply system at nuclear power plant
2006	Erwin		United States	Thirty-five liters of a highly enriched uranium solution leaked during transfer
2005	Sellafield	3	United Kingdom	Release of large quantity of radioactive material, contained within the installation
2005	Atucha	2	Argentina	Overexposure of a worker at a power reactor exceeding the annual limit
2005	Braidwood		United States	Nuclear material leak
2003	Paks	3	Hungary	Partially spent fuel rods undergoing cleaning in a tank of heavy water ruptured and spilled fuel pellets
1999	Tokaimura	4	Japan	Fatal exposure of workers following a criticality event at a nuclear facility
1999	Yanangio	3	Peru	Incident with radiography source resulting in severe radiation burns
1999	Ikitelli	3	Turkey	Loss of a highly radioactive Co-60 source
1999	Ishikawa	2	Japan	Control rod malfunction
1993	Tomsk	4	Russia	Pressure buildup led to an explosive mechanical failure
1993	Cadarache	2	France	Spread of contamination to an area not expected by design
1989	Vandellos	3	Spain	Near accident caused by fire resulting in loss of safety systems at the nuclear power station
1989	Greifswald		Germany	Excessive heating which damaged ten fuel rods
1986	Chernobyl	7	Ukraine (USSR)	Widespread health and environmental effects. External release of a significant fraction of reactor core inventory
1986	Hamm-Uentrop		Germany	Spherical fuel pebble became lodged in a pipe used to deliver fuel elements to the reactor

Year	Location	Level	Country	Description
1981	Tsuruga	2	Japan	More than 100 workers were exposed to doses of up to 155 mrem per day radiation
1980	Saint Laurent des Faux	4	France	Melting of one channel of fuel in the reactor with no release outside the site
1979	Three Mile Island	5	United States	Severe damage to the reactor core
1977	Jaslovske Bohunice	4	Czechoslovakia	Damaged fuel integrity, extensive corrosion damage of fuel cladding and release of radioactivity
1969	Lucens		Switzerland	Total loss of coolant led to a power excursion and explosion of experimental reactor
1967	Chapelcross		United Kingdom	Graphite debris partially blocked a fuel channel causing a fuel element to melt and catch fire
1966	Monroe		United States	Sodium cooling system malfunction
1964	Charlestown		United States	Error by a worker at a United Nuclear Corporation fuel facility led to an accidental criticality
1959	Santa Susana Field Laboratory		United States	Partial core meltdown
1958	Chalk River		Canada	Due to an inadequate cooling, a damaged uranium fuel rod caught fire and was torn in two
1958	Vinca		Yugoslavia	During a subcritical counting experiment, a power buildup went undetected – six scientists received high doses
1957	Kyshtym	6	Russia	Significant release of radioactive material to the environment from explosion of a high activity waste tank
1957	Windscale Pile	5	United Kingdom	Release of radioactive material to the environment following a fire in a reactor core
1952	Chalk River	5	Canada	A reactor shutoff rod failure, combined with several operator errors, led to a major power excursion of more than double the reactor's rated output at AECL's NRX reactor

TABLE 3.2
International Nuclear Events Scale (INES)

		International nuclear events scale (INES)			
Level	Definition	People and environment	Radiological barriers and control	Defense in depth	Example
7	Major accident	Major release of radioactive material with widespread health and environmental effects requiring implementation of planned and extended countermeasures			Chernobyl, Ukraine, 1986
6	Serious accident	Significant release of radioactive material likely to require implementation of some planned countermeasures			Kyshtym, Russia, 1957
5	Accident with wider consequences	Limited release of radioactive material likely to require implementation of some planned countermeasures. Several deaths from radiation	Severe damage to reactor core. Release of large quantities of radioactive material within an installation with a high probability of significant public exposure. This could arise from a major criticality accident or fire		Windscale, United Kingdom, 1957; Three Mile Island, 1979
4	Accident with local consequences	Minor release of radioactive material unlikely to result in implementation of planned countermeasures other than local food controls. At least one death from radiation.		Fuel melt or damage to fuel resulting in more than 0.1% release of core inventory. Release of significant quantities of radioactive material within an installation with a high probability of significant public exposure.	Fukushima 1, 2011

No.	Level				
3	Serious Incident	Exposure in excess of ten times the statutory annual limit for workers. Nonlethal deterministic health effect (e.g. burns) from radiation.	Exposure rates of more than 1 Sv h^{-1} in an operating area. Severe contamination in an area not expected by design, with a low probability of significant public exposure.	Near accident at a nuclear power plant with no safety provisions remaining. Lost or stolen highly radioactive sealed source. Misdelivered highly radioactive sealed source without adequate procedures in place to handle it.	Sellafield, United Kingdom, 2005
2	Incident	Exposure of a member of the public in excess of 10 mSv. Exposure of a worker in excess of the statutory annual limits.	Radiation levels in an operating area of more than 50 mSv h^{-1}. Significant contamination within the facility into an area not expected by design.	Significant failures in safety provisions but with no actual consequences. Found highly radioactive sealed orphan source, device or transport package with safety provisions intact. Inadequate packaging of a highly radioactive sealed source.	Atucha, Argentina, 2005
1	Anomaly			Overexposure of a member of the public in excess of statutory annual limits. Minor problems with safety components with significant defense-in-depth remaining. Low activity lost or stolen radioactive source, device or transport package.	

What does this mean? Well, driving is a very unsafe activity. Around 52 190 people are killed in the United States alone from driving-related accidents (6). People are generally more afraid of flying than of driving or riding in a car. If flying were as unsafe as driving, then the airline industry would go away. If nuclear power plants operated as unsafely as humans drive, the world would be highly contaminated with radionuclides.

On any given day, drivers can be observed committing some very risky driving maneuvers. They run red lights, they cut in front of other drivers, and they change lanes rapidly without looking. We now have the cellphone and texting distractions. A survey was conducted in 2014 by the National Safety Council for their annual injury and fatality report. The survey found that the use of cellphones causes 26% of the nation's car accidents. This was an increase from the previous year.

It was interesting to note that only 5% of cellphone-related crashes occur because the driver is texting. The majority of the accidents involved drivers that were distracted while they were talking on handheld or hands-free cellphones.

The NSC report, combined with Texas A&M research institute's "Voice-to-Text Driver Distraction Study," warns drivers that talking can be more dangerous than texting while operating a vehicle, and the use of talk-to-text applications is not a solution (7).

It appears as though they do not consider the consequences. The benefit of doing these maneuvers is the saving of seconds from their commute. In many cases, it is just seconds. In fact, in many communities the traffic lights are set up to control the rate of traffic flow, so even if a driver commits several risky maneuvers, they probably will only arrive at their destination at about the same time as the safe driver.

So, what are the consequences of risky driving?

- Drivers or passengers' deaths.
- Other drivers' or passengers' deaths.
- Permanent disabling injuries.
- Moderate injuries.
- Property damage (Most cars now cost well over $20 000.).
- Loss of driving privileges.
- Jail time.
- Tickets.
- Bad reputation.

Driving recklessly can have very severe consequences, yet people drive recklessly every day.

Consequences are a very important part of risk assessments. All of the consequences need to be considered and should help guide the risk assessment. For instance, if a risk assessment is being performed on a small chemical plant that produces very hazardous materials, then the following risks might be considered:

- Off-site chemical release (major and minor).
- On-site chemical release (major and minor).
- Fires and explosions (major and minor).
- Employee exposures (acute and chronic).
- Transportation accidents.
- Terrorist activities.
- Damage to company reputation.

These are not all the possible consequences, but they are a good start to help guide how the risk assessment should be conducted. The company or stakeholders and risk analysts need to bound the risk assessment by the consequences of interest.

Going back to the events of March 2011 in Japan, it is obvious that when the site planning of nuclear power plants was being conducted, the thought of a magnitude 9 earthquake and subsequent tsunami was not considered. The San Onofre nuclear power plant in California is designed to withstand a magnitude 7 earthquake (8). However, as seen in Japan and the Indonesian earthquake in December 2004, earthquakes with a magnitude of greater than 7 are not uncommon (9). Therefore, in future nuclear power plant construction, it should be obvious that consideration needs to be made on all the possible events and not those that are most likely. In hindsight, the Fukushima nuclear power plants should have been sited away from the coast so that any tsunami would not damage them to the point of meltdown. Also, the safety systems should have been designed to survive earthquake and tsunami. In the coming years, as the events in Japan play out, the full consequences of the decision to site the plants where they did will be understood.

The consequences of events like in Chernobyl and even the Lockerbie bombing are felt quite strongly today, over twenty years since either event occurred (10). This is further evidence to thoroughly consider all possible consequences when performing risk assessments.

3.3 CREDIBLE CONSEQUENCES

In a meeting in 1990, a group of new engineers had a roundtable discussion about the possible risks to a research reactor in the middle of a desert and the possible consequences. One possible scenario was a large commercial airliner crashing into the reactor. Upon discussing the scenario, there was laughter in the room. On the morning of 11 September 2001, that scenario became not just a "what if," but a real possibility. Therefore, when determining the possible set of credible consequences of which to consider risk ask at least the following questions:

1. What are all the known hazards?
2. What are all the possible events involving these hazards?
3. Have these events occurred before?
4. How frequently have these events occurred?
5. What happens when these events occur?
6. How severe are the consequences of these events?
7. How hard is it to clean up the mess left by the event?
8. Who ultimately pays for the cleanup?
9. What values does the organization consider important?

Always involve a team of subject matter experts when developing lists of possible consequences. The Delphi method discussed in Chapter 7 is a good tool to aid in these types of activities. Never dismiss a consequence until it is proven to not be credible. Then consider all the credible consequences when performing the risk assessment.

3.4 SUMMARY

As discussed above, all actions have consequences. Good eating habits and exercise have positive consequences, and reckless driving can have very negative consequences. Question all the possible consequences in the initial phases of a risk assessment, and consider those that are deemed to be credible. Don't toss our scenarios until they have been analyzed.

In summary, human's actions have consequences. The goal of risk analysts is to help determine which consequences are important enough to consider in risk assessments and to analyze realistically and appropriately.

Self-Check Questions

1. How is Newton's first law related to risk?
2. What are some consequences of risky driving?
3. Should you ever dismiss a consequence?

REFERENCES

1. CNCB (n.d.). Magnitude 6.1 earthquake kills three people, shutters factories in Japan's Osaka. https://search.cnbc.com/rs/search/view.html?source=CNBC.com&categories=exclude&partnerId=2000&keywords=MARCH%202011%209.0%20EARTHQUAKE (accessed 20 June 2018).

2. IAEA (n.d.). International nuclear and radiological event scale (INES). https://www.iaea.org/topics/emergency-preparedness-and-response-epr/international-nuclear-radiological-event-scale-ines (accessed 20 June 2018).
3. Gad-el-Hak, M. (2008). Engineering vs. disasters. https://www.researchgate.net/publication/292349698_Large-Scale_Disasters_Prediction_Control_and_Mitigation (accessed 20 June 2018).
4. US NRC 2010). Earthquake risk to nuclear power plants. http://www.nrc.gov/reading-rm/doc-collections/fact-sheets/fs-seismic-issues.html (accessed 20 June 2018).
5. French, A. (1971). *Newtonian Mechanics*, The M.I.T. Introductory Physics Series. New York: W.W. Norton & Company.
6. National Center for Statistics and Analysis (2018). *Passenger vehicles: 2016 data*, Traffic Safety Facts. Report No. DOT HS 812 537. Washington, DC: National Highway Traffic Safety Administration https://crashstats.nhtsa.dot.gov/Api/Public/ViewPublication/812537 (accessed 19 February 2019).
7. Mello, M. (2011). Edison: San Onofre could handle 7.0 quake. https://www.ocregister.com/2011/03/14/edison-san-onofre-could-handle-70-quake (accessed 20 June 2018).
8. Kratsas, G. (2014). Cellphone use causes over 1 in 4 car accidents. *USA Today* (28 March). https://www.usatoday.com/story/money/cars/2014/03/28/cellphone-use-1-in-4-car-crashes/7018505 (accessed 25 June 2018).
9. Wikipedia (2004). Indonesian earthquake. https://en.wikipedia.org/wiki/2004_Indian_Ocean_earthquake_and_tsunami (accessed 18 June 2018).
10. Wikipedia (n.d.). Lockerbie bombing. http://en.wikipedia.org/wiki/Lockerbie (accessed 20 June 2018).

CHAPTER 4

Ecological Risk Assessment

4.1 INTRODUCTION

Widespread ecological disasters are nothing new on Earth. The Earth has experienced countless natural disasters over its history. One of the first disasters that could most likely be the first step in progress toward current life forms on Earth is the advent of oxygen. Oxygen-producing organisms began spewing free oxygen into the atmosphere around 2.45 billion years ago (1). This free oxygen oxidizes iron, and the resulting iron oxide precipitated out to the floors of the oceans. It thus removed the dissolved iron from the oceans. These iron deposits are still mined today (2, 3). Maybe if those cyanobacteria or blue–green algae had not released free oxygen into the atmosphere, there would be some other forms of life on Earth.

The Earth has also had ice ages that have covered the entire globe or a major portion of the northern or southern hemisphere with ice, thus reducing the space that organisms could live to either in the depths of oceans or on land closer to the equator. The Earth has experienced hot, moist periods of time that allowed for the proliferation of new plant and animal species that drastically changed the environment. In fact, what is amazing is that the Earth remained habitable by some form of life for hundreds of millions of years.

Plants and animals, in general, have the tendency to destroy their living environment as they consume the resources they need to live. The production of alcohol by yeast in the manufacture of beer and wine is an excellent example of the cycle of an organism (4). Yeast converts sugars to alcohol. Once the alcohol level in a fermentation vat reaches somewhere between 4 and 12%, depending on the variety of the yeast, they shut down production due to the change in the environment. The growth

Risk Assessment: Tools, Techniques, and Their Applications, Second Edition. Lee T. Ostrom and Cheryl A. Wilhelmsen.
© 2019 John Wiley & Sons, Inc. Published 2019 by John Wiley & Sons, Inc.
Companion website: www.wiley.com/go/Ostrom/RiskAssessment_2e

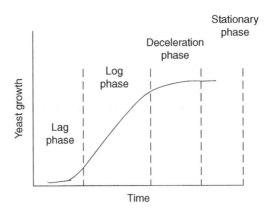

FIGURE 4.1   Yeast growth.

of the yeast has four distinct phases:

1. Lag Phase – yeast matures and acclimates to the environment.
2. Log or Expediential Phase – yeast rapidly multiplies, consuming resources and generating waste and alcohol.
3. Deceleration Phase – yeast growth slows due to lack of resources, sugar, and too much waste in the environment, alcohol.
4. Stationary Phase – limited growth occurs due to lack of resources and too much waste.

Yeasts literally eat themselves out of their environment. Figure 4.1 shows the four phases of yeast growth.

Algae blooms in aquatic environments, overpopulation of deer in protected areas, and the lemming and predator balance in the Arctic are all examples of similar observable ecological disasters in which animals or plants have a specific role.

Humans, of course, cause ecological disasters of their own. Those of us who grew up in the 1950s, 1960s, and 1970s still remember vividly the ecological problems that abounded in the United States during those decades. In north Idaho, for instance, the Silver Valley around Kellogg, ID, was under a constant layer of acrid smoke due to the lead smelters in the area. Driving from Coeur d'Alene, ID, to Kellogg, one would go from blue skies into a cloud of acrid smoke within a matter of 30 miles. Trees could not grow on the mountain sides due to the chemicals contained in the cloud of pollution. When vacationers ski in the Silver Valley today, they see very few remnants of the ecological disaster that once were blatantly obvious. The hillsides are once again tree covered and the air is relatively clean. However, lead levels remain high in the soil and on the bottom of Lake Coeur d'Alene (5). The primary operator of the lead smelters was Bunker Hill. The area around the Bunker Hill site was declared an Environmental Protection Agency (EPA) Superfund Site in the early 1980s and a great mount has been put into cleaning up the area.

More famous, or infamous, ecological disasters include:

- Chernobyl, Soviet Union (Belarus and Ukraine)
- Bhopal, India
- Deepwater Horizon, United States
- Love Canal, United States
- Minamata Methylmercury, Japan
- Agent Orange, Vietnam and the United States
- Seveso, Italy
- Jilin plant explosion

In this book we have discussed the Bhopal and Chernobyl events in several chapters. The following will discuss the attributes of the other events listed.

4.1.1 Deepwater Horizon

In the late spring and summer of 2010, the United States was dealing with one of the worst oil spills ever (6–8). It occurred in the Gulf of Mexico and involved the Deepwater Horizon drilling platform. The Deepwater Horizon was an offshore oil platform owned by Transocean Ltd. (Transocean). The rig was drilling within the Macondo Prospect oil field approximately 50 miles southeast of Mississippi in approximately 5000 ft of water. The Deepwater Horizon was considered an ultra-deepwater, dynamically positioned, column-stabilized, semi-submersible mobile offshore drilling unit (MODU) and could operate in waters of up to 8000 ft deep. Semi-submersibles are rigs that have platforms with hulls, columns, or pontoons that have sufficient buoyancy to cause the structure to float, but with weight sufficient enough to keep the structure upright. The Deepwater Horizon housed 126 workers, and its task was to drill wells and extract and process oil and natural gas from the Gulf of Mexico and export the products to shore. As with any large venture, there are several principle players in an oil exploration and development process. The rig was owned by Transocean, but the principal developer of the Macondo Prospect oil field was British Petroleum Company (BP). The Horizon had been leased to BP on a three-year contract for deployment in the Gulf of Mexico following its construction. On 20 April 2010, an explosion on the Deepwater Horizon, and its subsequent sinking two days later, resulted in the largest marine oil spill in the history of the United States and the deaths of 11 workers.

On 20 April 2010, at 10:45 p.m. EST, a sudden explosion rocked the Deepwater Horizon oil platform. The resulting fire traveled so fast that survivors stated they had less than five minutes to evacuate the platform after the first fire alarm. Most of the workers had to evacuate the platform by using the lifeboats from an auxiliary ship, the M/V Damon B Bankston. The Bankston had been hired to service the large platform oil rig. After the evacuation, eleven persons remained unaccounted for, and rescue procedures were put into place.

The US Coast Guard launched a rescue operation involving two cutters, four helicopters and a rescue plane. The Coast Guard conducted a three-day search covering approximately 5300 miles. They called off the search for the missing persons, concluding that the "reasonable expectations of survival" had passed. "Officials concluded that the missing workers may have been near the blast and not been able to escape the sudden explosion" (7). After many investigations, it has been suggested that the cause of the explosion and resulting fire was that a bubble of methane gas escaped from the well and rose up through the pipes, expanding and blowing out seals and barriers as it rose before exploding on the oil rig.

The Deepwater Horizon had been tethered to the ocean floor by a pipe used to extract oil called a riser. Because of the platform's sinking, the pipe was damaged. The damaged pipe began leaking a tremendous amount of oil in what is commonly known as a "gusher." Huge quantities of crude oil gushed from the riser pipe for approximately three months. A device called a blowout preventer (BOP) attached to the pipe at the ocean floor level to prevent such an occurrence failed to operate. Numerous attempts to manually operate the BOP also failed. The rate of oil that was released from the riser soon became a hotly debated issue. Real-time video feeds from the scene played out all over the United States and, in fact, the world to see. Eventually the resulting oil spill would cover almost $30\,000\,\text{miles}^2$ of ocean and an area, depending on weather conditions, larger than the state of South Carolina. The inaccuracies concerning the amount of oil release from government responders directly conflicted with the estimates of nongovernment scientists who suggested that the oil release figures were being grossly underreported. Though, in reality, the exact quantity of oil released was really not the issue. The real issue was how to clean up the oil that was there and, subsequently, preventing future occurrences. The most reliable estimate of the amount of oil that was released was "roughly five million barrels of oil were released by the Macondo well, with roughly 4.2 million barrels pouring into the waters of the Gulf of Mexico" (7).

BP's attempts to plug the leak had become a long and arduous task. BP engineers' initial plan was to use remotely operating underwater vehicles (ROV) to stop the leak by remotely activating the BOP, which was "a massive five story, 450 ton stack of shut-off valves, rams, housings, tanks and hydraulic tubing that sits on top of the well" (7, 8). As previously stated, the BOP failed to operate and speculation was that gas hydrates entered and formed in the BOP after a methane bubble rose up through the riser and blew out the seals and barriers in the pipes causing it to malfunction.

BP's next and subsequent attempts had become exercises in futility. On 7 May, BP engineers decided it would use a "top hat" or cofferdam to control the escaping oil from the broken riser. A top hat is a containment dome that is maneuvered over a blowout to collect the escaping oil so that it can be funneled through a pipe up to an awaiting drill ship on the surface. Except this top hat was 98 tons of steel. This project soon floundered because again, "the cofferdam containment system failed, becoming iced up with methane hydrates when hydrocarbons from the end of riser proved to have a higher gas content than anticipated" (9). A second smaller top hat that weighed a mere 2 tons and had the ability to be injected with alcohol to act as an antifreeze to reduce the formation of gas hydrates was the next course of action, but that plan was abandoned on 12 May, when engineers became unsure that plan would work either. "The first significant success at reducing the release of oil came on May 17, 2010 when robots inserted a four-inch diameter Riser Insertion Tube Tool (RITT) into the Horizon's riser, a twenty one-inch diameter pipe between the well and the broken end of the riser on the seafloor in five-thousand feet of water" (7). The RITT is supposed to work like a giant straw that siphons off the leaking oil and transports it to an awaiting tanker on the surface. This attempt brought some success.

The company's long-range plan was to initiate relief wells that would intercept the bored out well at approximately 13 000 ft below the ocean floor. After the relief wells were completed, heavy fluids and cement could be pumped down the damaged hole to kill the well; this is referred to as a "top kill." The only problem with this plan is it would take a minimum of 90 days to accomplish. Thus the reasoning for the stopgaps put into play was to reduce the oil leak at the broken riser early on after the explosion. On 25 May, the RITT was disabled for a "top kill" procedure scheduled for the following day.

> On May 26, 2010, the U.S. government gave BP the approval to proceed with a 'top kill' operation to stop the flow of oil from the damaged well. The procedure was intended to stop the flow of oil and gas from the damaged well and ultimately kill it by injecting heavy drilling fluids through the BOP. On May 29, 2010, BP engineers said that the 'top kill' technique had failed. Over thirty thousand barrels of heavy mud was injected into the well in three attempts at rates of up to 80 barrels a minute. Several different bridging materials had been tried and still the operation did not overcome the flow from the well (7).

After 86 days and several failed attempts and efforts to seal the leak, on 15 July, BP succeeded in stopping the flow of oil into the Gulf of Mexico.

Much of the work on oil platforms has become automated in its functions below the waves and on the ocean floor. But human error still manifests itself from time to time on these huge sea-going structures. These drilling rigs are some of the largest moveable man-made structures in the world; as such they have become virtual cities afloat that will always have minor equipment failure and human error, not to mention working in hurricane-prone environments. The Deepwater Horizon was no different; it had a long history of spills, fires, and other mishaps before the Gulf oil spill in April of 2010. There is even a collision documented in its recent history.

> Because vessels like the Deepwater Horizon operate 24 hours a day, Coast Guard officials said minor equipment problems appear frequently. If these problems are not corrected then such incidents could mushroom into bigger concerns (10).

The agency responsible for investigating the safety of offshore and gas operations is the US Department of the Interior's Minerals Management Service (MMS). The MMS had an extensive, detailed inspection program to help ensure the safety of offshore oil and gas operations. MMS inspectors are placed offshore on oil and gas drilling rigs and production platforms to audit operator compliance with extensive safety and environmental protection requirements.

The Deepwater Horizon had experienced many problems before.

- In 2005 the oil rig, still under contract with BP, "spilled 212 barrels of an oil based lubricant due to equipment failure and human error. That spill was probably caused by not screwing the pipe tightly enough and not adequately sealing the well with cement, as well as a possible poor alignment of the rig, according to records maintained by the federal Minerals Management Service" (10). Following that spill, MMS inspectors recommended increasing the amount of cement used during this process and applying more torque when screwing in its pipes.

- Also in 2005, a crane operator sparked a hazardous fire aboard Deepwater Horizon while refueling. His inattention caused diesel to overflow, and a spark initiated a fire on board. "In June 2003, the rig floated off course in high seas, resulting in the release of 944 barrels of oil. MMS blamed bad weather and poor judgment by the captain."

- "A month later, equipment failure and high currents led to the loss of an additional 74 barrels of oil" (10). These were just a few of the mishaps that were reported, and investigated by the MMS on the Deepwater Horizon before the blowout on April 2010.

The MMS, the caretaker of America's federal lands and oceans and watchdog of the oil and gas drilling industry, had come under increasing criticism in the years prior to the Deepwater Horizon mishap. "Investigators from the Interior Department's inspector general's office said more than a dozen employees, including the former director of the oil royalty-in-kind program, took meals, ski trips, sports tickets and golf outings from industry representatives. The report alleges that the former director, Gregory W. Smith, also netted more than $30,000 from improper outside work" (11). The collection of billions of dollars in royalties from oil and gas companies by government officials in their capacity were also alleged to have taken bribes, the steering of contracts to favored clients, and engaging in illicit sex with employees of the energy firms. In the report, investigators said they "discovered a culture of substance abuse and promiscuity" in which employees accepted gratuities "with prodigious frequency" (11).

The responsibility for the initial cleanup was assumed by BP oil corporation. Tony Haywood, Chief Executive Officer (CEO), formally verbalized to the American people that his company was taking full responsibility for the disaster "and where people can present legitimate claims for damages we will honor them." To augment the cost of the cleanup, under the Federal Water Pollution Control Act, the Oil Spill Liability Trust Fund (OSLTF), established in the Treasury helped defer expenses of a federal response to oil pollution and to help compensate claims for oil removal and damages as authorized by the Oil Pollution Act of 1990 (OPA). "The OPA requires that responsible parties pay the entire price tag for cleaning up after spills from off-shore drilling, including lost profits, destroyed property and lost tax revenue, but the statute caps their liability for economic damages at $75 million. Aggressive collection efforts are consistent with the 'polluter pays' public policy underlying the OPA. BP and Transocean have been named as responsible parties, although all claims are still being processed centrally through the BP Corporation" (7).

Many events led to the explosion on the Deepwater Horizon platform. Numerous events took place that contributed to the disaster. Working at great depths, 5000 ft or more, and pressures greater than 2000 lb in^{-2} (13 789 514.56 Pa) should be re-evaluated. Problems at these depths have very real dangers and are unfamiliar. Most equipment used to secure a well run amok has only been tested at depths half that of the Deepwater Horizon's. Guy Cantwell, a spokesman for the oil rig's owner, Transocean Ltd., said that the Swiss-based company planned to conduct its own investigation of what caused the explosion aboard the Deepwater Horizon. "The industry is going to learn a lot from this. That's what happens in these kinds of disasters," he said, citing a 1988 explosion of the Piper Alpha rig in the North Sea and a 1979 blowout of Mexico's IXTOC I in the Eastern Gulf (10). After the North Sea incident in which 167 men were killed, Great Britain revamped its safety requirements concerning deepwater drilling. There is no doubt that the same will happen in the United States. Numerous "well topping" devices and associated installation accessories have already been designed, built, and readied for future deployment, particularly where gas hydrates are concerned. It seems as though deepwater drilling is here to stay.

As of the writing of this book, we are only one year since this event occurred. There is still much controversy as to the long-term effects of the spill. Only time will tell what the long-term ecological effects of the spill are.

4.1.2 Love Canal

In 1920 Hooker Chemical (Hooker) had turned an area in Niagara Falls into a munic-ipal and chemical disposal site (12–14). In 1953 the site was filled and relatively modern methods were applied to cover the disposal site. A thick layer of impermeable red clay sealed the dump. The idea was that the clay would seal the site and prevent chemicals from leaching from the landfill.

A city near the chemical disposal site wanted to buy it for urban expansion. Despite the warnings made by Hooker, Niagara Falls School Board eventually bought the site for the amount of $1. Hooker could not sell for more, because they did not want to earn money off a project. As part of the development process, the city began to dig to develop a sewer. This damaged the red clay cap. Blocks of homes and a school were built, and the neighborhood was named Love Canal.

Love Canal seemed like a normal, regular neighborhood. The only thing that dis-tinguished this neighborhood from others was the strange chemical odors that often hung in the air and an unusual seepage noticed by inhabitants in their basements and yards. Children in the neighborhood often became sick. Love Canal families regularly experienced miscarriages and birth defects.

Lois Gibbs, an environmental activist, noticed the high occurrence of illness and birth defects in Love Canal and started documenting it. In 1978 newspapers revealed the existence of the chemical waste disposal site in the Love Canal area and Lois Gibbs started petitioning for closing the local school. In August 1978, the claim succeeded and the NYS Health Department ordered the closing of the school when a child fell ill from a chemical poisoning.

When the waste site at Love Canal was assessed, the researchers found over 130 lb of the highly toxic carcinogenic 2,3,7,8-tetrachlorodibenzo-p-dioxin (TCDD), a form of dioxin. A total of 20 000 tons of waste present in the landfill appeared to contain more than 248 different types of chemicals. The waste consisted of pesticide residues and chemical weapons research refuse, along with other organic and inorganic compounds.

Due to the breach of the clay cap, chemicals had entered homes, sewers, yards, and waterways and more than 900 families had to be relocated. President Carter provided federal funds to move all the families to a safer area. Hooker's parent company was sued and settled for $20 million.

Even though most of the chemicals were not removed from the chemical disposal site and despite protests by Gibbs's organization, some of the houses in Love Canal went up for sale some 20 years later. The site was resealed and the surrounding area was cleaned and declared safe. Today a barbed wire fence isolates the worst area of the site from the areas that are not as contaminated. Hooker's mother company paid an additional $230 million to finance this cleanup. They are now responsible for the management of the dumpsite. Today, the Love Canal dumpsite is known as one of the major environmental disasters of the century. Bacteria and other microbes might eventually break down the organic materials into safer compounds, but this could take hundreds, if not thousands, of years.

4.1.3 Minamata Methylmercury

The Chisso Corporation first opened a chemical factory in Minamata, Japan, in 1908 (15, 16). Initially producing fertilizers, the factory followed the nationwide expansion of Japan's chemical industry, branching out into production of acetylene, acetaldehyde, acetic acid, vinyl chloride, and other chemicals. The Minamata factory became the most advanced chemical company in Japan in the 1930s and 1940s. The waste products resulting from the manufacture of these chemicals were, as in many chemical and other process industries, released right into Minamata Bay. As with any chemical put into the environment, these pollutants had an impact, such as damaging the fisheries. In response, Chisso reached two separate compensation agreements with the fishery cooperative in 1926 and 1943 (16).

The Chisso Minamata factory was very successful and it had a very positive effect on the local economy (15). The area lacked other industry and, Chisso had great influence in Minamata. Over half of the tax revenue of Minamata City came from Chisso and its employees, and the company and its subsidiaries were responsible for creating a quarter of all jobs in Minamata.

The Chisso Minamata factory first started acetaldehyde production in 1932, producing 210 tons that year. Acetaldehyde is used as a chemical intermediary in the production of numerous products, for instance, vinyl. By 1951 production had jumped to 6000 tons per year and reached a peak of 45 245 tons in 1960 (16) The Chisso factory's output amounted to between a quarter and a third of Japan's total acetaldehyde production. The chemical reaction used to produce the acetaldehyde used mercury sulfate as a catalyst. A side reaction of the catalytic cycle led to the production of a small amount of an organic mercury compound, namely, methylmercury (17). This highly toxic compound was released into Minamata Bay from the start of production in 1932 to 1968. Interestingly enough, elemental mercury is poorly absorbed through the skin or through ingestion. However, the vapors of elemental mercury are much more hazardous. Methylmercury is very hazardous, as compared with elemental mercury. In 1968 the production process was modified and mercury was no longer used.

On 21 April 1956, a five-year-old girl was examined at the Chisso Corporation's factory hospital in Minamata, Japan, a town on the west coast of the southern island of Kyūshū. The physicians were puzzled by her symptoms: difficulty in walking, difficulty in speaking, and convulsions. Two days later her younger sister also began to exhibit the same symptoms and she too was hospitalized. The girls' mother informed the doctors that her neighbor's daughter was also experiencing similar problems. After a house-to-house investigation, eight further patients were discovered and hospitalized. On 1 May, the hospital director reported to the local public health office the discovery of an "epidemic of an unknown disease of the central nervous system," marking the official discovery of Minamata disease (15).

Researchers from Kumamoto University began to focus on the cause of the strange disease. They found that the victims, often members of the same family, were clustered in fishing hamlets along the shore of Minamata Bay. The staple food of victims was invariably fish and shellfish from Minamata Bay. The cats in the local area, who tended to eat scraps from the family table, had died with symptoms similar to those now discovered in humans. This led the researchers to believe that the outbreak was caused by some kind of food poisoning, with contaminated fish and shellfish being the prime suspects.

On 4 November, the research group announced its initial findings: "Minamata disease is rather considered to be poisoning by a heavy metal … presumably it enters the human body mainly through fish and shellfish" (16).

As soon as the investigation identified a heavy metal as the causal substance, the wastewater from the Chisso plant was immediately suspected as the origin. The company's own tests revealed that its wastewater contained many heavy metals in concentrations sufficiently high to bring about serious environmental degradation including lead, mercury, manganese, arsenic, selenium, thallium, and copper. Identifying which particular poison was responsible for the disease proved to be extremely difficult and time consuming. Between 1957 and 1958, many different theories were proposed for the cause of the ailments. Initially manganese was thought to be the causal substance due to the high concentrations found in fish and the organs of the deceased. A theory that there were multiple contaminants involving thallium and selenium was proposed. In March 1958, visiting British neurologist Douglas McAlpine suggested that the symptoms shown by victims in Minamata resembled those of organic mercury poisoning. From that point forward, the focus of the investigation centered on mercury.

In February 1959, the mercury distribution in Minamata Bay was investigated. The results showed that large quantities of mercury were detected in fish, shellfish, and sludge from the bay. At the mouth of the wastewater canal, there was approximately 2 kg of mercury per ton of sediment. This level would be economically viable to mine. Chisso did later set up a subsidiary to reclaim and sell the mercury recovered from the sludge (16).

Fifty years later, the legacy of Minamata Bay lives on. Victims still receive payment for their injuries and methylmercury still persists in the environment.

4.1.4 Agent Orange

Agent Orange was the code name for one of the herbicides and defoliants used by the US military as part of its herbicidal warfare program during the Vietnam War (18, 19). The campaign called Operation Ranch Hand involved spraying the countryside with the chemicals with the goal of defoliating the jungles and destroying crops.

Agent Orange was a 50:50 mixture of 2,4,5-T and 2,4-D (20). It was manufactured for the US Department of Defense primarily by Monsanto Corporation and Dow Chemical. The 2,4,5-T used to produce Agent Orange was later discovered to be contaminated with TCDD, an extremely toxic dioxin compound. It was given its name from the color of the orange-striped 55 US gal (200 l) barrels that were shipped in (18). It was the most widely used herbicide during the war.

During the Vietnam War, between 1962 and 1971, the US military sprayed nearly 20 000 000 US gal (75 700 000 l) of chemical herbicides and defoliants in Vietnam, eastern Laos, and parts of Cambodia, as part of the operation (18, 19).

Air Force records show that at least 6542 spraying missions took place over the course of the operation (18). Approximately 12% of the total area of South Vietnam had been sprayed with defoliating chemicals by the end of the war. It is estimated to be up to 13 times the recommended USDA application rate for domestic use (21). In South Vietnam an estimated 10 million ha of agricultural land was affected (19). In some areas TCDD concentrations in soil and water were hundreds of times greater than the levels considered "safe" by the US Environmental Protection Agency (EPA) (19, 21). Overall, more than 20% of South Vietnam's forests were sprayed at least once over a nine-year period (21).

The legacy of the spraying during the war lives on. Approximately 17% of the forested land had been sprayed, and to this day dioxins contained in the chemicals remain in the soil. In many places the natural foliage has been replaced by invasive plant species. Animal species diversity was also significantly impacted. For instance, a Harvard biologist found 24 species of birds and 5 species of mammals in a sprayed forest, while in two adjacent sections of unsprayed forest, there were 145 and 170 species of birds and 30 and 55 species of mammals (22).

Movement of dioxins through the food web has resulted in bioconcentration and biomagnification (23). The areas most heavily contaminated with dioxins are the sites of former US air bases (18). The Vietnam Red Cross reported as many as 3 million Vietnamese people have been affected by Agent Orange, including at least 150 000 children born with birth defects (24).

The problem with dioxins is that they are highly toxic, with no safe levels of exposure, and the chemicals take a tremendous amount of time to break down (25). This class of chemicals is produced as a by-product of producing chemicals for legitimate use or when chlorinated chemicals are burned. Vietnam is not the only place in the world where dioxins pose a threat. Seveso, Italy, experienced an environmental issue due to dioxin.

4.1.5 Seveso, Italy

During the middle of the day on 10 July 1976, an explosion occurred in a 2,4,5-trichlorophenol (TCP) reactor in the ICMESA chemical company in Meda, Italy (26, 27). A cloud containing toxins escaped into the atmosphere. The cloud contained high concentrations of TCDD. Downwind from the factory, the dioxin cloud polluted a densely populated area of 6 km long and 1 km wide, immediately killing many animals. Seveso is a neighboring municipality that was highly affected. The dioxin cloud affected a total of 11 communities.

Even though the media includes Seveso when other major disasters such as Bhopal and Chernobyl are discussed, the Seveso story is different when it comes to handling the toxins (26). Polluted areas were researched after the release and the most severely polluted soils were excavated and treated elsewhere. Health effects were immediately recognized as a consequence of the disaster and victims were compensated. A long-term plan of health monitoring was also put into place. Seveso victims suffered from a directly visible symptom known as chloracne and genetic impairments (27).

The Seveso accident and the immediate reaction of authorities led to the introduction of European regulation for the prevention and control of heavy accidents involving toxic substances. This regulation is now known as the Seveso Directive, which became a central guideline for European countries for managing industrial safety.

The most remarkable feature of the Seveso accident was that local and regional authorities had no idea the plant was a source of risk (26). The factory existed for more than 30 years and the public had no idea of the possibility of an accident until 1976. The European Directive was created to prevent such ignorance in the future and to enhance industrial safety. The Council of Ministers of the Council of Europe adopted the directive in 1982. It obligates appropriate safety measures and also public information on major industrial hazards, which is now known as the "need-to-know" principle (27).

4.1.6 Jilin Chemical Plant Explosion

This particular incident had impacts on both the environment and regional conflict. This account was adapted from the Environmental Justice Atlas (28). On 13 November 2005, an explosion at a petrochemical plant in China's northeastern Jilin province resulted in the release of 100 tons of toxins into the Songhua River. Much of the notoriety of the 2005 Songhua spill was derived from the flawed response capability exhibited by central and regional state authorities. On 26 November, representatives from China's State Environmental Protection Administration (SEPA) visited both the UN Environmental Program (UNEP) in Nairobi and the United Nations offices in Beijing to provide extensive data on the Songhua spill, after which SEPA continued to send regular updates to the United Nations. Although a decisive move on the part of the Chinese state, it approached the United Nations only after the pollution slick had reached Harbin, a full two weeks following the initial explosion in Jilin province. Harbin is an industrial city of 10 million people, with three major universities. While the central government went public largely because it could no longer keep information on the incident from its own citizens, it is perhaps safe to assume that this was the longest the Chinese government could wait without risking confrontation with Russia, whose border lay downriver (29).

Underlying China's delayed response at the international level was an almost paralyzing confusion at local and domestic levels. In the immediate aftermath of the explosion on 13 November, factory officials denied that pollutants had entered the Songhua River. It was five days before SEPA issued emergency monitoring instructions to its provincial counterpart, the Heilongjiang Province Environmental Protection Bureau (EPB). During those five days, it appears that factory officials together with local government officials attempted to manage the spill themselves without notifying Beijing, going so far as to drain reservoir water into Songhua River in an attempt to dilute the contaminants. Even after the central government intervened on 18 November, it failed to notify the general public of the danger until the slick reached Harbin. In the intervening period, public rumor regarding an undefined public emergency proliferated. Several days before the spill reached Harbin, city officials shut down municipal water services, citing a need to repair the system's facilities (29).

Following the central government's announcement, the domestic media was filled with material regarding the spill. Many of these articles traced the chain of responsibility and cover-up leading back to the explosion, in the process implicating high-ranking officials first in Jilin and later in Harbin. At this time the international media and international environmental groups also became closely involved, following the domestic media's lead in focusing on the dangerous lack of public information surrounding the spill. Even the UNEP team was invited to help evaluate the situation on the Songhua

River in December 2005. The initial field report generally approved of the Chinese government's response to the disaster, noted that the lack of public information management posed an avoidable danger, and indicated a lack of centralized emergency response procedures (29).

State media reported that five people were killed in the explosion only a few hundred meters from the river bank. Up to 10 000 people were temporarily evacuated. "We will be very clear about who's responsible," said Zhang Lijun, deputy director of SEPA, at a news conference in Beijing. "It is the chemical plant of the CNPC in Jilin Province." Zhang did not elaborate, but he said an investigation would be considered if there was any criminal liability for the spill. The official Xinhua news agency reported that the company had apologized for the contamination.

The company "deeply regrets" the spill and would take responsibility for handling the consequences, said CNPC's deputy general manager, Zeng Yukang. The vice governor of Jilin province, Jiao Zhengzhong, also apologized to the people of Harbin, according to a report Thursday in the newspaper Beijing News (30).

"Harbin's move to cut off the water supply was not a knee-jerk reaction," said Zhang Lanying, an environmental expert from Jilin University.

"If the contaminated water had been supplied to households, the result would have been unimaginable." Apart from the danger to the Chinese population and environment, the spill could also have diplomatic repercussions as it heads toward the point where the Songhua River joins the Heilongjiang River, which then crosses the Russian border in the region of Khabarovsk, about 550 km downriver from Harbin (29).

The longer-term environmental consequences of the Chinese spill are unknown. Environmental and other groups have suggested that the food chain in the river basin and corresponding region could be affected for some time. At the time, the Times (United Kingdom) reported on 21 December 2005 that fishing in the area could be banned for as long as four years. Other articles have suggested that the benzene contamination could present a long-term problem in that it can bioaccumulate in the basin's organisms, remain trapped in river ice that will melt and result in additional releases, and become trapped in the river's sediments (28).

4.1.7 Risk of Ecological Disasters

Unfortunately, the risk of another man-made ecological disaster looms just around the corner. The dumping of the reactor cores from nuclear-powered ships into the Barrett Sea by the Russian military and the other legacy nuclear sites in Russia and other former Soviet Union Republics pose a significant potential threat to the environment. In addition, many manufacturers in developing countries are displaying the same lack of environmental concern that developed countries had in the early to late 1900s. The resulting polluted bodies of water and land have the potential to cause widespread environmental problems. In addition, even well-run companies can have a process upset that results in a chemical spill that could cause catastrophic harm to the environment (31).

4.1.8 Ecological Risk Assessment

Ecological risk assessment (ERA) is a process that is used to help determine what risks are posed by an industry, government, or other entity. It further provides a logical process to help eliminate or mitigate such threats.

An ERA evaluates the potential adverse effects that human activities have on the living organisms (plants and animals) that make up ecosystems. The risk assessment process provides a way to develop, organize, and present scientific information so that it is relevant to future and present environmental decisions. When conducted for a particular place such as a forest or wetland, the ERA process can be used to identify vulnerable and valued resources, prioritize data collection activity, and link human activities to their potential effects. ERA results provide a basis for comparing different management options, enabling decision makers and the public to make better informed decisions about the management of ecological resources (32).

As EPA guidance states, ERAs are used to support many types of management actions, including the regulation of hazardous waste sites, industrial chemicals, and pesticides, or the management of watersheds or other ecosystems affected by multiple nonchemical and chemical stressors. The ERA process has several features that contribute to effective environmental decision making (32, 33):

- Through an iterative process, new information can be incorporated into risk assessments that then can be used to improve environmental decision making.

- Risk assessments can be used to express changes in ecological effects as a function of changes in exposure to stressors. This capability may be particularly useful to decision makers who must evaluate tradeoffs, examine different alternatives, or determine the extent to which stressors must be reduced to achieve a given outcome.

- Risk assessments explicitly evaluate uncertainty. Uncertainty analysis describes the degree of confidence in the assessment and can help the risk manager focus research on those areas that will lead to the greatest reductions in uncertainty.

- Risk assessments provide a basis for comparing, ranking, and prioritizing risks. The results can also be used in cost–benefit and cost-effectiveness analyses that offer additional interpretation of the effects of alternative management options.

- Risk assessments consider management goals and objectives as well as scientific issues in developing assessment endpoints and conceptual models during problem formulation. Such initial planning activities help ensure that results will be useful to risk managers.

The ERA approach contained in this book follows the EPA's guidelines (32). However, a search of the literature does present many other similar processes that can be used to assess ecological risk (33).

According to the EPA, the ERA process is based on two major elements:

1. Characterization of effects
2. Characterization of exposure

These provide the focus for conducting the three phases of risk assessment:

1. Problem formulation
2. Analysis
3. Risk characterization

Figure 4.2 shows the overall flow of an ERA.

The three phases of risk assessment are enclosed by a dark solid line.

Problem formulation is the first phase. During problem formulation, the purpose for the assessment is developed, the problem is defined, and a plan for analyzing and characterizing risk is also developed. Initial work on problem formulation includes the integration of available information on potential sources of contaminates, potential stressors to the environment, potential effects, and ecosystem and receptor characteristics. From this information two products are generated: assessment endpoints and conceptual models. Either product may be generated first (the order depends on the type of risk assessment), but both are needed to complete an analysis plan, the final product of problem formulation (32).

Analysis follows problem formulation. During the analysis phase, data are evaluated to determine how exposure to potential environmental stressors is likely to occur (characterization of exposure) and, if an exposure were to occur, the potential and type of ecological effects that could be expected (characterization of ecological effects). The first step in analysis is to determine the strengths and limitations of data on potential exposures, potential effects, and ecosystem and receptor characteristics. The products from these analyses are two profiles, one for exposure and one for stressor response. These products provide the basis for risk characterization (32, 33).

As part of the risk characterization phase, the potential exposure and stressor response profiles are integrated through the risk estimation process. Risk characterization includes a summary of assumptions, scientific uncertainties, and strengths and limitations of the analyses. The final product is a risk description in which the results of the integration are presented, including an interpretation of ecological adversity and descriptions of uncertainty and lines of evidence (33).

As with all risk assessments, ERAs are iterative in nature, as new data becomes available. For instance, in some cases, the health effects of chemicals are not fully elucidated until after the chemical or product has been in use.

4.1.9 Ecological Risk Assessment in Practice

The following will provide an example of how an ERA can be conducted. Note that all the information concerning organizations, locations, and scenarios have been

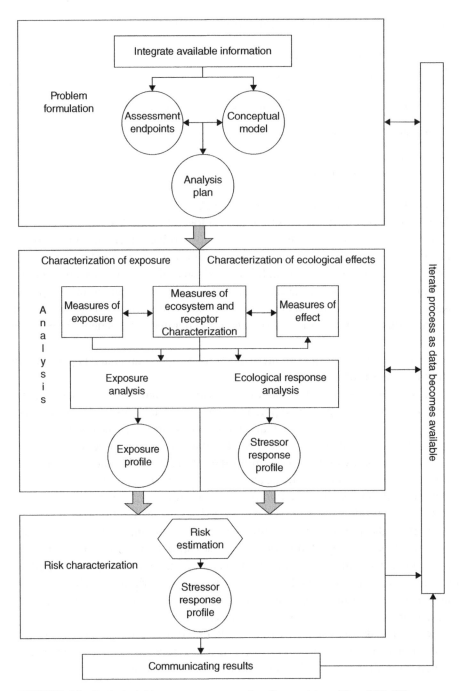

FIGURE 4.2 Ecological risk assessment process flow *Source*: Adapted from EPA (32).

fabricated. The nature of the chemicals are factual, but the information provided is for illustration only and not for reference. Also, this is not an exhaustive ERA, only an example one.

4.1.10 Production Plant, Inc.

Production Plant, Inc. (PPI) has been in the widget manufacturing business for 25 years. The product they manufacture is Widget A (WA). It is a stable product for the company and PPI plans on producing these widgets for the foreseeable future. In this regard, PPI is planning an expansion into a West Coast state so that they can serve the Pacific Rim countries better. The plot of land they are proposing to build on is approximately 20 acres and borders the Great River on one side, and the land contains a small creek that only flows during early winter until midsummer. Figure 4.3 shows the general layout and the proposed building location of the property. It also shows the flows of the waterways.

The proposed building site was used before as a farmland, and the farm-related chemicals found in the soil are shown in Table 4.1.

However, testing of Little Creek reveals that none of the farm chemicals are leaching into the waterway under current conditions.

The land generally slopes toward the Great River. There is a 10-ft elevation difference between the north end of the property and where it borders the river. At the

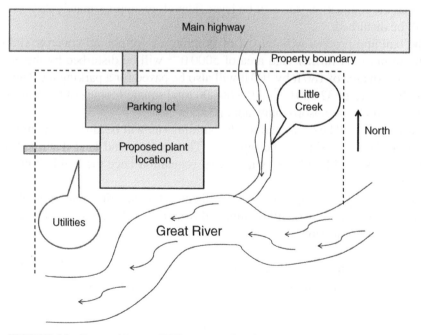

FIGURE 4.3 Proposed layout of PPI west coast location.

TABLE 4.1
Farm Chemicals Found in the Soil on Proposed Building Site

Product	Level detected	$LC_{50}{}^a$ (mg l^{-1})	$LD_{50}{}^b$ (mg kg^{-1})
Organophosphate A	50 ppb	0.1	10
Thiocarbamate C	1 ppm	20	25
Herbicide H	0.5 ppm	5	7
Fungicide X	10 ppb	100	120

[a] LC50 is lethal concentration for 50% of the population.
[b] LD50 is lethal dose to 50% of the population.

river's edge, there is a 10-ft drop off from the property to the river for most of the year. During spring runoff this decreases to 1–2 ft. The current vegetation consists of sage brush and nonnative grasses/weeds. Some bushy plants are found in Little Creek. The animals currently observed on the site are voles and gophers, raccoons near Little Creek, and several bird species, including great horned owls. The Great River contains a wide range of fishes, including sturgeon, white fish, rainbow trout, and largemouth bass. There are also several mollusks and crustaceans living in the river. Little Creek contains no fish because it is dry for most of the year. Some frog species enter the creek during wet periods.

The proposed land development process will entail digging up an area (140 ft by 100 ft) for the proposed building's foundation. The soil will remain on site. The building will not have a basement, but the soil will need to be removed to a depth of 2 ft so that the foundation slab will be level. Therefore, approximately 28 000 ft^{-3} of soil will be disturbed.

The building process will consist of bringing in pre-poured concrete slabs and erecting in place. A further land area of 5000 ft^{-2} will be disturbed by the erection equipment. An area of 180 ft by 100 ft will also be paved for a parking lot. Therefore, ultimately an area of 32 000 ft^{-2} of the 86 000 total square footage of the land will be no longer available to absorb rainwater and snowmelt.

The soil that will be excavated for the building site will be landscaped into rolling mounds on the south side of the building. The mounds and land disturbed by the building process will be planted with native vegetation once the building process is complete.

The utilities for the proposed factory will consist of electrical lines, potable water, natural gas, and ethanol that will be supplied from a neighboring plant. There will be three normal waste streams:

- Normal sewage waste from the toilets and shower areas – 20 000 gal per day
- Wash water containing biodegradable solvents and cleansers – 1000 gal per day
- Process water containing ethanol, methanol, and trace amounts of copper metal – 500 gal per day. This will be stored in a tank for pickup.

TABLE 4.2

Chemicals Used to Manufacture Widget A

Chemical	Proposed quantity on site	Usage volume	Hazardous nature	$LD_{50}{}^a$
Copper sulfate	1000 lb	10 lb per day	Moderately toxic	300–470 mg kg^{-1} (rat and mammals)
Aluminum Chloride	2000 lb	100 lb per day	Corrosive and moderately toxic	380 mg kg^{-1} (rat)
Ethanol	No onsite storage – supplied via pipeline	20 000 gal per day	Flammable, low toxicity	6300 mg kg^{-1} (rabbit)
Methanol	1000 gal stored on site	50 gal used per day	Flammable, moderate to low toxicity	5600 mg kg^{-1} (rat)

[a]These values are provided for example only and not for reference. Please consult the specific material safety data sheet (MSDS) for correct toxicity data.

It is proposed that the normal sewage will be discharged into the city system. It is proposed that the wash water be proposed on site by means of a digester. The process wastewater contains enough process chemicals that a local company might be interested in obtaining it and reprocessing it to remove the valuable contents.

The process for producing WA involves utilizing the chemicals shown in Table 4.2.

The Great River is 200 yards wide and 50 ft deep as it passes the proposed plant site. It discharges into the Pacific Ocean 100 miles below the proposed site. 100 000 people live below the plant site, along the Great River.

4.1.11 Problem Formulation

As per the EPA's ERA guidance, the first step in the process of assessing the impact of the proposed PPI plant is to develop the problem formulation. In this regard, the potential sources of contaminants need to be identified. The potential sources of containments from the proposed PPI plant include:

- Farm chemicals leaching out from the soil due to soil disturbance
- Spills from
  - Process upsets
  - Leaks from the ethanol supply line
  - Leaking process storage tanks and bins
  - Leaking process waste tank
  - Delivery trucks
  - Wastewater digester
  - Cleaning chemical spills

The assessment endpoints for this sample ERA that will be discussed are as follows:

- Reduction in species richness or abundance or increased frequency of gross pathologies in fish communities resulting from toxicity.
- Reduced richness or abundance of native plant species due to the establishment of the plant.
- Reduction in abundance or production of terrestrial wildlife populations resulting from toxicity.

The ecological assessment endpoints were selected based on meetings that included representatives of the local EPA, PPI, state department of environmental quality, and county planners.

1. The fish community in the Great River is considered to be an appropriate endpoint community because it is ecologically important and has a scale appropriate to the site.
2. The native plant life is considered to be an appropriate endpoint community because of its ecological importance and due to the amount of space the proposed PPI site will occupy.
3. The animal community is considered to be an appropriate endpoint community because it is ecologically important and due to the potential number of small mammals and birds that might be displaced by the PPI site.

This example ERA will consider the following measurement endpoints:

4.1.12 Fish

Single Chemical Toxicity Data – Chronic toxicity thresholds for freshwater fish are expressed as chronic EC20s or chronic values. These test endpoints correspond to the assessment endpoints for this community.

4.1.13 Terrestrial Plants

Biological Survey Data – Quantitative plant survey data does exist for the native and nonnative plants on the proposed plant site.

4.1.14 Terrestrial Animals

Single Chemical Toxicity Data – These include acute and chronic toxicity thresholds for contaminants of concern in birds and mammals with greater weight given to data from long-term feeding studies with wildlife species.

4.1.15 Conceptual Model

Conceptual models are graphical representations of the relationships among sources of contaminates, ambient media, and the endpoint biota. Suter (34, 35) has developed a complete guide on developing conceptual models for ERAs. Figure 4.4 is an example conceptual model that would be appropriate for the ERA for the proposed PPI plant.

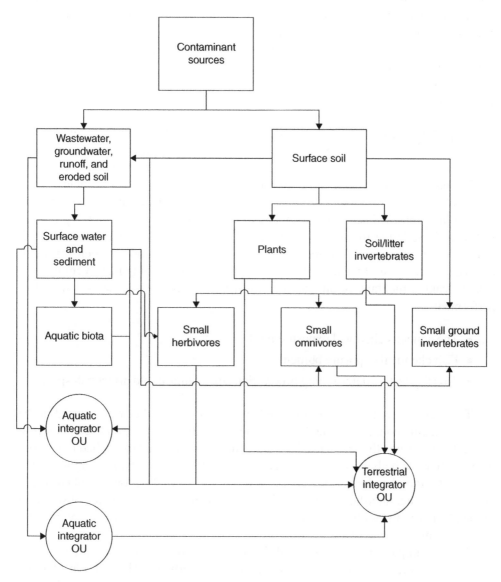

FIGURE 4.4   Example conceptual model for proposed PPI plant *Source*: Adapted from Suter (35).

In this ERA the three areas that will be assessed are the impact on fish in the Great River, the native plants growing on the proposed PPI plant site and the area that will be disturbed by the construction of the building and parking lot, and the animals inhabiting the PPI property.

4.1.16 Analysis Plan

The analysis plan for the PPI ERA follows the problem formulation. Therefore, in this example the analysis plan will concern the fish in the Great River, the terrestrial plants, and the terrestrial animals. These analyses can be very involved and demanding. In this example only a sampling of the types of analyses will be presented. In this regard, a partial analysis of the risk to fish will be presented.

4.1.17 Risks to Fish

In this partial analysis, the risk to fish will be examined to a limited degree. The fish are potentially exposed to contaminants in their natural environment, water. The contaminants in the water could come from upstream of the proposed plant, from the 30 communities that discharge treated water from sewage disposal plants, from farm runoff, from permitted and nonpermitted industrial plant discharge locations, and from military bases. The contaminants that potentially harm fish from the proposed PPI plant include the runoff from the ground that contained the farm chemicals. It could come from a process upset and a resulting spill of the industrial chemicals. It could come from a spill from a delivery truck and from something like an employee's vehicle leaking fuel tank or oil leak. In a full ERA, the following analyses would be performed:

- The aqueous chemical exposure model.
- Fish chemical exposure burdens.
- Toxicity tests – determining how various exposures could affect fish species.

These analyses can be quite involved and result in data that will be used in the decision making concerning the design of the plant. Because the proposed plant is to be located in an area that has had a large amount of farm chemicals applied to the soil, the potential fish body burden of farm chemicals in the fish population might already be high. It might then be considered that water runoff from the proposed plant would not pose a further significant health risk to the fish. However, since the construction of the plant will disrupt the soil, there could be a much higher potential to release the chemicals into the waterways via runoff.

As an example, the analysis found that the fish in the Great River already carry a high body burden of farm chemicals, especially Herbicide H. Addition of more chemicals into the river could harm not only the fish but also birds who feed on the fish.

4.1.18 Risk Characterization

Risk characterization is the final phase of an ERA and is the culmination of the planning, problem formulation, and analysis of predicted or observed adverse ecological effects related to the assessment endpoints. Completing risk characterization allows decision makers to clarify the relationships between stressors, effects, and ecological entities and to reach conclusions regarding the occurrence of exposure and the adversity of existing or anticipated effects. The analysts performing the ERA use the results of the analysis phase to develop an estimate of the risk posed to the ecological entities included in the assessment endpoints identified in problem formulation, in this case, fish in the Great River, terrestrial plants, and terrestrial animals. After estimating the risk, the assessor describes the risk estimate in the context of the significance of any adverse effects and lines of evidence supporting their likelihood. Finally, the analysts identify and summarize the uncertainties, assumptions, and qualifiers in the risk assessment and reports the findings to decision makers.

Since this is a proposed plant, the ERA could be used to help design the plant site to avoid or eliminate any threat found. For instance, it is apparent that disrupting the soil has the potential of releasing farm chemicals from soil. The analysis found that the fish already carry a high body burden of these chemicals. Therefore, something needs to be done to eliminate or reduce to the level possible the amount of chemicals that could be released from the soil. How can this be accomplished? Several things could be done:

- Remove the soil from the site and placing it in an approved landfill.
- Lay a clay pad first and then putting the excavated soil on the pad and then placing a clay berm around the site to contain the water runoff from the soil.
- Place a dike below the proposed site and collecting all the water runoff and processing it to remove the pesticides.

A cost–benefit analysis would need to be performed to help choose the best solution.

The ERA would then be updated to reflect the changes in the plant design.

For instance, if the proposed chemicals were changed so that they posed less of an ecological risk, then that aspect of the ERA would be modified. An example of this would be the elimination of the copper sulfate as a process chemical. A safer process using something other than copper sulfate would significantly reduce the ecological risk posed by this plant. The goal, as with every risk assessment, is to reduce the risk, in this case, the ecological risk.

4.2 SUMMARY

An ERA is a very systematic and very scientifically intensive analysis. In this chapter an abbreviated example was presented that provides an example of the types of analyses that are performed as a part of an ERA. Experienced ERA analysts should be employed to perform such an analysis to ensure every aspect of the ERA is performed appropriately.

Self-Check Questions

1. Ecological risk assessment is an analysis more difficult to conduct than other types of risk assessments. Provide five reasons why it is more difficult to conduct.

2. Why is the most sensitive organism used as the basis of an ecological risk assessment?

3. Pick an ecological disaster and explain how it could have been prevented.

4. Pick an ecological disaster and explain the long lasting effects of the disaster. These might be political or ecological.

5. What happened to the oil from the Deepwater Horizon event?

6. Heavy metal contamination in many ways is more hazardous to the environment than hydrocarbon contamination. Why?

REFERENCES

1. Scientific American (2009). The origin of oxygen in Earth's atmosphere. http://www.scientificamerican.com/article.cfm?id=origin-of-oxygen-in-atmosphere (accessed 22 August 2011).
2. Wikipedia. History of Earth. http://en.wikipedia.org/wiki/History_of_the_Earth (August 2018).
3. Raynolds, R.G. (ed.) (2008). *Roaming the Rocky Mountains and Environs: Geological Field Trips.* Boulder, CO: Geological Society of America.
4. Wikipedia (2018). Yeast. http://en.wikipedia.org/wiki/Yeast (accessed August 2018).
5. Environmental Protection Agency. Bunker Hill mining and metallurgical complex. http://www.class.uidaho.edu/kpgeorge/issues/bunkerhill/bunker.htm (accessed February 2019).
6. Wikipedia. Deepwater horizon. http://en.wikipedia.org/wiki/Deepwater_Horizon (accessed August 2011).
7. Cleveland, C. Posts by Cutler Cleveland. http://www.theenergywatch.com/author/cutler (accessed August 2010).
8. BP (2010). Deepwater Horizon Accident Investigation Report, September 2010.
9. Hagerty, C.L. and Ramseur, J.L. (2010). Deepwater horizon oil spill: selected issues for congress. http://www.energy.gov/open/oilspilldata.htm (accessed 4 March 2019).
10. Nola.com. Deepwater horizon. http://www.nola.com/news/gulf-oil-pill/index.ssf/2010/04/deepwater_horizon_rig_had_hist.html (accessed August 2011).
11. Washington Post. Report Says Oil Agency Ran Amok. http://www.washingtonpost.com/wp-dyn/content/article/2008/09/10/AR2008091001829.html (accessed August 2011).
12. Wikipedia. Love Canal. http://en.wikipedia.org/wiki/Love_Canal (accessed August 2011).
13. Smith, R.J. (1982). The risks of living near Love Canal. *Science* 212: 808–809, 811. "Controversy and confusion follow a report that the Love Canal area is no more hazardous than areas elsewhere in Niagara Falls."

14. Gibbs, L.M. (1998). *Love Canal: The Story Continues … .* Stony Creek, CT: New Society Publishers; Anv edition.
15. Wikipedia. Minamata disease. http://en.wikipedia.org/wiki/Minamata_disease (accessed August 2011).
16. Mishima, A. and Brown, L.R. (1992). *Bitter Sea: The Human Cost of Minamata Disease.* Tokyo: Kosei Publishing Company.
17. US Geological Survey. Mercury concentrations in streams found to go through daily cycles. http://toxics.usgs.gov/definitions/methylmercury.html (accessed August 2011).
18. Wikipedia. Agent Orange. http://en.wikipedia.org/wiki/Agent_Orange (accessed August 2011).
19. Wilcox, F.A. (2011). *Waiting for an Army to Die: The Tragedy of Agent Orange.* New York: Seven Stories Press.
20. Fund for Reconciliation and Development. Voices for Agent Orange victims. http://ffrd.org/Voices/AgentOrange.htm (accessed August 2011).
21. Stellman, J.M., Stellman, S.D., Christian, R. et al. *The extent and patterns of usage of Agent Orange and other herbicides in Vietnam. Nature* 422: 681.
22. Chiras, D.D. (2010). *Environmental Science*, 8e, 499. Burlington, MA: Jones & Bartlett Learning.
23. Vallero, D.A. (2007). *Biomedical Ethics for Engineers: Ethics and Decision Making in Biomedical and Biosystem Engineering*, 73. Academic Press http://books.google.com/books?id=AeT56Pi8LFYC&pg=PA73 (accessed 4 March 2019).
24. Department of Veterans Affairs Office of Public Health and Environmental Hazards (2010). Agent Orange: Diseases Associated with Agent Orange Exposure, 25 March 2010. http://www.publichealth.va.gov/exposures/agentorange/diseases.asp (accessed August 2011).
25. US Environmental Protection Agency. Dioxin. http://cfpub.epa.gov/ncea/CFM/nceaQFind.cfm?keyword=Dioxin (accessed September 2011).
26. Lenntech. Environmental disasters. http://www.lenntech.com/environmental-disasters.htm#ixzz1VtBe3lKH (accessed August 2011).
27. Wikipedia. Seveso disaster. http://en.wikipedia.org/wiki/Seveso_disaster (accessed August 2011).
28. Environmental Justice Atlas. The Jilin chemical plant explosions and Songhua River Pollution Incident, China. https://ejatlas.org/conflict/the-jilin-chemical-plant-explosions-songhua-river-pollution-incident (accessed August 2011)
29. Nat Green (2011). Positive Spillover? Impact of the Songhua River Benzene Incident on China's Environmental Policy. https://www.wilsoncenter.org/publication/positive-spillover-impact-the-songhua-river-benzene-incident-china-s-environmental (accessed August 2011).
30. Parsons Behle & Latimer. The Songhua River Spill: China's pollution crisis. https://parsonsbehle.com/publications/the-songhua-river-spill-china-s-pollution-crisis (accessed August 2011).
31. Hernan, R.E., Graham Nash (Preface), Bill McKibben (Foreword) (2010). *This Borrowed Earth: Lessons from the Fifteen Worst Environmental Disasters Around the World.* New York: Palgrave Macmillan.
32. Environmental Protection Agency (1998). Guidelines for Ecological Risk Assessment PA/630/R-95/002F, April 1998.
33. US Environmental Protection Agency. Get the latest news on Superfund. http://www.epa.gov/superfund/programs/nrd/era.htm (accessed 22 August 2011).
34. Suter, G.W. II (2006). *Ecological Risk Assessment*, 2e. CRC Press.
35. Suter, G. (1996). *ES/ER/TM-186, Guide for Developing Conceptual Models for Ecological Risk Assessments.* Washington, DC: United States Department of Energy.

CHAPTER 5

Task Analysis Techniques

5.1 WHAT IS TASK ANALYSIS?

A task analysis is any process for assessing what a user does (task), how the task is organized, and why it is done in a particular way and using this information to design a new system or analyze an existing system. Task analysis is an investigative process of the interaction between operators and the equipment and/or machines they utilize. It is the process of assessing and evaluating all observable tasks and then breaking those tasks into functional units. These units allow for the evaluators to develop design elements and appropriate training procedures and identify potential hazards and risks.

Task analysis has been defined as "the study of what an operator (or team of operators) is required to do, in terms of actions and/or cognitive processes, to achieve a system goal" (1). Approaches to task analysis have been classified into three categories: normative, descriptive, and formative. According to Vicente (2), normative approaches "prescribe how a system should behave," descriptive approaches "describe how a system actually works in practice," and formative (also called predictive) approaches "specify the requirements that must be satisfied so that the system could behave in a new, desired way." The nature of the work domain and the task influences the type of analysis that is appropriate. Both the normative and descriptive approaches are applicable to analyzing existing systems, while the formative approach can be applied to developing and designing a new system that will support work that has not previously been done or to allow the work to be done in a new way. The normative approach is appropriate for a very mechanical and predictable work environment; as the work becomes more complex and unpredictable, requiring more judgment, the formative approach becomes a better analysis tool (3).

Risk Assessment: Tools, Techniques, and Their Applications, Second Edition. Lee T. Ostrom and Cheryl A. Wilhelmsen.
© 2019 John Wiley & Sons, Inc. Published 2019 by John Wiley & Sons, Inc.
Companion website: www.wiley.com/go/Ostrom/RiskAssessment_2e

5.2 WHY A TASK ANALYSIS?

Evaluation and design of a system using task analysis effectively integrate the human element into the system design and/or operation. In utilizing this analysis in system design, you effectively consider the human as a component of the system to ensure efficient and safe operation. To be effective, the task analysis must look at the system as comprising three interrelated components: human operator, equipment, and environment. This systematic analysis of the tasks results in equipment that is safer to use, easier to maintain, and operated using effective procedures. Through a task analysis, you build a concrete and thorough description of the task and can attain a clear definition of what other factors contribute to the system.

Task analysis provides a great deal of information and insight into possible areas for performance improvement in systems, equipment, and workers. It can also aid in identifying areas of waste, in terms of time, resources, production, and dollars. A task analysis identifies weaknesses in the production process and in human performance. It identifies both what is working well and what is not working within the current organization (4). And a task analysis can provide knowledge of the tasks that the user wishes to perform. Therefore, it can be a reference in which the value of the system functions and features can be tested (3).

5.3 WHEN TO USE TASK ANALYSIS?

Ideally, a task analysis should be used when designing the system, procedures, and training. Conducting a task analysis as part of the design of any system or process should be one of the focuses. Those systems that include humans as operators, maintainers, or support personnel should have each of the areas represented in the analysis phase in order to incorporate the capabilities and limitations of the system. This should be done in the analysis phase. Performing a task analysis early in the design and with the user's input helps to eliminate rework down the road. It saves dollars and time.

The overall design process, as well as the task analysis, is considered iterative processes. In other words, design specs are presented, and the tasks needed to complete the job are incorporated into the design process with continual changes until the final design documents are approved. The user's input is valuable, because they are the ones performing the tasks and have the knowledge as to the best ways and resources to complete those tasks. There is always compromise, but the fact remains that the task analysis feeds into the design for a successful overall system.

After the results of the task analysis are incorporated into the system design, it is necessary to perform the analysis again to ensure that the changes do not produce an unforeseen consequence. In addition to providing useful information to incorporate into the design of the system, task analysis information can be used to develop and improve the personnel and training requirements. Task analysis can be used to evaluate an existing system. If a new piece of equipment is added or a problem has been identified, a task analysis can be used to improve the system.

Task analyses are useful in many industries such as the auto industry along with the maintenance of the automobiles, manufacturing and assembly processes, and business and financial processes and medical and aviation industries.

5.4 TASK ANALYSIS PROCESS

There are hundreds of task analysis techniques with advantages and disadvantages; therefore, you may want to use a mix of task analysis methods during your assessment process.

5.4.1 Step 1: Data Collection Information

Identifying the focus of the analysis helps in deciding what type of information should be collected and how it should be gathered. Focus on the system and what the results will be used for, such as in the design of a new system, modifying an existing system or developing training.

Develop a list of tasks and positions associated with the overall job. Provide descriptions of each task that characterize the task according to issues of potential importance to the system, process, or job. How complex the system may be and the eventual application of the results will determine the level of analysis to be performed.

A system is composed of a job or jobs that ultimately lead to a common goal. Those jobs can be broken into tasks that must be performed in order to complete the job. The tasks can then be broken down into steps or what is referred to as subtasks. Those steps need to be performed in order to accomplish the overall task. Each task or subtask can be defined differently depending on the application. It is very important that consistency be applied within a given analysis.

Table 5.1 lists types of information that might be required in order to complete a job.

Earlier we talked about the design and the analysis process as being iterative. The data collected concerning the job and the tasks are broken down into subtasks and once reviewed, it might be necessary to then collect more details and information.

5.4.2 Step 2: Recording the Data

Some of the methods for collecting information in a task analysis can be shown in Table 5.2.

Document the information gathered utilizing the methods suggested in Table 5.2. The data needs to be compiled in a format that could be used to analyze and process the information. A simple and straightforward format that can be used to organize and record the collected data is a column format (see Table 5.3).

TABLE 5.1
Required Types of Information

Task information
Identify tasks of involved with the process, system, or training
Identify subtasks involved with the tasks
Group and organize the subtasks involved in a task
Identify which subtasks have commonalities and are linked to each other
Identify the priority based on criticality of the subtasks
Identify how frequently the subtasks occur under different conditions
Identify the sequence of the subtasks, depending on the occurrence
Identify which subtasks should be executed based on the event or a decision made in a previous subtask or even task
Identify the goals or objectives for each subtask
Identify the knowledge the user has on how the system functions (i.e. tasks, subtasks)

TABLE 5.2
Methods of Collecting Information

Data collection methods
Observe and record information on the worker(s) performing the job. This can be done as job shadowing or just taking notes as the worker(s) complete their duties
Interview the worker(s) by asking questions about their job. These questions can be open-ended questions or specific questions related to the specific task. The best way to collect this information is to allow the worker(s) to remain anonymous. This helps them to be more honest knowing the management will not be able to identify them specifically
Review existing documentation such as standard operating procedures (SOPs), safety and injury reports, training manuals, and any other previous surveys or analyses
Checklists may be used to identify workplace concerns, human system interfaces, ergonomics issues, environmental issues, and any other workplace concerns
Surveys or questionnaires may be used to collect the worker(s) views of the system or tasks
Videotape or record the worker(s) performing the job or specific tasks involved in completing the job

Other examples of formats are hierarchical diagrams, operational sequence diagrams, and timelines. Breaking down the tasks into subtasks can be accomplished by the use of a hierarchical format. The general tasks are listed at the top of the diagram. Detailed subtasks that comprise each task are illustrated by branching off from the appropriate task box (Figure 5.1).

Timelines are used to define not only the sequence of steps that make up the task but also the time that they occur and duration. This presentation is particularly useful when there are several workers and machines interacting and can help to identify when the worker and/or machines are being overloaded or underloaded during the completion of a task (Figure 5.2).

The use of operational sequence diagrams shows the sequence of steps and the relationships between them in completing the task. This method requires making a

TABLE 5.3

Examples of Data Recording

Subtasks	Comments	Tools required	Feedback and displays
Remove panel A	Panel A is 40 in from the ground and is unobstructed. The panel is removed by unscrewing six sheet metal screws	Phillips screwdriver	Panel is clearly labeled on both sides
Close fuel feed valves B1 and B2	Fuel feed valves B1 and B2 are brass, quarter turn valves. They are clearly visible and located on either side of the fuel filter	None	The valves are open when the stem are in-line with the fuel line and are closed when the stem is perpendicular to the fuel line. The valves are clearly labeled
Remove fuel filter B	The fuel filter is clear and is mounted on the structure by a spring clamp. The fuel hoses slip over the nozzles of the filter and are held by spring clamps	None	The spring clamps are hand operated and snap either open or closed

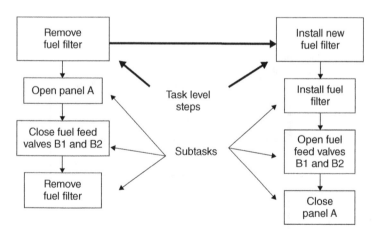

FIGURE 5.1   Hierarchical task analysis diagram.

flowchart of the task using standard symbols to present the information (Figure 5.3). Note that the example provided is very basic.

5.4.3 Step 3: Data Analysis

There are many ways to use the data obtained from task analyses. Table 5.4 below lists some of the techniques that can be used with task analysis data that can also aid in

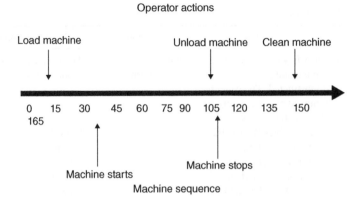

FIGURE 5.2   Timeline.

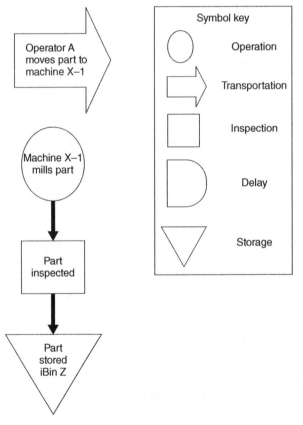

FIGURE 5.3   Operational sequence diagram.

TABLE 5.4
Analysis Techniques

Technique

Hierarchical task analysis is a broad approach used to represent the relationship between the tasks and the subtasks. This is a useful approach in documenting the system requirements as well as the ordering of the tasks

Link analysis identifies the relationships between the components of a system and represents the links between those components

Operations sequence diagrams identify the order in which the tasks are performed and identify the relations between the person, equipment, and the time

Timeline analysis is used to match up the process performance over time, which includes the task frequency, interactions with the other tasks, the worker(s), and the duration of the task

risk analyses. It is beyond the scope of this guide to discuss these techniques beyond this table.

Self-Check Questions

1. What are the three categories described in task analysis?

2. What must the task analysis look at to be effective?

3. When should a task analysis be used?

4. What are some of the types of information that might be required in order to complete a job?

REFERENCES

1. Ainsworth, B.K. (1992). *A Guide to Task Analysis*. London: Taylor & Francis.
2. Vicente, K.J. (1999). *Cognitive Work Analysis: Toward Safe, Productive and Healthy Computer-Based Work*. Mahwah, NJ: Lawrence Erlbaum Associates.
3. Salvendy, G. (1997). *Handbook of Human Factors and Ergonomics*, 2e. New York: Wiley.
4. Kirwan, B. and Ainsworth, L.K. (eds.) (1992). *A Guide to Task Analysis: The Task Analysis Working Group*, 1e. CRC Press.

CHAPTER 6

Preliminary Hazard Analysis

According to the Consumer Product Safety Commission (CPSC), there are over 21 000 deaths and approximately 28 million injuries associated with the 15 000 products under the commission's jurisdiction (1). The death of any child is an almost unbearable event for a family and an infant death is even more so. occurrence of an unexpected release of hazardous occurrence of an unexpected release of hazardous Infants die each year due to poor product designs. Deaths associated with the designs of cribs are one of the issues the CPSC deals with. One of the most common causes of infant death due to crib design is when infants get their head through the bars of the crib and then becomes strangled when they cannot remove their head. This happens because of two reasons: (i) an infant's head is small and (ii) an infant's skull is still soft and can deform, which allows the head to pass through the bars. Imagine the shock and horror of young parents when they discover that the infant they love has died because his/her head passed through the bars of the crib and was subsequently strangled to death.

Manufacturers do not deliberately produce products that can hurt children. However, when a product designer does not consider all the possible consequences of their design, fatal accidents can and will happen.

Preliminary hazard analysis (PHA) will be the focus of this chapter. Several applications of this tool will be explored. The partial analysis of a piece of infant furniture will be one example.

6.1 DESCRIPTION

The PHA technique was developed by the US Army and is listed in the US Military standard system safety program requirements (MIL-STD-882B). PHAs have been

Risk Assessment: Tools, Techniques, and Their Applications, Second Edition. Lee T. Ostrom and Cheryl A. Wilhelmsen.
© 2019 John Wiley & Sons, Inc. Published 2019 by John Wiley & Sons, Inc.
Companion website: www.wiley.com/go/Ostrom/RiskAssessment_2e

proven to be cost-effective in identifying hazards in the beginning of a conceptual design phase. Because of its military legacy, the PHA technique is sometimes used to review process areas where energy can be released in an uncontrolled manner.

The main purpose of a PHA is to identify the hazardous states of a system and its implications. In order to obtain maximum benefit of a PHA, it should occur as early as possible in the system life cycle. Tasks and requirements involved in preparing a PHA should include the following:

- **Establish for purpose of the analysis:**
  - Boundaries between the system, any system with which it interacts, and the domain.
  - Overall system structure and functionality.
- **Identify:**
  - Detailed list of hazards of system based on preliminary hazards list report and the requirements.
  - Update hazard list.
  - Accidents to the most practicable extent.
  - Events of accident sequence and those that can be discounted.
  - Record in hazards list.
- **Assign:**
  - Each accident a severity categorization and each accident sequence a predicted qualitative/quantitative probability.
  - Each hazard a preliminary random and systematic probability target.
- **Document:**
  - Any safety features that are to be implemented during the design and development phase.

6.1.1 Process of Preliminary Hazard Analysis

Hazard analysis is usually performed during the early stages of design, but it also can take place in different stages of the life cycle of processes and facilities.

- For new processes and facilities, at the conceptual (work) design phase, PHAs are performed to identify opportunities to eliminate or reduce hazards, before resources are committed to engineering design and construction.
- During engineering design and construction, design hazard analyses are performed to identify needed system changes or process controls not identified at the conceptual design phase.
- Before initial start-up of a new system or process, pre-start-up or operational readiness reviews are conducted to ensure that systems are in place to control all identified hazards.

- Process hazard analysis are conducted periodically during the lifetime of operating facilities/processes and every time a process/facility undergoes significant modification, to identify any new hazards that may have occurred from a process change and to also ensure that all hazards are adequately controlled.

- When dealing with the decontamination and decommissioning (D&D) and environmental restoration projects, hazard characterizations are conducted to characterize hazards and, when possible, to develop and define prioritized rankings for hazard elimination.

- As in any process and especially during D&D, hazards can change constantly, which requires hazards to be identified and controlled.

The major goal of a preliminary hazards analysis is the ability to identify and characterize possible known hazards in the beginning of a design phase. Partial lists of those hazards are listed below:

- Raw material, intermediate and final products, and their reactivity.
- Plant equipment.
- Facility layout.
- Operating environment.
- Operational activities (testing, maintenance, etc.).
- Interface among system components.

PHA identifies such known hazards as explosions, radioactive sources, pressure vessels or lines, toxic materials, high voltage, machinery, etc. It specifies where each hazard will occur, their significance, and the method that will be used to eliminate the hazards or how the associated risk will be controlled.

The probability of occurrence of an unexpected release of hazardous energy or material (an accident) that determines its credibility for the purpose of PHA is as follows:

Frequency of Occurrence (Per Year)

Less than 10^{-6} $\begin{cases} \text{Less than credible: Events are expected not to occur} \\ \quad \text{during the life cycle of the facility.} \end{cases}$

10^{-6} to 10^{-4} $\begin{cases} \text{Credible and extremely unlikely: Events will probably} \\ \quad \text{not occur during the life cycle of the facility.} \end{cases}$

10^{-4} to 10^{-2} $\begin{cases} \text{Credible and unlikely: Events may occur once during} \\ \text{the life cycle of the facility,} \\ \text{e.g. natural phenomena and trained worker error.} \end{cases}$

$$10^{-2} \text{ to } 10^{-1} \begin{cases} \text{Likely: Events may occur several times during the life cycle} \\ \qquad \text{of the facility, e.g. general worker error.} \\ \text{Very likely: Events may often occur,} \\ \qquad \text{e.g. back strains and abrasions.} \end{cases}$$

For most nonnuclear facilities, qualitative determinations of credibility are all that is necessary, while nuclear facilities require quantitative determinations of credibility. Quantitative measures:

$$0.1 \text{ to } 10 \text{ per year}$$

$$0.001 \text{ to } 1 \text{ per year}$$

$$\left. \begin{array}{l} 0.000\ 01 \text{ to } 0.001 \text{ per year} \\ 0.000\ 001 \text{ to } 0.000\ 01 \text{ per year} \\ 0.000\ 000\ 1 \text{ to } 0.000\ 001 \text{ per year} \end{array} \right\} \text{Rare events}$$

A PHA seeks to rank hazards in a qualitative measurement of the worst potential consequence resulting either from personnel error, environmental conditions, design inadequacies, procedural deficiencies, and system, subsystem, and component failure or malfunction. The categories are defined as follows:

- Class I Hazards (Negligible): a hazardous occurrence in which the worst-case effects could cause less than minor injury, occupational illness, or system damage.
- Class II Hazards (Marginal Effects): a hazardous occurrence in which the worst-case effects could cause minor occupational illness or system damage.
- Class III Hazards (Critical): a hazardous occurrence in which the worst-case effects will cause severe (non-disabling) personnel injury, severe occupational illness, or major property or system damage.
- Class IV Hazards (Catastrophic): a hazardous occurrence in which the worst-case effects will cause death, disabling personnel injury, or facility or system loss.

The most reliable solution when identifying PHAs is to eliminate the source or cause of the hazards. If the source or cause cannot be eliminated, the hazard should be reduced or redesigned as much as possible such as:

- Redesigning, changing, or substituting equipment to remove the sources (i.e. excessive temperature, noise, or pressure).
- Redesigning a process to possibly use less toxic chemicals.

- Redesigning a workstation to relieve physical stress and remove ergonomic hazards.

- Designing general ventilation with sufficient fresh outdoor air to improve air quality and generally to provide a safe, healthful atmosphere.

Once a PHA has been established, a preliminary hazard list needs to include details of the hazards and serves as the central reference that documents the safety characteristics of the system being analyzed. A hazard list/matrix should be kept for the entire life cycle of the system. Table 6.1 shows an abbreviated example of a hazard list. The structure of a hazard list should be as follows:

- Complete description of system and scope of use. This should also include references that identify unique system identifiers.

- Reference list to the systems safety requirements.

- Accident severity categories, probability categories, equivalent numerical probabilities, and accident risk classification scheme for the system.

There are numerous software programs specifically designed for conducting PHAs. Standard word processors and spreadsheets can assist with documenting the PHA results. Uses of flowchart diagrams and process modeling as discussed in Chapter 5 are also beneficial when developing PHAs. Table 6.2 presents a template for a PHA. Table 6.3 shows common hazards to be considered.

Consider a design concept that feeds phosgene ($COCl_2$) from a cylinder to a process unit. At this stage of the design, the analyst knows only that this material will be used in the process, nothing more. The analyst recognizes phosgene has toxic properties, and the analyst identifies the potential release of phosgene as a hazardous situation. The analyst lists the following causes for such a release:

- The pressurized storage cylinder leaks or ruptures.

- The process does not consume all of the Phosgene.

- The phosgene process supply lines leak/rupture.

- A leak occurs during connection of a cylinder to the process.

The analyst then determines the effects of these causes. In this case, fatalities could result from large releases. The next task is to provide guidance and design criteria by describing corrective/preventive measures for each possible release. For example, the analyst might suggest that the designer:

- Consider a process that stores alternative, less toxic materials that can generate phosgene as needed.

- Consider developing a system to collect and destroy excess phosgene from the process.

TABLE 6.1
Example Hazard List for a Home Propane Grill

System	Sub system	Reference list	Codes and standards	Accident categories	Relative likelihood of occurrence	Accident risk classification
Propane grill	Propane tank	Model 231	National Fire Protection Association's Pamphlet 58 - LP-Gas Code, 1998 Edition (Code National Fire Protection Association (NFPA) 58, *LP-Gas Code* (2001 edition), §§2.1.5 [DOT cylinders]; 2.2.1.6 [Cylinder in fire]; 2.2.6.4 [Label]	Explosion	Moderate	Moderate
				Valve failure	Moderate	Low
			29 Code of Federal Regulations §1910.1200, Hazard Communication			
			49 Code of Federal Regulations §173.301 [Specification, Inspection]			
	Burner assembly		American National Standards Institute (ANSI) standard for outdoor cooking gas appliances Z21.58-1995 83EK	Uncontrolled flame	Low	Low

TABLE 6.2
Example Table for a Preliminary Hazard Analysis

Subsystem	Mode	Hazardous element	Event causing hazardous condition	Hazardous condition	Event causing potential accident	Potential accident	Effect	Hazard class	Accident prevention measure
Hardware or functional element being analyzed	Applicable system phase or mode of operation	Hazardous energy system	Can be personnel error, deficient or inadequate design, or malfunction	Situation that occurs because of a combination of events	Undesired event or faults that could cause the hazardous condition to become the identified potential accident	Any potential accident that could result	Possible effects of the potential accident (list everything possible and weed out non-credible items)	Qualitative measure of significance for the potential effect. Class I – Negligible Class II – Marginal effects Class III – Critical Class IV – Catastrophic	Recommended preventative measures (i.e. hardware, procedures, personnel)

- Provides a plant warning system for phosgene releases.
- Minimize on-site storage of phosgene without requiring excessive delivery/handling.
- Develop a procedure using human factors engineering expertise for storage cylinder connection.
- Consider a cylinder enclosure with a water deluge system that is triggered by phosgene leak detectors.
- Locate the storage cylinder for easy delivery access but away from other plant traffic.
- Develop a training program to be presented to all employees before start-up (and subsequently to all new employees) on phosgene effects and emergency procedures.

6.1.2 Examples of Hazardous Energy Sources

TABLE 6.3
Examples of Hazardous Energy Sources

Chemical energy	Electrical energy	Thermal energy
Corrosive materials Flammable materials Toxic materials Reactive material Oxygen deficiency Carcinogens	Capacitors Transformers Batteries Exposed conductors Static electricity	Steam Fire Friction Chemical reaction Spontaneous combustion Cryogenic materials Ice, snow, wind, rain
Radiant energy	**Kinetic energy**	**Pressure energy**
Intense light Lasers Ultraviolet X-rays, Infrared sources Electron beam Magnetic fields RF fields Nuclear criticality High energy particles	Pulley, belts, gears Shears, sharp edges Pinch points Vehicles Mass in motion	Confined gases Explosives Noise
	Potential energy	**Biological energy**
	Falling Falling objects Lifting Tripping, slipping Earthquakes	Pathogens (virus, bacteria, etc.) Allergens

6.2 USING PHA FOR PROCEDURE DESIGN

PHA is most effectively used during the initial development of a process and the procedures for performing that process. Also, it has utility when updating/changing a process and its procedure. The following example illustrates its use for analyzing a maintenance process/procedure.

6.2.1 Purpose of Process

Flushing an automobile cooling system.

6.2.2 Initial Basic Procedure Steps

1. Begin with the engine cold and ignition off. Remove the radiator pressure cap.
2. Open the petcock at the bottom of the radiator and drain the coolant into a bucket.
3. Close the petcock and fill the radiator with water.
4. Start the engine and turn the heater control to Hot. Add cooling system cleaner, and idle the engine for 30 minutes (or as per the instructions on container).
5. Stop the engine and allow it to cool for five minutes. Drain the system.
6. Close the petcock, fill the radiator with water, and let the engine idle for five minutes.
7. Repeat step No.5. Close the petcock.
8. Install new 50/50 mixture of water and ethylene glycol antifreeze/coolant.

6.2.3 Analysis

Table 6.4 shows the analysis of the procedure. Note that in a PHA the idea is to list the hazards and not to analyze each step of the procedure. Failure mode and effect analysis and failure mode, effect, and criticality analysis are better suited for analyzing the steps of the procedure.

6.2.4 Using the Results of the Analysis

The procedure has now been revised based on the findings from the analysis. The new procedure reads:

Warning: Cooling system must be below 100 °F prior to draining.
1. Begin with the engine cold and ignition off. Remove the radiator pressure cap.

TABLE 6.4
Analysis of Procedure

Subsystem	Mode	Hazardous element	Event causing hazardous condition	Hazardous condition	Event causing potential accident	Potential accident	Effect	Hazard class	Accident prevention measure
Cooling system	Drain, flush, and refill	Heat and pressure	Opening radiator cap or petcock before cooling system has sufficiently cooled	Release of hot coolant, water, and steam	Mechanic contacting hot coolant, water, and steam	Mechanic in contact with hot coolant, water, or steam	Severe burns	Class III	Ensure cooling system is less than 100 degrees F prior to removing radiator cap in step 1
		Ethylene glycol coolant	Not properly storing coolant	Uncontrolled toxic material	Child or animal could consume coolant	Child or animal consuming coolant	Illness due to ethylene glycol poisoning or death	Class IV	Replace ethylene glycol coolant with safer chemicals
			Not properly disposing of coolant						

Note that there are two primary hazards associated with flushing the cooling system: the heat and pressure associated with the hot coolant and the toxic nature of the coolant itself. The accident preventive measures in respect to the procedure will result in the addition of two warnings and the substitution of ethylene glycol with a nontoxic coolant.

Warning: Ethylene glycol coolant is toxic and must be disposed of in an appropriate manner.

2. Open the petcock at the bottom of the radiator and drain the coolant into a bucket.

3. Close the petcock and fill the radiator with water.

4. Start the engine and turn the heater control to Hot. Add cooling system cleaner, and idle the engine for 30 minutes (or as per the instructions on container).

Warning: Cooling system must be below 100 °F prior to draining.

5. Stop the engine and allow it to cool for five minutes. Drain the system.

6. Close the petcock, fill the radiator with water, and let the engine idle for five minutes.

7. Repeat step No.5. Close the petcock.

8. Install new 50/50 mixture of water and nontoxic antifreeze/coolant.

6.3 USING PHA FOR PRELIMINARY PRODUCT DESIGN

At the beginning of the chapter, we discussed the injuries that can occur with poor design of infant furniture. We will now develop a partial PHA for a proposed piece of infant furniture, a crib, using the PHA technique. Figure 6.1 shows an example crib.

The steps we will use to do the analysis are as follows:

1. Determine function(s) of the crib.
2. Determine required specifications for the crib.
3. Determine major systems/subsystems of the crib.
4. Determine important components of each system/subsystem of the crib.
5. Determine operating modes of the crib.
6. Determine hazards associated with each system/subsystem of the crib in the identified operating modes.
7. Determine accident conditions.
8. Determine potential consequences.
9. Determine hazard class, if appropriate.
10. Determine preventative/mitigating measures.

To accomplish the PHA, we do the steps in sequence.

Step 1: Determine Functions of the Crib

The primary function of the crib is a place for the infant to sleep. There are several secondary functions and/or unintended functions that still can impact the safety of the crib:

Changing table
Storage area of baby toys
Step 2: Determine Required/Critical Specifications
The required/critical specifications of the crib can come from regulations, case law, quality standards, previous accidents, manufacturing requirements, style requirements, and company standards. These types of requirements include the following:

- Hold up to 100 lb of bedding, infant, toys, and weight of caregiver as they lean against the device.
- Safe.
- Nontoxic coatings.
- Nonflammable.
- Waterproof mattress.

Step 3: Determine Major Systems/Subsystems of the Crib
The major system is the crib itself. The subsystems include:

- Legs
- Rails/end pieces
- Mattress
- Mattress support
- Side rail lowering mechanism
- Casters

Step 4: Determine Important Components of Each System/Subsystem
This can be a very tedious task and is where many errors can occur in the design or safety analysis of a system. Just what is important? It can be a critical bolt, screw, line of code, indicator light, etc. That is one of the reasons for the analysis in the first place. In a crib it can be every part. The plastic sleeve on the top rail of the sides can be critical, the side rail locking mechanism, the casters, the mattress, and so on. In fact, the CPSC recently issued a bulletin on the hazards of crib mattress supports (2). Therefore, in a device that has been associated with so many deaths, a thorough analysis of all components should be performed.
Step 5: Determine Operating Modes of the Crib
The operating modes, per se, are listed below:

- Infant in crib
- Placing infant into crib
- Removing infant from crib

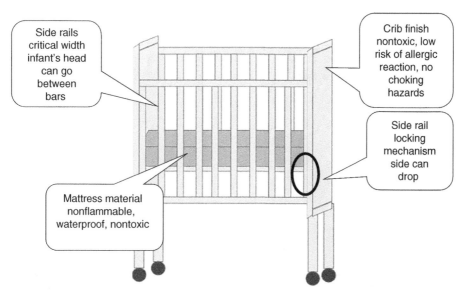

FIGURE 6.1   Crib design.

- Cleaning
- Assembly/disassembly

The remaining steps are performed in sequence in relation to the system, required/critical specification, subsystem, component, and operating mode as shown in Table 6.5 below. What we feel is always the most critical step is doing something about the hazards that are found. This is step 10 in this process – determine preventative/mitigating measures. In the case of a crib design, one would want to eliminate any potential hazards. However, in other industries mitigating measures might be all that is needed.

The following is the hierarchy of hazard control technologies (best to least desirable):

1. Elimination of the hazard by design or the substitution of a less hazardous process/chemical.
2. Engineering controls. Installing a barricade, machine guarding, lifting device, etc.
3. Administrative control. These are procedures, warnings, stay times in radiation zones, and signage.
4. Personnel protective equipment (PPE). Examples are safety glasses, boots, anti-contamination suits, etc.

TABLE 6.5
Partial PHA of Infant Crib

Subsystem	Mode	Hazardous element	Event causing hazardous condition	Hazardous condition	Event causing potential accident	Potential accident	Effect	Hazard class	Accident prevention measure
Side rails	Infant in crib	Infant head through rails	The infant has inadvertently slipped his/her head through the rails	Potential strangulation	Infant unable to remove head from between rails and becomes exhausted from struggle	Infant is strangled	Death	Class IV	Ensure distance between rails is less than 1 percentile infant head width at narrowest point
		Infant chews on crib	Infant ingests crib finish	Poisoning, allergic reaction, or Choking	Infant chews on finish of rails and ingests finish or material	Infant is poisoned by the finish	Death		Use nontoxic finish on the crib that has a low potential for allergic reaction
						Infant suffers allergic reaction from the finish	Rash		Construct crib from materials that do not create choking hazards when chewed on by infants
						Infant chokes on pieces of the crib	Death		

6.4 SUMMARY

PHA is a great tool for beginning to understand the hazards of a system. In some cases, a PHA is all that is needed to analyze a simple system. It is also the first step in the hazard analysis of more complicated systems. PHAs are usually done during the design phase of a system. However, there are no rules that say it cannot be used after a system has been put in place and used to eliminate hazards in existing systems. As with any of the tools we discuss in this book, determine the hazards and then do something about them. The job is not complete until the hazards are eliminated or mitigated. The analysis tools are a means to an end, not the end themselves.

Case Study

Mt. Everest is the highest mountain on planet Earth. However, some consider Mauna Kea to be a taller mountain in that it rises approximately 32 808 from the bottom of the ocean to its summit. On the other hand, Mt. Everest 29 029 ft from mean sea level into the atmosphere. Figure 6.2 shows Mt. Everest in the Himalayan Mountains.

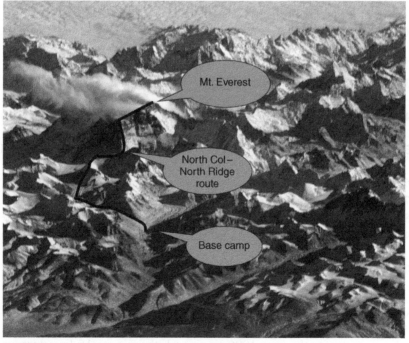

FIGURE 6.2   Mt. Everest in the Himalayan Mountains.

An article on the Independent web page listed 60 facts about Mt. Everest (3). Here is an abbreviated list:

1. The officially recognized height of Mount Everest is 29 029 ft (8 848 m), based on a 1954 ground-based measurement. A disputed satellite-based measurement in 1999 suggested it was 6 ft taller.
2. There are two main routes to the summit: the south-east ridge from Nepal and the north ridge from Tibet.
3. At Everest's highest point, you are breathing in a third of the amount of oxygen you would normally breathe due to the atmospheric pressure.
4. At least one person has died on Everest every year since 1969, except in 1977.
5. Sir Edmund Hillary's son, Peter, has climbed Everest five times. His first summit was in 1990.
6. Tenzing Norgay unsuccessfully tried to get to the top of Everest six times before reaching it with Hillary.
7. Winds on the mountain have been recorded at more than 200 mph.
8. Comparatively, the safest year on Everest was 1993, when 129 reached the summit and eight died (a ratio of 16 : 1).
9. The deadliest year for climbers of Everest was 1996, when 15 died.
10. One in 10 successful climbs to the summit ends in death.
11. The fastest descent was made in 11 minutes: Frenchman Jean-Marc Boivin paraglided down in 1988.
12. There are estimated to be 120 dead bodies on the mountain.
13. The youngest person to reach the top is Jordan Romero, who made it, aged 13, in 2010.
14. The Nepalese call it Sagarmatha, meaning "Forehead (or Goddess) of the Sky." In Tibet it is known as Chomolungma, "Mother Goddess of the Universe."
15. In 1865 the mountain was renamed in honour of the Surveyor General of India George Everest, from its original name of Peak 15.
16. The first woman to climb Everest was Junko Tabei, from Japan, in 1975.
17. A Nepalese government permit to climb Everest can cost up to £17 000.
18. There are 18 named climbing routes on Everest.
19. The most dangerous area on the mountain is Khumbu icefall, which is thought to have claimed 19 lives.
20. Climbers burn 20 000 calories on the day of the summit climb and an average of 10 000 a day on the rest of the climb.

10 May 1996 is remembered for the deadly storm that took the lives of eight mountain climbers (4–6). It is recorded as the worst loss of lives in a single day in Mt. Everest. As for the year 1996 in general, there were 15 lives lost on this mountain over spring as a result of 17 uncommon expeditions. Despite this tragedy,

98 mountain climbers made it to the peak safely. Mt. Everest is the world's tallest mountain, making it the most sought-out peak for mountain climbers.

The death toll has reached as high as 91 in 1980–2002. This glorious peak was first overcome by two men in 1953 by the names of Edmund Hilary and Tenzing Norgay. Mt. Everest peaks at approximately 29 029 ft high; at 26 000 ft oxygen levels drop drastically. With a success rate of 67%, only 8 000 of the 12 000 attempts up the mountain have been successful.

In 10 May, there were four separate expeditions scheduled to take place. As the groups were preparing to make their way back down the Himalayan Mountain, a dangerous powerful blizzard began and made it difficult for even experienced mountain climbers to climb down. This made them vulnerable with depleted oxygen supply and conditions were frigid at a −30° temperature. Typically when a blizzard is bad enough where it's hard to see and becomes a "whiteout," it is best to stay, but staying put wasn't much of an option on Mt. Everest because of the cold temperature capable of freezing you to death. Strong wind conditions were also a factor; winds up to 100 mph made it even more difficult to return to base camp. Despite the dangerous inclement weather, the teams of mountain climbers persisted on reaching the top of the Himalayan peak. It was said that the disastrous journey was a result of expeditionary teams competing among each other. One team by the name of Mountain Madness was headed by American Scott Fischer. They were competing with Adventure Consultants, led by Rob Hall from New Zealand.

In a 1995 expedition, Hall turned around as a precautionary measure to ensure the safety of his team. In this 1996 expedition, he continued climbing and refused to turn around when he saw Fischer taking his team up the summit. It was reported that Rob Hall wanted to remain as the world leader in professionally organized expeditions and felt Scott Fischer was a threatening competition. Besides the pressure of seeing the opposite team on the mountain, there was also pressure to continue due to expedition cost. This journey alone costed a little over $59 000, motivating members to keep climbing.

The fatalities of the 1996 Mt. Everest expedition could have certainly been avoided. As the team leaders, both Rob Hall and Scott Fischer could have avoided taking their team into the storm. The decision to turn around for safety was disregarded as both teams continued up Mt. Everest. As shown in this situation, competition and ego driven decisions in dangerous environments can lead to disaster.

Assignment

From the information provided and what you can find, develop a PHA for climbing Mt Everest. Develop the hazards, relative ranking of the hazards, and potential mitigation efforts.

REFERENCES

1. Center for Chemical Process Safety (CCPS) (1992). *Guidelines for Hazard Evaluation Procedures, Second Edition with Worked Examples*, Publication G18. New York: American Institute of Chemical Engineers.
2. Mil Std 882B Preliminary hazard analysis (Internet Access). https://www.system-safety.org/Documents/MIL-STD-882B.pdf (accessed August 2018).
3. Conquering Everest: 60 facts about the world's tallest mountain. https://www.independent.co.uk/environment/nature/conquering-everest-60-facts-about-the-worlds-tallest-mountain-8632372.html (accessed August 2018).
4. 1996 Death on Mount Everest. http://www.history.com/this-day-in-history/death-on-mount-everest (accessed August 2018).
5. Rollings, G. (2015). Truth of Everest tragedy is even more horrifying than film shows. 17 September 2015. https://www.thesun.co.uk/archives/news/123362/truth-of-everest-tragedy-is-even-more-horrifying-than-film-shows (accessed August 2018).
6. Plan a Mt Everest Base Camp Tour in Tibet. https://www.tibettravel.org/tibet-travel-advice/trip-to-everest-base-camp.html (accessed August 2018).

CHAPTER 7

Primer on Probability and Statistics

7.1 INTRODUCTION

Probability theory was developed around 1654 by the French mathematicians Blaise Pascal and Pierre de Fermat. The theory was created in order to help gamblers in games of chance, like roulette, dice, craps, and cards (1). This chapter provides only a brief overview of the concepts of probability. Consult a text on probability and statistics for an in-depth discussion.

Probability theory is an integral part of risk assessment. In fact, they go hand in hand. Whether it is gambling at a casino or the stock market or gambling with one's life on the freeway, people act based on some knowledge or perception of the probability of a successful/unsuccessful outcome. Many times risks are taken and successful outcomes occur and the common method of explaining this is that the person was lucky or conversely people will say they or someone else was unlucky if a negative outcome occurred. Luck, of course, is defined as a chance happening. Mostly everyone from every culture has some feeling or perception of luck. Those entering a casino hope they would be lucky. They want a successful outcome. If they are successful, then they were lucky, if not, well karma or bad luck or some other supernatural force caused them to be unlucky. This superstitious view of luck or chance is different than viewing events based on the probability they will occur.

All games of chance (luck) are based on some predefined probability of success and failure. All betting games at a casino have a house edge. Card games, for instance, Blackjack or 21, have a house edge of between 0.17 and 0.44% (2). A card game like "Let it Ride," called a carnival game, has a house edge of approximately 3.5%. Slot machines and other computer-based gambling devices can have a house edge of 10%. This house edge can be governed by state law.

Risk Assessment: Tools, Techniques, and Their Applications, Second Edition. Lee T. Ostrom and Cheryl A. Wilhelmsen.
© 2019 John Wiley & Sons, Inc. Published 2019 by John Wiley & Sons, Inc.
Companion website: www.wiley.com/go/Ostrom/RiskAssessment_2e

The dice game craps is a good way of explaining the very basic principles of probability and risk. In the game of craps, a participant rolls two dice. He/she is called the thrower. The casino ensures that each die is fair, meaning the probability of any of the numbers of the die coming up is equal. The die has six sides and, as we all know, each is numbered from one to six. If the die is loaded or one side has a higher probability of coming up, then the casino can be at a disadvantage. In Chapter 8 (Bayesian Update), there is a discussion on loaded dice. The craps table is designed so the dice tumble when they hit the table. This ensures randomness. Randomness plays in favor of the casino. Throwers could manipulate the dice by sliding them if the table did not contribute to the tumbling of the dice.

When a player walks up to a craps table, he/she places a bet. There are many bets that can be placed. Assume the player walks up to the table when there is no number that is "On." This will be explained soon. The player lays down a bet of $10 on the pass line. At this point the player is betting that one of two events will occur. Either a seven or an eleven will be thrown, and the player will be paid even money ($10) or that the thrower will throw a 4, 5, 6, 8, 9, or 10. If the thrower throws one of these numbers, no money is exchanged and a point number is determined. The point number then becomes the "On" number. If the thrower throws a 2, 3, or 12, the player loses his/her bet. This is a "Craps." The thrower, whether it's our player or it's another player, throws the dice. The probabilities of the 36 combinations of numbers coming up are shown in Table 7.1. These probabilities or odds cannot be changed if the dice are fair and the table provides a surface that ensures the dice will tumble.

Table 7.2 shows the probabilities of the possible events at this point in the game. For the simulation, assume on the first throw a six comes up. As explained above, then six becomes the point number and it is said to be "On." The probability that a six would come up is 5/36 or approximately 14%. The probability that either a 7 or an 11 would come up is approximately 22%. The probability that a 2, 3, or 12 would come up is approximately 12%.

So, what was the risk the player was taking? He risked his bet of $10. He had a 12% probability he would lose his $10. He had an approximately 22% probability of winning $10. There was an approximately 66% probability of a neutral event. Was there a risk beyond this bet? No, the player can make a decision at this point to walk away from the table. However, our player chooses to play one more time.

There are several betting options that can now occur, such as placing a side bet on the point number behind the pass line, betting one other numbers, and so on. Our player keeps his simple bet on the pass line. He is now betting that a six will come up. Table 7.3 shows the probabilities of the possible events that our player is possibly subjected to. At this point the house has a 3% edge. To make a long story short, at this point a seven is thrown and the player loses his $10.

TABLE 7.1
Dice Probabilities

Dice 1	Dice 2	Outcome	Probability of event	Outcome group	Combined probability
1	1	2	1/36	2	1/36
1	2	3	1/36	3	2/36 or 1/18
2	1	3	1/36		
2	2	4	1/36	4	3/36 or 1/12
1	3	4	1/36		
3	1	4	1/36		
2	3	5	1/36	5	4/36 or 1/9
3	2	5	1/36		
1	4	5	1/36		
4	1	5	1/36		
2	4	6	1/36	6	5/36
4	2	6	1/36		
1	5	6	1/36		
5	1	6	1/36		
3	3	6	1/36		
1	6	7	1/36	7	6/36 or 1/6
6	1	7	1/36		
2	5	7	1/36		
5	2	7	1/36		
3	4	7	1/36		
4	3	7	1/36		
2	6	8	1/36	8	5/36
6	2	8	1/36		
3	5	8	1/36		
5	3	8	1/36		
4	4	8	1/36		
3	6	9	1/36	9	4/36 or 1/9
6	3	9	1/36		
4	5	9	1/36		
5	4	9	1/36		
4	6	10	1/36	10	3/36 or 1/12
6	4	10	1/36		
5	5	10	1/36		
5	6	11	1/36	11	2/36 or 1/18
6	5	11	1/36		
6	6	12	1/36	12	1/36

TABLE 7.2
Probabilities of Dice Events

Event	Type of event	Approximate individual probability	Approximate combined probability (%)
7	Advantageous to player	6/36 or 1/6 or 17%	22
11	Advantageous to player	2/36 or 1/18 or 6%	
4	Neutral event	3/36 or 1/12 or 8%	66
5	Neutral event	4/36 or 1/9 or 11%	
6	Neutral event	5/36 or 14%	
8	Neutral event	5/36 or 14%	
9	Neutral event	4/36 or 1/9 or 11%	
10	Neutral event	3/36 or 1/12 or 8%	
2	Advantageous to house	1/36 or 3%	12
3	Advantageous to house	2/36 or 1/18 or 6%	
12	Advantageous to house	1/36 or 3%	

TABLE 7.3
Probabilities of Second Round

Event	Type of event	Approximate individual probability	Approximate combined probability (%)
6	Advantageous to player	5/36 or 14%	14
2	Neutral event	1/36 or 3%	69
3	Neutral event	2/36 or 1/18 or 6%	
4	Neutral event	3/36 or 1/12 or 8%	
5	Neutral event	4/36 or 1/9 or 11%	
8	Neutral event	5/36 or 14%	
9	Neutral event	4/36 or 1/9 or 11%	
10	Neutral event	3/36 or 1/12 or 8%	
11	Neutral event	2/36 or 1/18 or 6%	
12	Neutral event	1/36 or 3%	
7	Advantageous to house	6/36 or 1/6 or 17%	17

7.2 PROBABILITY THEORY

Probability is defined as the likelihood that the event will occur. Probability measures the uncertainty associated with the outcomes of a random experiment. Some other terms or words used in place of probability are chance, likelihood, uncertainty, and odds. Probability is usually expressed as a fraction with the denominator representing the total number of ways things can occur and the numerator representing the number of things that you are hoping will occur. Probability is always a number between 0 and 1 or between 0 and 100%. Zero means that something cannot happen (impossible), and 1 or 100% means it is sure to happen. Another way to express this is $0 \leq P(A) \leq 1$ where A is the event. This expression is the first basic rule of probability (3).

There is also a rule that applies to two events, A and B, which are mutually exclusive, that is, the two events cannot occur at the same time. In this case we express this as $P(A \text{ or } B) = P(A) + P(B)$. Some textbooks will use mathematical symbols for the words "and" and "or" and the expression would look like $P(A \cup B) = P(A) \cap P(B)$.

Although these rules of probability are extremely few and simple, they are incredibly powerful in application. In order to understand probability, you must know how many possible ways a thing can happen. For instance, if you flip a coin, there are two possible ways it can land, either heads or tails. If we want to calculate the probability of the coin landing on a head, we see that the head is one of two possible ways so the probability is ½ or 0.5. The probabilities do not change on the second or subsequent coin tosses. This is because the events are independent. One event is not tied to a prior or future event.

As with the dice game, craps, discussed above, every time one tosses a coin or throws the dice, the probability of that individual event occurring is the same. This concept is sometimes very difficult for people, whether engineers, politicians, or gamblers, to comprehend. If a risk analysis of an airplane is performed and a probability of the airplane breaking apart in mid-flight is calculated as one event in every 10 000 flights, it does not mean the event will not occur on its first flight. It has a one in 10 000 chance of occurring on the first and every subsequent flight. The only time the probability changes is if the assumptions on which the probability estimate were made changes.

Let's reconsider the dice game. If a thrower throws two sixes, what is the probability on the next throw that double sixes will come up again? The probability is the same, one out of 36. The two events are independent. However, the probabilities change significantly if one is predicting the probability in advance that double sixes will be thrown twice in a row. This will be explained in another section.

The probability of winning a major lottery is one out of 10 000 000. A one dollar ($1) ticket is purchased by our player. Our player wins. Next week our player considers playing again. What is the probability our player will win next week's lottery? It is the same, one out of 10 000 000. However, the probability of our player winning both lottery events is:

$$\frac{1}{10\,000\,000} \times \frac{1}{10\,000\,000}$$

or

$$(1 \times 10^{-7}) \times (1 \times 10^{-7}) = 1 \times 10^{-14}$$

This would be a very rare event, but since it is not "0," it could occur.

7.3 COMBINING PROBABILITIES

When probabilities are combined, as with our lottery example above, the probabilities are multiplied together to determine the final probability for the combined events. The rules of Boolean algebra are used to combine probabilities (4).

The two most commonly used Boolean algebra terms are the logical "AND" and "OR." When two probabilities are combined using AND logic, as with our lottery example, the two probabilities are multiplied together:

Winning the first lottery AND winning the second lottery.

$$(1 \times 10^{-7}) \times (1 \times 10^{-7}) = 1 \times 10^{-14}$$

In an accident analysis, our events might appear something like,
A driver runs a red light AND the driver in the cross street cannot stop.

Another example would be
A terrorist gains access to a building AND is undetected.

A probability can be associated with the event and then combined to develop an overall probability for the event.

The probability associated with a driver running a red light might be 1/1000 red lights or 0.001. The probability associated with the cross street driver not being able to stop might be 1/100 or 0.01. Combining these two probabilities yields the overall probability of an accident occurring because of the two events:

$$0.001 \times 0.01 = 0.000\,01$$

The logical "OR" is used when several events can occur, but only one of these events can lead to an outcome. For instance, an accident can occur if any of the following events occur:

Fail to stop soon enough behind another vehicle – probability = 0.001.
Skid on ice and slide into another vehicle – probability = 0.0001.
Fail to yield when turning and another vehicle collides – probability = 0.005.

In this case the probabilities are added instead of multiplied.

$$0.001 + 0.0001 + 0.005 = 0.0061$$

The structure of fault trees use these basic logical terms in their construction. This will be explained in detail in Chapter 14.

7.4 CONDITIONAL PROBABILITY

Conditional probability is a probability whose sample space has been limited to only those outcomes that fulfill a certain condition. It is expressed as:

$$P(A \mid B) = \frac{P(A \cap B)}{PB},$$ which is the probability of event A given event B.

Events can be categorized into two main categories, single events and compound events. A single event is when you only expect one event to occur such as flipping coins or rolling dice. It could even be drawing a certain card, such as the ace of spades, from a deck of 52 cards.

A compound event is when two or more events or things are happening. Frequently, we want to know the probability of two things happening. In other words, one thing happens and then the other thing happens. Since "and" means multiply, the probability of one event must be multiplied by the probability of another event happening. An example of this would be flipping two heads in a row. The expression would be $\frac{1}{2} \times \frac{1}{2} = \frac{1}{4}$.

Generally speaking when working with probability, "and" means to multiply and "or" means to add. However, you must be careful. Here are three important rules for compound probabilities:

1. If A and B are independent, or the occurrence of one does not affect the other, then $P(A \text{ AND } B) = P(A)*(B)$.
2. If A and B are dependent, or the occurrence of one does affect the other, then $P(A \text{ AND } B) = P(A)*P(B)$ given that A has occurred.
3. $P(A \text{ OR } B) = P(A) + P(B) - P(A \text{ AND } B)$. [$P(A \text{ AND } B)$ may equal zero if A and B are "mutually exclusive" or that they cannot happen simultaneously.]

There are three basic terms of probability: experiment, sample space, and event. An experiment is a process by which an observation, or outcome, is obtained. A sample space means the set (S) of all possible outcomes of an experiment. And an event means any subset (E) of the sample space. The events are subsets of S, but they are not outcomes. When an event is certain the $E = S$, like rolling a number between 1 and 6 [$E = \{1, 2, 3, 4, 5, 6\} = S$], this is called a certain event. If you were to say, the event to roll a 7, since the events are subsets of the sample space and 7 is not contained in the sample set, we say this is an impossible event. Obviously, only one die is used for this event.

7.5 PROBABILITY DISTRIBUTIONS

The probabilities of events fall within one of several probability distributions. Using the craps example again and using a bar chart to display the distribution of events and their corresponding probabilities, we see a very discrete distribution (Figure 7.1).

A discrete distribution is just as it sounds, the values of the variable are discrete. For instance, there is no such thing as a 7.5 on a pair of dice. It is a value of 7 or 8. Same as with the probabilities of heads or tails on a coin, they are discrete values.

Another example of a discrete distribution would be the probabilities associated with a certain type of car being in a wreck. These are discrete numbers associated with a certain type of vehicle. Common types of discrete variables would be

- Number of times an electrical switch is thrown before failure.
- Number of cycles of an airplane before failure of a component.
- Number of demands on a fire suppression system before failure.
- Number of demands on a backup diesel generator before failure.

However, in some cases, depending on the range of values, a discrete-type distribution can appear continuous.

The alternative to discrete distributions is continuous distributions. The number of hours drivers operate a vehicle between accidents can range from some fraction of

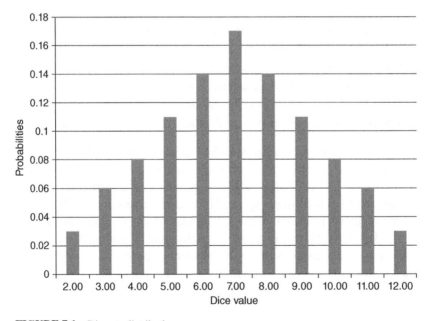

FIGURE 7.1 Discrete distribution.

one to thousands of hours. The values of the variable are continuous. Some common types of continuous variables would be

- Hours of operation of a pump, valve, piping, controller, or other mechanical piece of equipment before failure or between repairs.
- Hours of operation of an airplane, ship, car, or train between accidents or repairs.

Common types of continuous distributions used for risk assessment are:

- Normal distribution
- Uniform distribution
- Chi-square distribution
- Weibull distribution
- Poisson distribution
- Exponential distribution

Continuous probability distributions possess a probability density function. As discussed above, a continuous random variable is the one that can take a continuous range of values. For example, the time between pump failures can have an infinite number of values; however, it has probability zero because there are infinitely many other potential values.

The most important continuous probability distribution for risk analysis and reliability engineering is the exponential distribution. The exponential distribution is considered to be memoryless and is, therefore, a good distribution to model the constant hazard rate of a reliability bathtub curve (3).

The expected value or mean of a variable X that is exponentially distributed is:

$$E[X] = \frac{1}{\lambda}$$

Lambda (λ)
The variance of X:

$$\text{Var}[X] = \frac{1}{\lambda^2}$$

The median value is

$$m[X] = \ln \frac{2}{\lambda}$$

When determining failure rates for components, the exponential distribution is one of the preferred continuous distributions because an error factor can be calculated. The error factor is the ratio of the 95th percentile to the median or 50th percentile

failure rate (www.nrc.gov). For example, if the median value for a pump failure is
2E-3 hours and the 95th percentile failure rate is 1.5E-2, the error factor is

$$\frac{1.5\text{E-2}}{2\text{E-3}} = \text{error rate} = 7.5$$

Other continuous probability distributions are used in risk assessment but are
much more difficult to apply for most common types of applications. The Weibull
distribution is the most widely used continuous distribution used in modeling reliability. In many cases, what appears to be a continuous variable can be converted to a
discrete variable and can, therefore, be much more easily dealt with. Run time hours
on a pump can be binned into a set of predefined bins, for instance, low, medium, and
high. As with any concept, the simpler way information is presented, and the more
people will understand it and be able to do something with it.

7.6 USING PROBABILITY

The following examples will explore more applications of probability. Let's start by
looking at flipping a coin, which is a single event. The total number of possible outcomes when flipping a coin is 2, either heads or tails. The two would be placed in the
denominator. If we wanted to know the probability that the outcome would be heads,
then the expression would look like ½, or a 0.5% chance it would be heads.

Frequently we want to know the probability of two things happening; in other
words, one thing happens, AND then another thing happens (AND means multiply).
You multiply the probability of one thing happening by the probability of the other
thing happening. What is the probability of the following:

1. Flipping two heads in a row? [Answer: ½ × ½ = ¼]
2. Flipping three heads in a row? [Answer: ½ × ½ × ½ = 1/6]
3. Flipping six heads in a row? [Answer: ½ × ½ × ½ × ½ × ½ × ½ = 1/64]
   (Rolling one dice twice is the same as rolling two dice together once.)
4. Drawing two aces? It depends on if you put the first one back before drawing
   the second. If you did put it back, then it is 4/52 × 4/52 = 16/2704 or 1/169. If
   you didn't put it back, then it is 4/52 × 3/51 = 12/2652 or 1/221. The second
   probability is 3/51 because there are only 3 aces left and 51 cards left to choose
   from after you successfully take out the first ace. You have a better chance of
   getting 2 aces if you put the first one back before drawing again.
5. Throwing an 11 using 2 dice? [Answer: 2/36 because there are two ways of
   throwing an 11 {5 + 6 and 6 + 5}]

Now look at one that is a little more complicated, like rolling two dice. There are
36 possible outcomes when rolling two dice. Using Table 7.4 below, the probability
that a three would be rolled could be determined.

TABLE 7.4
Combinations Using Two Dice

	1	2	3	4	5	6
1	2	3	4	5	6	7
2	3	4	5	6	7	8
3	4	5	6	7	8	9
4	5	6	7	8	9	10
5	6	7	8	9	10	11
6	7	8	9	10	11	12

There are two possible ways of rolling the two dice so that they add up to three, and therefore the probability is 2/36 or 1/18 chances.

Using the matrix above, one could determine the possibility of rolling two dice and rolling a three 3 times in a row. The probability of rolling a three was 1/9 as determined above. To find the probability of rolling a three 3 times in a row, multiply 1/9 by itself 3 times or $1/9 \times 1/9 \times 1/9 = 1/729$.

Suppose you have a bag containing 14 red marbles, 12 blue marbles, and 18 green marbles.

1. What is the probability that if you pull out a marble at random you get either a red or a blue marble? [P(red or blue)] There are 44 total marbles and 26 chances to pull either a red or a blue marble, so the probability is 26/44 or 13/22.
2. What is the probability of not getting a blue marble? [P(not blue)] There are 44 total marbles and 32 are not blue, so the probability is 32/44 or 8/11.

Suppose you roll two dice, a red one and a green one.

1. What is P(a sum of 4)? There are three ways of making a sum of 4 (see matrix above) out of 36 possible sums, so the probability is 3/36 or 1/12.
2. What is P(a sum of 5 or 6 or 7)? There are four ways of making a sum of 5, five ways of making a sum of 6, and six ways of making a sum of 7, so the probability of making a sum of 5 or 6 or 7 is $4/36 + 5/36 + 6/36 = 15/36$ or 5/12.

When a polling agency says its confidence level is 95%, it is saying that the probability of its numbers being correct is 0.95. What is the probability their numbers are wrong? $(100 - 95\%)$ or $(1 - 95) = 5\%$ or 0.05.

An art class has 13 right-handers and 7 left-handers.

1. What is the probability that a student chosen at random from the class is right-handed? There are 20 total students, so the probability that the student would be a right-hander is 13/20.

2. If three students are chosen, and the first two are right-handers, what is the probability that the third is also a right-hander? Two have already been picked and that leaves 18 students to choose the third student from and only 11 are right-handed since two have been chosen, so the probability that the student would be a right-hander is 11/18.

Tree diagrams can help you visualize a probability problem such as flipping a coin or rolling dice. Suppose you are going to flip a coin three times. You could use a tree diagram to visualize all of the possibilities. An example of this is in Figure 7.2.

So, one of the possibilities would be T, H, H. There are eight possible combinations when flipping a coin three times in a row.

1. What is the probability of getting three heads in a row (H, H, H)? There is only one branch that produces three heads in a row (H, H, H), so the probability is 1/8.

2. What is the probability of getting two heads and one tail (in any order)? There are three branches that have H, H, T (in any order), so the probability is 3/8.

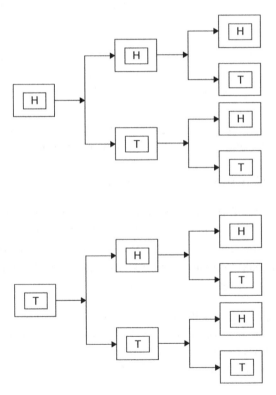

FIGURE 7.2   Tree diagrams of coin flip.

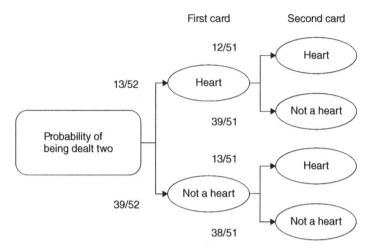

First card   Second card

12/51
Heart

13/52   Heart

39/51
Not a heart

Probability of
being dealt two

13/51
Heart

39/52   Not a heart

38/51
Not a heart

FIGURE 7.3   Tree diagram of two hearts.

What is the probability of being dealt two hearts? Solve this problem by using Figure 7.3: a tree diagram as follows.

To solve, multiply the probabilities along one whole "limb." So multiplying along the top limb, we have $13/52 \times 12/51 = 0.06$ or 6%. With a tree diagram we can see other possible outcomes. For example, looking along the very bottom "limb," the probability of not getting a heart at all is $39/52 \times 38/51 = 0.56$ or 56%.

To summarize tree diagrams for probability,

(a) Conditional probabilities start at their condition.
(b) Nonconditional probabilities start at the beginning of the tree.
(c) Multiply when moving horizontally across a limb.
(d) Add when moving vertically from limb to limb.

7.7 SUMMARY

Probability theory is an integral part of risk assessment. Games of chance like craps and card games that have discrete probabilities associated with various outcomes are a good way of explaining probability. Real-life events like car accident potentials might have a discrete probability either associated with them or continuous. For most common types of risk assessments, though, the use of a discrete probability is probably advised so that it is more understandable. Please consult a text on probability and statistics for a more in-depth understanding on these topics.

Self-Check Questions

1. These questions and activities are all applied.
   Get a standard deck of cards and shuffle them well.

   (a) Deal two cards. Calculate the odds of getting that particular hand.

   (b) Deal three cards. Calculate the odds of getting that particular hand.

   (c) Deal five cards. Calculate the odds of getting that particular hand.

   (d) Deal seven cards. Calculate the odds of getting that particular hand.

2. With a fellow student or friend, try playing craps and calculate the odds of each role. Do they align with the predicted values?

3. Sit at a stop sign, and observe the cars as they stop or if they don't stop and run the sign. What did you observe? Can you calculate the odds of a car stopping/not stopping at the stop sign? What are the odds of a near miss? If you watch long enough, you might observe an accident. If so, what are the odds of a collision?

4. Put skittles or colored marbles in a container. While not looking, draw out at separate times 10, 50, and 100, and calculate the odds of the various colors.

5. Using the data you collected in Question 4, do the same experiment and compare the results.

REFERENCES

1. de Laplace, P.S. (1812). *Analytical Theory of Probability*. Paris: Courcier.
2. Wizard of Odds Wizardofodds.com (accessed August 2018).
3. Ross, S.M. (2009). *First Course in Probability*, 8e. Upper Saddle River, NJ: Prentice Hall.
4. NRC. Guidance on the Treatment of Uncertainties Associated with PRAs in Risk-Informed Decision Making. NUREG-1855, Vol. 1. http://www.nrc.gov/reading-rm/doc-collections/nuregs/staff/sr1855/v1/sr1855v1.pdf (accessed 4 March 2019).

CHAPTER 8

Mathematical Tools for Updating Probabilities

8.1 INTRODUCTION

This chapter discusses two common tools used to update and to help generate failure probabilities. These are Bayesian update and Monte Carlo analysis. Both of these techniques are used widely in the probabilistic risk assessment (PRA) community and can range from very simple mathematical manipulation of data to very complex algorithms. This chapter will focus on relatively simple versions of these techniques. More complex methods can be found in the references provided.

8.2 BAYESIAN UPDATE

This section will discuss Bayesian update and how it is used in PRA to aid in further developing probabilities.

8.2.1 Symbols and Terms Used in Bayesian Update

The common mathematical symbols and mathematical terms associated with Bayesian update are as follows:

∩ – This symbol means that two probabilistic events occur together. For example, $P(A \cap B)$ is the probability that A and B occur together or intersect. If the events are mutually exclusive, then $P(A \cap B) = 0$.

∪ – This is union symbol. The probability that events A or B occur is the probability of the union of A and B. The probability of the union of events A and B is denoted by $P(A \cup B)$.

Risk Assessment: Tools, Techniques, and Their Applications, Second Edition. Lee T. Ostrom and Cheryl A. Wilhelmsen.
© 2019 John Wiley & Sons, Inc. Published 2019 by John Wiley & Sons, Inc.
Companion website: www.wiley.com/go/Ostrom/RiskAssessment_2e

$P(A|B)$ – The probability that event A occurs, given that event B has occurred, is called a conditional probability. The conditional probability of event A, given event B, is denoted by the symbol $P(A|B)$. Conditional probability was discussed in Chapter 7.

$\sum$ – This is the common symbol for sum.

$\prod$ – This symbol means the product of all values in range of series. For example,

$$\prod_{N=1}^{4} 3n = 3 \times 6 \times 9 \times 12 = 1944$$

8.2.2 Bayes' Theorem

Bayes' theorem or law was developed over 250 years ago by Reverend Thomas Bayes (1). The theorem was updated by Pierre-Simon Laplace of Laplace transforms published in 1812 (2). The basic premise of Bayes' theorem describes the probability of an event that is based on prior knowledge of conditions related to the event. For instance, using the Bayes approach, the probability of a certain illness can be better predicted within a cohort or for an individual knowing attributes like age, weight, and ethnic origin.

Mathematically stated, Bayes' theorem is as follows:

$$P(A|B) = \frac{P(B|A)P(A)}{P(B)}$$

where

A and B are events and $P(B) \neq 0$.

$P(A|B)$ is a conditional probability. The probability of event A occurring given that B is true.

$P(B|A)$ is a conditional probability as well. The probability of event B occurring given that A is true.

$P(A)$ and $P(B)$ are the likelihoods or probabilities of observing A and B independently of each other or marginal probability.

Take, for instance, a population of individuals who might or might not be taking a certain drug. If people take the drug, there is an 80% chance that a test will provide positive results. If they are a nonuser, then there is a 99% chance the test results will be negative. It is found that 10% of the population uses the drug. What is the probability that a randomly selected individual with a positive drug test is a user?

Therefore: *Person is a user is abbreviated PU. Nonuser is NU.*

$$P(\text{PU}| + \text{test}) = \frac{P(+\text{test}|\text{PU})P(\text{PU})}{P(+\text{test})}$$

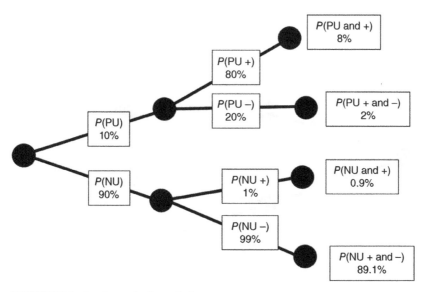

FIGURE 8.1 Logic tree for Scenario 1.

$$P(\text{PU}|+\text{test}) = \frac{P(+\text{test}|\text{PU})P(\text{PU})}{P(+\text{test}|\text{PU})P(\text{PU}) + P(+\text{test}|\text{NU})P(\text{NU})}$$

$$P(\text{PU}|+\text{test}) = \frac{0.80 \times 0.10}{0.80 \times 0.10 + 0.01 \times 0.90} = 0.898 \text{ or approximately } 90\% \text{ chance}$$

This is because there is a large number of drug users in the population, and it is more likely a person selected for testing, who is a user of the drug, will test positive even though the test is not specific. The number of users in a population of 1000 is 100.

Reversing these numbers so that the user population is small (0.1%), the false positive rate in the nonuser group is 1%, and the test is not as specific (80%) and produces the following results:

$$P(\text{PU}|+\text{test}) = \frac{0.80 \times 0.001}{0.80 \times 0.001 + 0.01 \times 0.999} = 0.008 \text{ or } 0.8\%$$

Diagrams of the logic trees for the two scenarios are shown in Figures 8.1 and 8.2.

Examples of Bayes' theorem use can get much more complicated. The following is an example using the chemical industry as a backdrop. Take a chemical manufacturer with four production lines. The total production capacity of the plant is 200 000 gal of product a day. Line A is the oldest and makes 40 000 gal; Line B is the largest producer and makes 60 000 gal; Lines C and D make 50 000 gal each. However, the quality of the product varies. Line A has the lowest quality level with 5% material out of spec. Line B the next worst at 3%. Lines C and D produce the best quality at 1% and 0.5%, respectively, out of spec material.

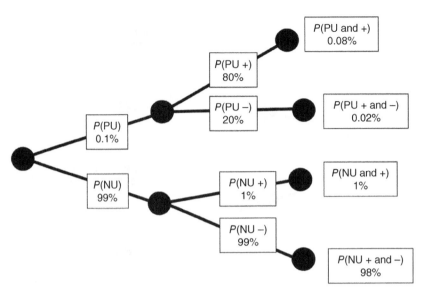

FIGURE 8.2 Logic tree for Scenario 2.

Therefore, $P(A) = 0.2$ (20%); $P(B) = 0.30$ (30%); $P(C) = 0.25$ (25%); $P(D) = 0.25$ (25%)

The defect rate (DF) for each of the lines is:

$P(DF|A) = 0.05$ (5%); $P(DF|B) = 0.03$ (3%); $P(DF|C) = 0.01$ (1%); $P(DF|D) = 0.005$ (0.5%)

The total amount of out of spec product is $(40\,000 \times 0.05) + (60\,000 \times 0.03) + (50\,000 \times 0.01) + (50\,000 \times 0.005) = 4\,550\,\text{gal}$

Overall, 2.27% (DFR) of the plant's production is out of spec.

A quality control staff member pulls a sample and it is out of spec. What is the probability it came from production Line C? Please note the question. It is already understood what the defect rate for Line C is. This test provides us insight into the probability that a random sample, which is defective, is from Line C.

Therefore,

$$P(C|DF) = \frac{P(DF|C)P(C)}{DFR}$$

Or

$$\frac{(0.01) \times (0.25)}{0.0227} = 0.11 \text{ or } 11\%$$

A tree diagram is shown in Figure 8.3.

8.2.3 Frequentist Theory vs. Bayesian Theory

The differences between classical probability theory, frequentist probability theory, and Bayesian probability are discussed in almost all books and papers about Bayesian

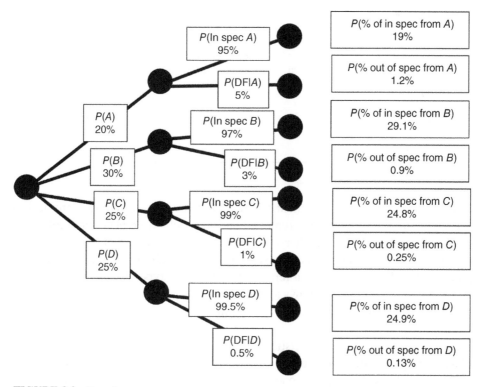

FIGURE 8.3   Tree diagram.

statistical inference (BSI). This book will be no different. However, the effort here is to bring it down to a usable level. Chapter 7 of this book discussed classical probability theory.

8.2.4 Frequentist Probability Theory

In risk assessment, statistics are used to help answer a question. Frequentist statistical tests are used to determine whether an event occurs or not. It is more of a yes/no type test. It calculates the probability of an event in the long run of the experiment. Frequentist probability theory is an interpretation of probability (3). Mathematically frequentist probability is defined as:

$$P(x) \approx \frac{n_x}{n_t}$$

where n_x is the number of times an event of interest is observed and n_t is the number of trials. As the number of trials approaches infinity, the true probability is determined:

$$P(x) = \lim \lim_{n_t \to \infty} \left(\frac{n_x}{n_t} \right)$$

TABLE 8.1
Card Experiment

Number of draws	Number of red cards drawn	Deviation from predicted	Calculated probability
20	12	+2	0.6 or 60%
50	23	−3	0.46 or 46%
100	53	+3	0.53 or 53%
1 000	489	−11	0.489 or 48.9%
10 000	5 015	+15	0.5015 or 50.15%

For example, take an experiment where a card is drawn from a random deck of cards where the jokers have been removed and the decks have been randomly shuffled. The event of interest is whether a randomly drawn card is part of red suit of cards. Since 50% of a deck contains red cards, the overall probability should at some point approach 50%. Table 8.1 shows a representation of this type of experiment. As this table shows, as the number of draws approaches infinity, the true probability of drawing a red card is predicted. This approach requires a great number of trials to confirm a probability of an event.

The concept of frequentist statistical inference has been used to a high degree over the last century. However, frequentist statistics has some great flaws in its design and interpretation. These flaws pose concerns as real problems. Examples of these problems include that p-values measured for experiments can vary depending on when the experimenters have different stopping intentions. If one experimenter stops at 10 samples and another at 10 000, the results can vary tremendously. These types of differences are found commonly in medical experiments where small changes in trial sizes can have big impacts.

Confidence intervals (CIs) are also heavily dependent on the sample size. The more samples, the more narrow the CI. Therefore, once again, if one experimenter uses a large sample size and another a small sample size, the CI will be different. Also, CIs are not probability distributions and do not provide the most probable value for a parameter.

For reliability models, frequentist methods treat model parameters as unknown, fixed constants and employ only observed data to estimate the values of parameters. In the case of failure time data, one might assume the data is exponentially distributed.

Bayesian models treat parameters as unknown random variable whose distribution, or what is called the prior, represents the current belief about the parameter.

8.2.5 Bayesian Theory

BSI is an important tool risk assessment professionals, forensic scientists, and others use to provide important information about some event that has a probabilistic component to it. BSI requires the establishment of parameters and models as part of the process. Models are the mathematical formulation of the observed events. Parameters are the factors in the models affecting the observed data.

The simplest example of this that has been discussed in many sources is the fairness of a coin (4–6). Fairness of a coin may be defined as the parameter of a coin and is symbolized as Θ. The outcome of the events associated with the coin is symbolized as D. Using the Bayesian theory concept, we are not interested in the absolute outcome of coin tosses, for instance. We are interested in knowing whether the outcome of coin tosses is fair. Therefore, given (D), what is the probability that a coin is fair ($\Theta = 0.5$)? Putting this concept into Bayes' theorem, we get:

$$P(\Theta|D) = \frac{P(D|\Theta) \times P(\Theta)}{P(D)}$$

In this case $P(\Theta)$ is the prior. The prior is the strength in the belief that the coin is fair before it is tossed. For a coin the obvious prior is 0.5 (50%) for heads or tails. However, the bias/fairness can range from 0 to 1 (100%) heads or tails. If the coin deviates a great amount from 0.5 for either possible outcome, then the coin is not fair.

The likelihood of observing our result given our distribution of Θ is $P(D|\Theta)$. If we knew the coin was fair, then this gives the probability of observing the number of heads or tails given a number of tosses of the coin.

The evidence is given by $P(D)$. This is calculated by summing across all possible values of Θ. These values are weighted by how strongly we believe in particular values of Θ.

$P(\Theta|D)$ is what is called the posterior. The posterior is the belief of our parameters after observing the evidence.

The literature likes to use biased coins and horse races where one horse has an advantage on a muddy track over another horse to first help explain BIS. Here we will look at a biased coin. The following example was adapted from MIT OpenCourseWare (4).

If a coin is tossed or flipped the same way each time and it is unbiased, the results will converge at 50% heads and 50% tails. However, a coin might have a bias. We will assume a coin has bias toward tails.

$$\text{Bias} = P(\text{Tails}) \text{ OR Bias} = P(\text{Heads})$$

As an example, there are three types of coins that have different probabilities of landing heads when tossed.

- Type A coins are fair, with probability 0.5 of heads or tails.
- Type B coins are not weighted correctly and have probability 0.75 of tails.
- Type C coins are not weighted correctly and have probability 0.9 of heads.

Suppose a container has 10 coins: 5 of type A, 3 of type B, and 2 of type C. One coin is selected. Without revealing the type of category of the coin, the results of the flip is heads. What is the probability it is type A? Type B? Type C?

Let A, B, and C be the event that the chosen coin was type A, type B, and type C. Let D be the event that the toss is heads. The problem asks us to find

$$P(A|D), P(B|D), P(C|D).$$

In this case,
$$P(A) = 0.5, P(B) = 0.3, P(C) = 0.2.$$

The likelihood function is $P(D|H)$. This is the probability of the data assuming that the hypothesis is true. In this case, likelihood and probability are synonyms. Most often we will consider the data as fixed and let the hypothesis vary. For example, $P(D|A) = $ probability of heads if the coin is type A. In our case the likelihoods are:

$$P(D|A) = 0.5, P(D|B) = 0.25, P(D|C) = 0.9.$$

Note that since we are interested in heads rather than tails for event "B," the likelihood is $1 - 0.75$ or 0.25. Please note that likelihood and probability are synonyms.

The probability (posterior) of each hypothesis given the data from tossing the coin is:
$$P(A|D), P(B|D), P(C|D).$$

These posterior probabilities are what is of interest.

Bayes' theorem is now used to compute each of the posterior probabilities. We are going to write this out in complete detail so we can pick out each of the parts. (Remember that the data D is that the toss was heads.)

Figure 8.4 shows the tree for these possible events.

Bayes' theorem says
$$P(A|D) = \frac{P(D|A)P(A)}{P(D)}$$

The denominator $P(D)$ is computed by using the law of total probability:
$$P(D) = P(D|A)P(A) + P(D|B)P(B) + P(D|C)P(C) = (0.5 \times 0.5) + (0.25 \times 0.3) + (0.9 \times 0.20) = 0.505$$

Now each of the three posterior probabilities can be computed:

$$P(A|D) = \frac{P(D|A)P(A)}{P(D)} = \frac{0.5 \times 0.5}{0.505} = \frac{0.25}{0.505}$$

$$P(B|D) = \frac{P(D|B)P(B)}{P(D)} = \frac{0.25 \times 0.3}{0.505} = \frac{0.075}{0.505}$$

$$P(C|D) = \frac{P(D|C)P(C)}{P(D)} = \frac{0.9 \times 0.2}{0.505} = \frac{0.18}{0.505}$$

Notice that the total probability $P(D)$ is the same in each of the denominators and that it is the sum of the three numerators. We can organize all of this very neatly in a Bayesian update table (Table 8.2).

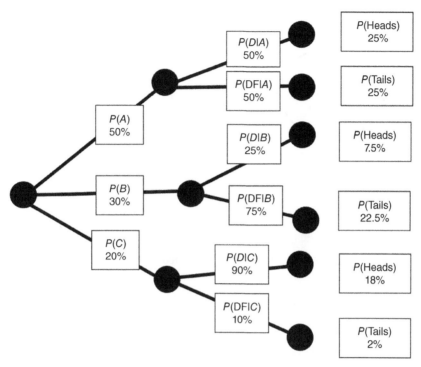

FIGURE 8.4   Logic tree showing possible events.

TABLE 8.2
Bayesian Update Table

Hypothesis	Prior	Likelihood	Bayes numerator	Posterior
H	$P(H)$	$P(D\|H)$	$P(D\|H)P(H)$	$P(H\|D)$
A	0.5	0.5	0.25	0.495
B	0.3	0.25	0.075	0.148
C	0.2	0.9	0.18	0.356
Total	1		0.505	1

The Bayes numerator is the product of the prior and the likelihood. We see in each of the Bayes' formula computations above that the posterior probability is obtained by dividing the Bayes numerator by $P(D) = 0.505$. We also see that the law of total probability says that $P(D)$ is the sum of the entries in the Bayes numerator column.

Please note that the probabilities for each of the hypothesis have changed for the posterior. "A" changed the least. However, "B" and "C" have changed significantly. The Bayes numerator column determines the posterior probability column. The likelihood column does not necessarily sum to 1.0; however, the prior and posterior columns do because they represent probabilities.

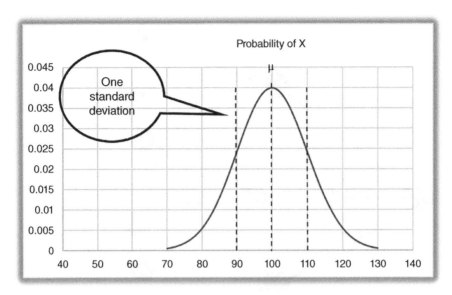

FIGURE 8.5 Probability density function for $\mu = 100$ and $\sigma = 10$.

The maximum likelihood estimation (MLE) is used when you want to find the parameter values that best fit the dataset using a specified distribution (6). The likelihood term represents this type of information. The difference is that the likelihood and prior are inputs to Bayesian analysis, not the output. The critical point in Bayesian analysis is that the posterior is a probability distribution function (pdf) of the parameter given the data set, not simply a point estimate. This enables all the properties of a pdf to be employed in the analysis. Figures 8.5 and 8.6 show pdf's for a $\mu = 100$ and $\sigma = 10$ and $\mu = 500$ and $\sigma = 25$, respectively.

As stated previously, the differences between frequentist and Bayesian methods become negligible as the sample size increases. However, when the data sets are small, these differences can be significant, with Bayesian interval estimates often narrower than the frequentist methods (7).

8.2.6 Steps to Implementing Bayesian Analysis

The following is adapted with permission from Practical Bayesian Analysis for Failure Time Data Best Practice (7). This process focuses on mean time between failure (MTBF) data:

1. Choose a prior distribution that describes our belief of the MTBF parameter.
2. Collect failure time data and determine the likelihood distribution function.
3. Use Bayes' rule to obtain the posterior distribution.
4. Use the posterior distribution to evaluate the data.

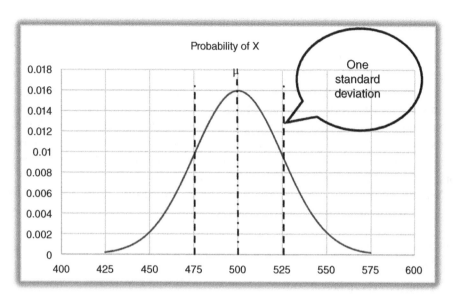

FIGURE 8.6   Probability density function for $\mu = 500$ and $\sigma = 25$.

Choose a Prior Distribution That Describes Our Belief of the MTBF Parameter
Any distribution can be chosen as a prior so long as it accurately describes the parameter information known and is determined *before* collecting any new data. In the case of failure times, we choose an inverse gamma distribution as it relates to the specific example below.

The inverse Gamma distribution is defined as:

$$f(\lambda; \alpha, \beta) = \frac{\beta\alpha}{\Gamma(\alpha)} \lambda^{-\alpha-1} e^{\frac{-\beta}{\lambda}} \text{ for } \alpha > 0, \beta > 0$$

where λ is the exponential distribution parameter, α is the shape parameter, β is the rate parameter, and λ is the parameter of interest, for instance MTBF.

The parameters that feed into this equation can be developed by several different ways. If the MLE is known, then the values can be determined directly. If a subject matter expert's opinion, engineering knowledge, or a failure database is used, then the mean and standard deviation of MTBF, the expected mean (μ), and variance (σ^2) can be used to determine the values of α and β:

$$\mu = \frac{\beta}{\alpha - 1} \text{ for } \alpha > 1, \quad \sigma^2 = \frac{\beta^2}{(\alpha - 1)^2(\alpha - 2)} \text{ for } \alpha > 2$$

The report by Michael Hamada (7) contains Matlab code for these equations.

Harmon references a NIST report that states that most defense systems' failure times are assumed to be exponentially distributed (7). However, the data should be used to select the appropriate likelihood distribution. As stated prior, the posterior is based on the prior and likelihood distributions.

Choosing the prior is an important but challenging task. The data used for the priors need to be defensible and as objective as possible. Priors can be classified broadly as either informative or noninformative. Informative priors, as the name suggests, contain substantive information about the possible values of the unknown parameter θ. Noninformative priors, on the other hand, are intended to let the data dominate the posterior distribution; thus, they contain little substantive information about the parameter of interest. Other terms for noninformative priors are diffuse priors, vague priors, flat priors, formal priors, and reference priors (8).

Chapter 9 discusses many sources to obtain probability data. These sources include the following:

- SME opinion.
- Delphi sessions.
- Industry experience.
- Maintenance log data.
- Past experience with similar equipment.
- Reliability databases.

Examples:

The following are two examples that demonstrate how to develop a posterior from failure data.

Example 8.1 A Valve Fails to Open This valve has two possible modes – open or closed. The failure mode is if the valve fails to open on demand. An industry database shows a similar valve has a beta distribution failure profile:

$$\alpha_{prior} = 1.75$$

$$\beta_{prior} = 156\,000$$

Operating data from Facility B shows three failures in 350 demands. What is the posterior probability that the valve will fail to open? What is a credible 90% CI for the probability?

The mean of the beta prior distribution is calculated by:

$$\mu = \frac{\alpha}{(\alpha + \beta)} \text{ or } 1.12 \times 10^{-5} = \frac{1.75}{(1.75 + 156\,000)}$$

To calculate the 5th and 95th percentiles of the prior distribution, the Microsoft Excel (2016) (Excel) EXCEL BETA.INV function can be used. The syntax for this function is

BETA.INV(probability, α, β)

Therefore,

5th percentile = BETA.INV(0.05, 1.75, 156 000) = 1.66×10^{-6}

95th percentile = BETA.INV(0.95, 1.75, 156 000) = 2.77×10^{-5}

The posterior distribution can also be described adequately with a beta distribution. The posterior distribution is calculated by:

$$\alpha_{\text{post}} = \alpha_{\text{prior}} + \text{failures}; 1.75 + 3 = 4.75$$

$$\beta_{\text{post}} = \beta_{\text{prior}} + \text{demands}; 156\,000 + 350 = 156\,350$$

The posterior mean is calculated by:

$$3.03 \times 10^{-5} = \frac{4.75}{(4.75 + 156\,350)}$$

The 5th and 95th percentiles of the posterior distribution can again be calculated using the BETA.INV function.

5th percentile = BETA.INV(0.05, 4.75, 156 350) = 1.16×10^{-5}

95th percentile = BETA.INV(0.95, 4.75, 156 350) = 5.63×10^{-5}

The probability of failure has increased with the addition of the new data.

Figure 8.7 shows a graph of the prior and posterior distributions.

Example 8.2 Pump Fails to Operate A feedwater pump fails during a run of 140 days of continuous operation. The next big push for the plant will involve a run of 240 days. What will be the posterior mean and 5th percentile to 95th percentile operating envelope? The pump parameters are:

$$\alpha_{\text{prior}} = 1.3$$

$$\beta_{\text{prior}} = 250\,000 \text{ hours}$$

$$140\,\text{days} = 3360\,\text{hours}$$

$$240\,\text{days} = 5760\,\text{hours}$$

The original mean is $\mu = \frac{r}{\beta_{\text{prior}}}$ $5.2 \times 10^{-6} = \frac{1.3}{250\,000}$ or 5.2×10^{-6}

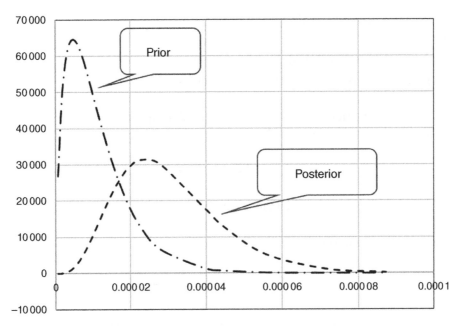

FIGURE 8.7 Prior and posterior distributions.

The 5th and 95th percentiles can be calculated using the EXCEL GAMMA.INV function:

$$\text{percentile} = [\text{GAMMA.INV (probability, alpha, 1)}]/\text{beta}$$

In this case beta = 1/beta calculated or the entire function divided by the hours.

$$\text{5th percentile} = [\text{GAMMA.INV (0.05, 1.3, 1)}]/250\,000 = 4.7 \times 10^{-7}$$

$$\text{95th percentile} = [\text{GAMMA.INV (0.95, 1.3, 1)}]/250\,000 = 1.42 \times 10^{-5}$$

It appears reasonable because the mean falls between the 5th and 95th percentiles. A Poisson model is selected because this is a continuous distribution.

$$\alpha_{\text{post}} = \alpha_{\text{prior}} + \text{failures}; \; 1.3 + 1 = 2.3$$

$$\beta_{\text{post}} = \beta_{\text{prior}} + \text{demands}; \; 250\,000 + 3\,360 = 253\,360 \text{ hours}$$

$$\text{Posterior mean} = \mu = \frac{\alpha_{\text{post}}}{\beta_{\text{post}}} 9.07 \times 10^{-6} = \frac{2.3}{253\,360} \text{ or } 9.07 \times 10^{-6} \text{ failures per hour}$$

$$\text{5th percentile} = [\text{GAMMA.INV (0.05, 2.3, 1)}]/253\,360 = 1.9 \times 10^{-6}$$

$$\text{95th percentile} = [\text{GAMMA.INV (0.95, 2.3, 1)}]/253\,360 = 2.06 \times 10^{-5}$$

Using the posterior mean failure rate of 9.07×10^{-6} failures per hour, the probability that the pump will not fail during a run of 5760 hours is calculated by

$$\text{Probability of failure} = \exp[-(9.07 \times 10^{-6} \times 5760)] = 0.949 \text{ or } 95\%$$

8.2.7 Conclusion

Bayesian analysis is a very useful tool for updating data for use in risk assessments. As Harmon states (7), "Bayesian analysis is essentially a weighted solution where the prior effect varies with the size of the data set. Posteriors from a small data set will typically not move significantly from the prior. Conversely, increasingly larger data sets will begin to overwhelm the prior and reduce its effect. If you chose Bayesian analysis because of the benefits it provides to small data sets, you may unintentionally impact the results with an unrealistic prior. A prior with a small variance (optimistic) implies a high degree of confidence the parameter only exists over a small range of values and is fair only if the data supports it. Conversely, a large variance (vague) used in an attempt to 'be fair' imparts little knowledge to the posterior and may result in unnecessarily wide and uninformative intervals. So, an overly optimistic or needlessly vague prior does not serve the analysis well. Ultimately, the analyst must be able to defend the choice of a prior and allow the data to tell the story."

8.3 MONTE CARLO ANALYSIS

The Monte Carlo method (or Monte Carlo simulation) can be used to describe any technique that approximates solutions to quantitative problems through statistical sampling. As used here, "Monte Carlo simulation" is more specifically used to describe a method for propagating (translating) uncertainties in model inputs into uncertainties in model outputs (results). Hence, it is a type of simulation that explicitly and quantitatively represents uncertainties. Monte Carlo simulation relies on the process of explicitly representing uncertainties by specifying inputs as *probability distributions*. If the inputs describing a system are uncertain, the prediction of future performance is necessarily uncertain. That is, the result of any analysis based on inputs represented by probability distributions is itself a probability distribution.

There are thousands of articles written on the Monte Carlo method. There are also good texts on the subject (9–11). Here we will keep it relatively simple. Excel can be used to develop simple Monte Carlo simulations. We will explain this method using Excel and some straightforward examples. Monte Carlo simulation is stochastic/probabilistic and as such uses probabilistic distribution curves as part of the model. Here we will focus on the normal and Poisson distributions.

Example 8.3 Normal Distribution   In this first example, we will assume the data are normally distributed. A normal distribution requires three variables: probability, mean, and standard deviation. The analyst has to have a starting point for these data,

TABLE 8.3
Input Data

Item of interest	Mean time between failures	Standard deviation
Pump fails to run	15 000 hours	2300

as with Bayesian. The method will not generate the starting point. MTBF will be the focus of this first example. Table 8.3 shows the inputdata.

The Excel function that will be used to generate the data is

NORM.INT(probability, Mean, SD)

However, since we are interested in simulating this, a random number generator will be used. The function now becomes

NORM.INT(RAND[], Mean, SD)

This will provide a random result based on the mean and standard deviation.

Figure 8.8 shows the result from one iteration. Next we want to run a set of iterations. We can select between two and a million. For this example we will use 25. Set up the Excel spreadsheet with 25 cells as per Figure 8.9.

Next, select the "Data" tab and then the "What-If Analysis" from the ribbon. From that pulldown select "Data Table." Put a blank cell into the "Column input

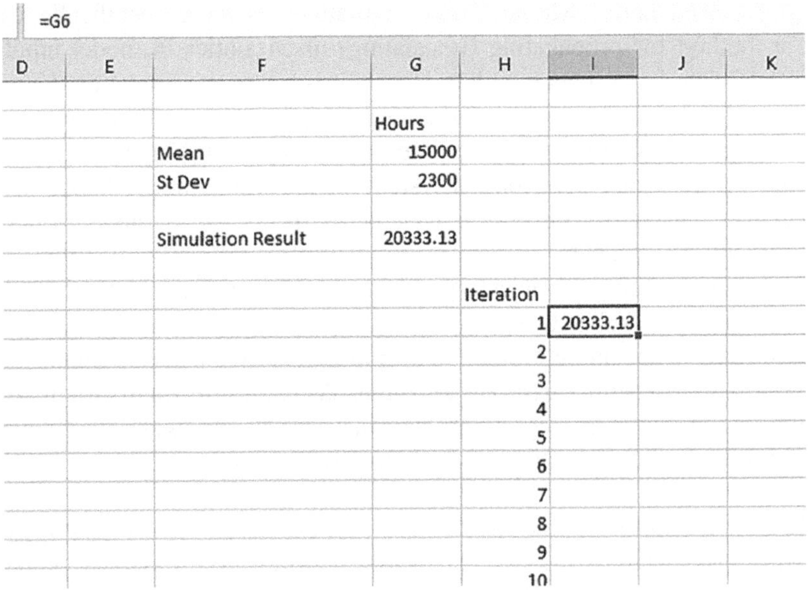

	=G6							
D	E	F	G	H	I	J	K
			Hours				
		Mean	15000				
		St Dev	2300				
		Simulation Result	20333.13				
				Iteration			
				1	20333.13		
				2			
				3			
				4			
				5			
				6			
				7			
				8			
				9			
				10			

FIGURE 8.8   Results from one iteration.

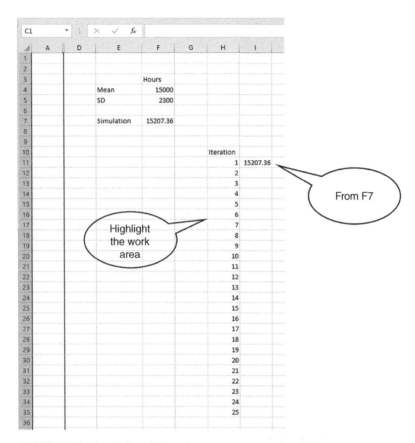

FIGURE 8.9   Results from 25 iterations.

cell." See Figure 8.10. A populated data table is inserted when "OK" is clicked. Note, that the simulation might change when a cell is clicked in the spreadsheet.

Statistics can now be calculated from the data that were developed. Table 8.4 shows a standard descriptive statistic table developed from the data. The resulting mean from the simulation is 14 954 and the standard deviation is 1 708.

Example 8.4 Switch Decision   Table 8.5 contains data on four different switches. The cost of labor to replace this critical switch and the associated downtime is $5000. Which switch should you order?

First, develop a spreadsheet with the variable and fixed costs. One way of setting this up is shown in Figure 8.11. Next, the following Excel equation is used to calculate how many switch failures will occur for 10 000 attempts:

$$= \text{RAND()}^* \text{ probability of failure for SWA}^*10\,000$$

A set number of iterations of this simulation are run. This can be two iterations to a million. The results are tabulated and the mean and standard deviation are calculated. A set of 20 iterations are run for each switch type. The results are shown in Figure 8.12.

FIGURE 8.10 Column input cell example.

TABLE 8.4
Statistical Data for Monte Carlo
Analysis using Normal Distribution

	Statistics
Mean	14 954.43
Standard error	341.75
Median	15 022.50

TABLE 8.5
Switch Data

Switch model	Unit cost ($)	Probability of failure
SWA	543.00	1/5 000 attempts
SWB	1750.00	1/12 000 attempts
SWC	2200.00	1/14 000 attempts
SWD	4200.00	1/20 000

Switch Model	Unit Cost	Probability of Failure	Fixed Cost	Demand on the Switch
SWA	$543	0.0002000	$5,000	10,000
SWB	$1,750	0.0000833		20,000
SWC	$2,200	0.0000714		30,000
SWD	$4,200	0.0000500		40,000
				50,000
				60,000
				70,000
				80,000
				90,000
				100,000

FIGURE 8.11 Spreadsheet example.

	SWA	SWB	SWC	SWD
1	1.516606	0.383962	0.077733	0.463184
2	0.306288	0.078528	0.349591	0.134942
3	0.751756	0.800709	0.178807	0.149484
4	1.13362	0.805178	0.276596	0.204489
5	1.761223	0.434585	0.63759	0.480649
6	1.442202	0.651354	0.702562	0.283158
7	0.24836	0.731052	0.260662	0.491316
8	1.952009	0.385593	0.671681	0.289489
9	0.509178	0.6865	0.160328	0.194664
10	0.275108	0.768709	0.464016	0.031599
11	0.172066	0.357836	0.642201	0.143906
12	1.009096	0.808381	0.276683	0.167027
13	1.067182	0.164775	0.521594	0.437526
14	1.744066	0.734256	0.642642	0.389462
15	1.549908	0.386964	0.642315	0.448494
16	1.099462	0.586713	0.174923	0.306136
17	1.742884	0.048292	0.266912	0.302651
18	1.632028	0.458787	0.337143	0.467587
19	0.19317	0.113588	0.018768	0.183579
20	1.772942	0.148717	0.14635	0.243045
Mean	1.093958	0.476724	0.372455	0.290619
Standard Deviation	0.624898	0.266141	0.22318	0.140037

FIGURE 8.12 Results from 20 iterations.

TABLE 8.6
Cost Comparison for Switches

Switch model	Unit cost ($)	Probability of failure	Mean failures per 10 000 attempts	Total cost ($)
SWA	543.00	1/5 000 attempts	0.97	5376.71
SWB	1750.00	1/12 000 attempts	0.46	3105.00
SWC	2200.00	1/14 000 attempts	0.36	2599.20
SWD	4200.00	1/20 000	0.23	2116

One cost factor might be the cost of a certain type of switch for 10 000 attempts of its use. This can be calculated by:

Cost per 10 000 attempts = (Unit cost + Fixed cost of repair) × Mean failures

Table 8.6 shows examples of these data. In this case, Switch D is the most cost-effective.

These are very simple examples of Monte Carlo simulation, but they do illustrate how the techniques are used in risk assessment.

Self-Check Questions

1. At an airport all passengers are checked carefully. Let T with $t \in \{0, 1\}$ be the random variable indicating whether somebody is a terrorist ($t = 1$) or not ($t = 0$) and A with $a \in \{0, 1\}$ be the variable indicating arrest. A terrorist shall be arrested with probability $P(A = 1|T = 1) = 0.98$, and a non-terrorist with probability $P(A = 1|T = 0) = 0.001$. One in ten thousand passengers is a terrorist: $P(T = 1) = 0.0001$. What is the probability that an arrested person actually is a terrorist?

2. A drug test (random variable T) has 3% false positives (i.e. 3% of those not taking drugs show positive in the test) and 10% false negatives (i.e. 10% of those taking drugs test negative). Suppose that 3% of those tested are taking drugs. Determine the probability that somebody who tests positive is actually taking drugs (random variable D).

3. A coin may or may not have a bias. It is thought that the coin has a 10% bias toward heads. A student is tasked with flipping the coin a thousand times to test this. The results come back as 575 times heads and 425 times tails. Is the coin biased? What is the true bias of the coin using a Bayesian update technique?

4. A certain pump has a manufacturers reliability value of one failure every 11 000 hours. A plant has three of the same pumps. Pump A failed at 12 500 hours, Pump B at 8 000 hours, and Pump C at 14 000. Using a Bayesian update technique, determine an updated reliability value.

5. A switch has a failure rating of one failure for every 5000 attempts. A switch fails after 2300 attempts. Use Bayesian update to determine a better failure rate.

6. Perform a simple Monte Carlo simulation to determine a failure rate for a pump that has a mean time between failures of 12 000 hours. The standard deviation is 1500. Run at least 100 iterations.

7. Using the same mean time between failures from (6), change the standard deviation to 3500. Run at least 100 iterations.

8. What benefit is there in using Monte Carlo simulation to help determine failure data?

REFERENCES

1. Bernardo, J.M. and Smith, A.F.M. (2000). *Bayesian Theory*, 1e. Wiley.
2. Laplace, P.-S. (1815). *Essai Philosophique Sur Les*. Paris: Courcier.
3. Discussed inGilles, D. (2000). The frequency theory. In: *Philosophical Theories of Probability*, 88. Psychology Press.
4. Orloff, J. and Bloom, J. (2018). *Bayesian Updating with Discrete Priors*. MIT OpenCourseWare https://ocw.mit.edu.
5. Nolan, D. (2002). Teacher's corner: you can load a die, but you can't bias a coin. *American Statistician* 56: 308–311. https://doi.org/10.1198/000313002605.
6. Cox, D.R. and Hinkley, D.V. (1974). *Theoretical Statistics (Example 11.7)*. Chapman & Hall.
7. Hamada, M., Wilson, A., Shane Reese, C., and Martz, H. (2008). *Bayesian Reliability*. New York: Springer Print.
8. Dezfuli, H. and United States National Aeronautics and Space Administration (2009). *Bayesian Inference for NASA Probabilistic Risk and Reliability Analysis*, NASA/SP-2009-569. Washington, DC: National Aeronautics and Space Administration.
9. Fishman, G. (2003). *Monte Carlo: Concepts, Algorithms, and Applications*, Springer Series in Operations Research and Financial Engineering. Springer.
10. Kroese, D.P., Taimre, T., and Botev, Z.I. (2011). *Handbook of Monte Carlo Methods*, 1e. Wiley.
11. Graham, C. and Talay, D. (2013). *Stochastic Simulation and Monte Carlo Methods: Mathematical Foundations of Stochastic Simulation*, Stochastic Modelling and Applied Probability. Springer.

CHAPTER 9

Developing Probabilities

9.1 RISK ASSESSMENT DATA

9.1.1 Introduction

Finding data to populate risk assessments can sometimes be problematic. For example, an analyst is in the process of performing a failure mode and effects analysis on a new, complicated system. When the analyst begins to populate the data tables with failure rates for various failure modes, he/she hits a brick wall. No failure rate data. The question becomes what to do next? Do a "scientific" wild-ass guess (SWAG), make it up (worse than a SWAG), or find a subject matter expert (SME) and have that person do a SWAG?

Data are available in many forms for hardware failure rates and for human error probabilities (HEPs), and this chapter will provide examples on how to find data and/or how to develop it.

9.1.2 Hardware Failure Rate Data

In the truest sense, hardware failure data is much easier to obtain than is HEPs. In many cases, failure rate data is available at the system as well as the subsystem and component levels. Hardware failure rate data can be obtained from the manufacture, from historical data, government and military handbooks, accident data or generated from testing by the user (1).

9.1.3 Manufacture

Failure rate data can be obtained from the manufacturer for certain pieces of industrial equipment, for instance, pumps, valves, motors, electrical panels, controllers, and even for components such as chips, diodes, and resistors. These data are usually

Risk Assessment: Tools, Techniques, and Their Applications, Second Edition. Lee T. Ostrom and Cheryl A. Wilhelmsen.
© 2019 John Wiley & Sons, Inc. Published 2019 by John Wiley & Sons, Inc.
Companion website: www.wiley.com/go/Ostrom/RiskAssessment_2e

supplied on a product data sheet or can be requested from the manufacture. In addition, the product data sheets will sometimes supply failure modes for the equipment. For instance, does a valve fail to open or closed, and does a control panel fail so that no spurious signal is sent?

9.1.4 Historical Data

Many organizations maintain internal databases of failure information on the devices or systems that they produce that can be used to calculate failure rates for those devices or systems. For new devices or systems that are similar in design and manufacture, the historical data for similar devices or systems can serve as a useful estimate.

9.1.5 Government and Military Handbooks

The Reliability Information Analysis Center (RIAC) compiles data and develops products for use in assessing the reliability of components and their failure modes (2). Handbooks of failure rate data for various components are available from government and commercial sources. MIL-HDBK-217F, *Reliability Prediction of Electronic Equipment*, is a military standard that provides failure rate data for many military electronic components (3).

National laboratories have also done studies on the reliability and failure modes of various systems and components in specialized applications. For instance, the Idaho National Laboratory has conducted numerous studies on the reliability of components for use in fusion test reactors (4). Table 9.1 shows the types of data available from these studies.

The error factor was discussed in Chapter 6.

TABLE 9.1
Example Failure Rate Data

Component	Failure mode	Failure rate	Error factor
Blow out panels	Fail on demand	1E−03/demand	Upper bound
Blow out panels	Leakage	1E−07/h	10
Blow out panels	Rupture	1E−09/h	10
Vent duct	Leakage	1E−11/m-h	10
Wet gas scrubber	Fail to operate	1E−02/demand	Upper bound
Wire mesh scrubber	Fail to function	1E−05/h	10
Gravel bed scrubber	Plugging	3E−06/h	10
Gravel bed scrubber	Internal rupture	5E−07/h	10
Gravel bed scrubber	Internal leakage	3E−06/h	10
Air dryer (refrigeration)	Fail to start	1E−02/demand	10
Air dryer (refrigeration)	Fail to operate	3E−05/h	10
HEPA filter	Plugging	3E−06/h	10
HEPA filter	Leakage	3E−06/h	10
HEPA filter	Rupture	5E−07/h	10

Table excerpted from EXT-98-00892 (MIL-HDBK-217F).

9.1.6 Commercial Data Sources

There are many commercially available failure rate data sources. Loss prevention handbooks, insurance companies, data mining organizations, and trade organizations can be sources of data for use in inclusion in risk assessments (5).

9.1.7 Operational Data and Testing

Within an organization failure rate data can be calculated from failures of components within a facility or multiple facilities. Accurate records as to the failure need to be kept for this data to be useful. The types of data that aids in a risk assessment include the following:

- How many hours, demands, or length of travel was on the device when it failed?
- What other factors were involved?
  - Environment – hot, cold, wet, dry.
  - Periodic maintenance performed or not.
  - How was the system operated?

Failures that lead to or are caused by accidents can aid in determining failure rates as well.

The most accurate source of data is to test samples of the actual devices or systems in order to generate failure data. This is often prohibitively expensive or impractical, so the previous data sources are often used instead.

9.1.8 Failure Rate Calculations

The failure rate of a system usually depends on time, with the rate varying over the life cycle of the system. Aircraft structure is an excellent example of a system that fails over time. A new aircraft structure, whether aluminum or composite, has not been stressed. However, over time and many thousands of cycles, an aircraft structure can become weakened and can develop fatigue cracks.

Aloha Airlines Flight 243 had a catastrophic decompression, and a portion of the fuselage ripped off after 23 minutes during a flight from Hilo to Honolulu, HI, on 28 April 1988 (6). The aircraft had over 89 000 flight cycles. The estimated life of the aircraft skin was approximately 75 000 cycles.

Table 9.2 shows failure data for a system. Failure rates can be easily calculated from such data. In fact, several types of failure rates can be calculated. A more qualitative rate can be determined by binning the failures into one of several bins, high number of hours (long life components), medium number of hours (medium life components), and low number of hours (short life components). In this case the analyst decides what constitutes the cutoffs for these values. These types of rates are useful for analyses such as a failure mode and effects analysis that are discussed in

TABLE 9.2
Hardware Failure Data

Component	Hours	Failure	Operating temperature (°F)
1	12 000	Yes	70–100
2	8 000	No	70–100
3	3 000	Yes	121–130
4	11 000	Yes	101–120
5	7 000	Yes	121–130
6	6 500	Yes	121–130
7	9 300	Yes	70–100
8	12 400	No	70–100
9	1 200	Yes	140+
10	4 600	Yes	101–120
11	8 900	No	70–100
12	7 100	Yes	70–100
13	4 300	Yes	121–130
14	6 900	Yes	101–120
15	8 500	Yes	70–100
16	2 800	Yes	121–130
17	4 100	Yes	121–130
18	9 800	Yes	70–100
19	3 900	Yes	101–120
20	8 700	Yes	70–100
21	9 200	Yes	70–100
22	9 600	Yes	70–100
23	800	Yes	140+

Chapter 8 or to provide management a rough order of magnitude estimate of the life of a component.

The first step in this process is to construct the bins. The bins for this example are shown in Figure 9.1. The components are binned according to their failures. Those components that did not fail are excluded from the bins. Using this method provides a great amount of information, such as that eight components had a long life, nine had a medium life, and three had a short life. Added to this can be performance data. Maybe the components that had a short life were subjected to an adverse environment or were maintained better. Table 9.3 shows the binning of the data by operating temperature and the associated failure rates. Workers, management, and lay people can sometimes understand a qualitative analysis such as this, rather than a hard, cold number. In this example it is evident that reducing the operating temperature increases the life of the component.

The operating hours for the components are totaled and then divided by the number of components to calculate a quantitative estimate of the failure rate. In this

Component	Hours
3	3000
5	7000
6	6500
10	4600
12	7100
13	4300
14	6900
17	4100
19	3900

Component	Hours
1	12000
4	11000
7	9300
15	8500
18	9800
20	8700
21	9200
22	9600

Component	Hours
9	1200
16	2800
23	800

Short-life components less than 3000 hours

Medium-life components 3000 – 7999 hours

Long-life components 8000 + hours

FIGURE 9.1   Binning failures.

TABLE 9.3
Binning of the Data.

Operating temperatures (°F)	70–100	101–120	121–140	140+
Failure rate (per h)	1.08E−4	1.5E−4	2.1E−4	1E−3
MTBF (h)	9275	6600	4616	1000
N	8	4	6	2

example there are 20 failures in 130 300 hours, so the estimated failure rate is:

$$\frac{20}{130\,300} = \lambda = 1.54\text{E} - 4 \text{ failures per hour.}$$

The inverse of this is the mean time between failures (MTBF):

$$\text{MTBF} = \frac{1}{\lambda} = \frac{1}{1.5\text{E} - 4} = 6494 \text{ hours.}$$

MTBF is often used instead of the failure rate. The MTBF is an important system parameter in systems where failure rate needs to be managed, in particular for safety systems. The MTBF appears frequently in the engineering design requirements and governs frequency of required system maintenance and inspections (1). MTBF assumes the unit can be repaired. MTBF is calculated directly by dividing the total operating hours by the total number of failures, as shown below:

$$\text{MTBF} = \theta = \frac{T}{N},$$

where θ is the MTBF, T is the total time, and N is the number of units.

Another useful term is mean time to failure (MTTF). MTTF assumes that all systems fail in the same way. MTTF is calculated by dividing the total amount of time of all components and dividing by the number of units under test.

$$\text{MTTF} = \gamma = \frac{T}{N},$$

where γ is the MTTF, T is the total time, and N is the number of units under test.

9.1.9 Accident Data

On 28 January 1986, the Space Shuttle Challenger lifted off from Cape Canaveral, and 73 seconds into the flight, the spacecraft blew up. This accident has been analyzed in great depth, and it is not the intention here to provide an analysis of the event. However, Table 9.4 lists some interesting facts that help calibrate accident probabilities before data about the shuttle program is presented.

Accident facts concerning the Space Shuttle program:

Some studies place the probability that an accident, similar to the Challenger accident where the vehicle is destroyed on launch, occurring is 1 in 556.

The probability of an accident occurring during the entire mission is 1 in 265. However, a study in the 1970s placed the odds of losing a Space Shuttle as 1 in 50.

TABLE 9.4
Accident Probabilities

Accident type	Probability
Injury from fireworks	5.1E−5
Injury from shaving	1.5E−4
Injury from using a chain saw	2.1E−4
Injury from mowing the lawn	2.7E−4
Fatally slipping in bath or shower	8.8E−4
Drowning in a bathtub	1.4E−6
Being killed sometime in the next year in any sort of transportation accident	0.012
Being killed in any sort of non-transportation accident	0.014
Being killed by lightning	4.3E−7
Being murdered	5.5E−5
Getting away with murder	0.5
Dying from any kind of fall	5.0E−5
Dying from accidental drowning	1.2E−5
Dying from exposure to smoke, fire, and flames	1.2E−5
Dying in an explosion	9.3E−6
Dying in an airplane accident	2.8E−6

The launch of the Challenger on 28 January was the 25th Shuttle mission. After that accident the actual probability of losing a launch vehicle was one in 25 missions. Subsequent to the accident, there were 87 successful launches. At this point, the risk of losing a Space Shuttle fell to one in 113 launches. On 16 January 2003 the Space Shuttle Columbia failed on re-entry. The probability for a shuttle accident then rose to two in 113 missions. However, the probability of losing a Space Shuttle on launch remained one in 113 missions, and the probability of losing a Space Shuttle on re-entry became one in 113 missions. As of this writing, there have been 132 Space Shuttle flights and two failures. However, this does not tell the entire story. A failure rate can be calculated from the flight hours, which is probably a better way to describe the reliability of the system. Table 9.5 contains flight statistics for Space Shuttles (7, 8).

There were two failures in 30 946 flight hours for a failure rate of:

$$\frac{2 \text{ failures}}{30\,946 \text{ hours}} = 6.5\text{E-5 failures per hour.}$$

A daily failure rate can also be calculated:

$$\frac{2 \text{ failures}}{1289.4 \text{ days}} = 1.5\text{E-3 failures per operating day.}$$

Finally, a yearly failure rate can be calculated:

$$\frac{1289.4 \text{ operating days}}{365 \text{ days per year}} = 3.53 \text{ operating years,}$$

$$\frac{2 \text{ failures}}{3.53 \text{ years}} = 0.56 \text{ failures per operating year.}$$

Using these rates puts the accidents in a better perspective. One accident in over 15 000 operating hours is a reasonable rate for a transportation vehicle. In 15 000 operating hours, a fleet of five cars would drive approximately 750 000 miles at 50 miles per hour.

TABLE 9.5
Space Shuttle Flight Stats

Orbiter	Year of first flight	Number of spaceflights	Flight time
Enterprise	1977 landing tests only	0	0
Columbia	1981	28	300 d 17 h 46 m 42 s
Challenger	1983	10	62 d 07 h 56 m 15 s
Discovery	1984	38	351 d 17 h 50 m 41 s
Atlantis	1985	32	293 d 18 h 29 m 37 s
Endeavor	1992	24	280 d 09 h 39 m 44 s
Totals		132	1289 d 09 h 52 m 48 s
Total hours			30 946 h

Rates similar to this can be calculated for all accident types, as long as the numerator and denominator are valid or as valid as possible.

Injury and illness rates are usually calculated using standardized formulas (9, 10). However, in general these rates are not as important for risk assessments; there are times they do have utility. Injury frequency rate can be calculated in several different ways. One way is based on 1 000 000 labor hours and is calculated using the following formula:

$$\text{Accident frequency rate} = \frac{\text{number of accidents} \times 1\,000\,000}{\text{actual labor hours}}.$$

For example, the accident frequency for an organization with 10 accidents and 250 000 labor hours is:

$$\text{Accident frequency rate} = \frac{10\,\text{accidents} \times 1\,000\,000}{250\,000\,\text{labor hours}}$$
$$= 40\,\text{accidents per million labor hours}.$$

The Bureau of Labor Statistics (BLS) uses a base of 100 full-time workers instead of 1 000 000 labor hours. Therefore, the rate using the BLS formula the rate becomes

$$\text{BLS accident rate} = \frac{10\,\text{accidents} \times 200\,000}{250\,000}$$
$$= 8 \ \text{accidents per 200 000 labor hours}.$$

Accident severity rates can be calculated using a variety of methods. For instance, the BLS disabling injury rate is calculated using the formula:

$$\text{BLS disabling injury severity rate} = \frac{\text{total days charged} \times 1\,000\,000}{\text{labor hours of exposure}}.$$

Illness rates are commonly calculated using two formulas (11). The first rate is an incidence rate. Incidence rate for a disease is the number of new cases for a population. As an example, there is a town with a population of 10 000. A flu virus invades the town. The town's medical staff begins to monitor the town, and during week one 150 people catch the flu. During week one, the incident rate for the flu becomes:

$$\text{Incident rate} = \frac{150\,\text{new cases}}{10\,000 \ \text{persons}} \times 100 = 1.5\%.$$

During week two, another 400 new cases of the flu are reported. The incident rate for week two becomes:

$$\text{Incident rate} = \frac{400\,\text{new cases}}{10\,000 \ \text{persons}} \times 100 = 4\%.$$

The prevalence rate is the number of active cases of disease in a population. In our mythical town of 10 000, there were 150 cases of flu during week one and 400 cases of flu during the week two. However, 80 of the people who caught the flu during week one got well. So, during week two, there are 470 active cases of the flu. The prevalence rate is calculated using the formula:

$$\text{Prevalence rate} = \frac{470 \text{ cases}}{10\,000 \text{ persons}} \times 100 = 4.7\%.$$

Many different versions of these calculations are available, and this is just a sample of the types of formulas that are used.

9.1.10 Monte Carlo Simulation

Monte Carlo simulation techniques are also used to aid in developing and updating failure rates. A general overview of Monte Carlo simulation is in Chapter 8. Other good sources of information on this topic are David Vose's *Risk Analysis: A Quantitative Guide* and Douglas Hubbard's *How to Measure Anything: Finding the Value of Intangibles in Business* (12, 13).

9.1.11 Human Error Probabilities

Chapter 12 discusses human reliability analysis in detail, along with the technique for human error-rate prediction (THERP) and other techniques. In this chapter methods are discussed on how to develop HEPs and/or rates from other sources.

Developing reliable data on any human behavior is problematic because humans do not always do what we want them to do. Therefore, the goal is to develop the most reliable HEPs possible. Similar methods are used to develop these data as were used to develop hardware data. However, humans do not like to be monitored 24/7, so the number of errors humans commit, but then discover and correct prior to an accident, cannot be absolutely determined. Also, most hardware systems are designed to work in certain environments, and tests can be conducted to determine the changes in the reliability or life of the system when the environment is altered. Humans operate in a wide range of environments, and how one human responds to a set of environmental conditions and stress levels varies widely from how another human might respond.

The first American astronauts were said to have the right stuff because they could handle environmental and psychological stress much better than the average person (14). Some people handle cold conditions better, while some people hot conditions. Some people can work in noisy environments, while some people have to have if possible completely quiet environments. Even with this, knowing where there is a higher probability for a human to commit an error provides analysts with information they can use to direct improvements to the system and make it less susceptible to human error.

Modifiers to HEPs are called performance shaping factors (PSFs) (15). PSFs allow risk analysts to help account for human response to environmental and psychological stressors.

Common PSFs include:

- Hotness/coldness.
- Noise level.
- Light (too much or too little and/or quality of the light).
- Vibration.
- Physical ergonomics.
- Experience.
- Training.
- Management.
- Time (too much or not enough to perform the task).
- Stress (too much or too little).
- Equipment design.
- Human machine interface.
- Sequence of tasks.

In a recent study, Ostrom and Wilhelmsen (16) developed risk models, HEPs, and PSFs for aircraft maintenance and inspection tasks. This work involves visiting airlines and third-party repair stations around the world, collecting data on aircraft maintenance and inspection, and, from these data, developing parameters on the risk of these operations.

Two methods of developing HEPs are discussed. These are the Delphi method, which is a formal approach to HEP development, and using performance data, which is a more informal approach to developing HEPs.

9.1.12 Delphi Method

The Delphi approach to decision making has been in use for many years (17, 18). It is a relatively simple technique to use but does require a commitment on the part of the panel who participates in the Delphi sessions. The technique utilizes a panel of experts and a facilitator to work on the consensus to develop HEP estimates and estimates of hardware failure rates. It is systematic in nature and has been used to help develop sales forecasts, to speculate on technology (military and civilian), and even to aid in national policy making. The focus of the following discussion is on developing HEP estimates using an approach Ostrom et al. used to develop HEP estimates for aircraft maintenance tasks (19). However, it can be applied to hardware failure rates as well, with the correct panel. The room where the Delphi process is conducted should be comfortable so that the panel members feel at ease.

The steps of the Delphi process for developing HEPs are as follows:

1. Select the panel.
2. Introduce the topic and calibrate the panel on the topic.
3. Present the parameters of the task to evaluate.
4. Perform the initial round of discussions on the error probability and voting.
5. Discuss the results of the first-round and second-round voting.
6. Discuss the results of the second round and repeat until consensus is reached.
7. Develop PSFs.

These steps are discussed below in the context of an example. This example will be to develop HEPs for an aircraft inspection task. This inspection task will be to determine the HEPs for visually inspecting a composite aircraft panel for small anomalies. The determination of the HEPs will be important in the development of a risk model for a next-generation, primarily composite aircraft.

Select the Panel

The panel should consist of SMEs in the field of interest, along with an expert in risk assessment and a facilitator. The SMEs should provide representation for the complete process. For this example, our SMEs should include:

- Qualified inspectors from an airline.
- Representation from a structures group from an aircraft manufacturer.
- Composite repair person.

The qualified inspectors are on the panel to provide reliable estimates of detection rates for composite anomalies. The representative from the structure group of an aircraft manufacturer provides information on the materials of interest, the types of anomalies that can develop, and data from the airline community as to the anomalies and the relative rates of detection. The composite repair person provides information on the anomalies seen in the field and the types of anomalies that are/are not detected.

The risk analyst on the panel ensures the panel discussions are pertinent to developing the HEPs of interest and the HEPs and PSFs are reasonable. The facilitator should understand the process, but does not need to be an SME. Instead, the facilitator should be good at eliciting a response on the topic, keeping the group focused on the goal, and moving things along.

It is always best to select the panel so that the least biased data can be developed. Therefore, the SMEs should, in our example, represent several airlines. Generally, avoid selecting people who commonly work with each other to break up, to the degree possible, the interdependences that can develop between colleagues. On our fictitious panel there are six inspectors, an aircraft manufacturer representative, a composite repair person, and a risk analyst.

Introduce the Topic and Calibrate the Panel

The facilitator begins the process by briefly introducing the panel. Each panel member should introduce themselves and briefly discuss their expertise. The facilitator next introduces the risk analyst. The risk analyst discusses the task at hand, in this case developing HEPs for the visual inspection of an aircraft composite panel. Next, the risk analyst presents data on other types of similar tasks. In this example, data on the probability of detection (POD) for fatigue cracks from the visual inspection of aluminum aircraft panels might be presented (20):

- 0.005 Probability of detecting a 0.1-in crack on a detailed inspection.
- 0.05 Probability of detecting a 0.5-in crack on a detailed inspection.
- 0.99 Probability of detecting a 2-in crack on a detailed inspection.

The probability of not finding the cracks is then:

- 0.1-in crack HEP is approximately $1.0 - 0.005 = 0.995$.
- 0.5-in crack HEP is approximately $1.0 - 0.05 = 0.95$.
- 2-in crack HEP is approximately $1.0 - 0.99 = 0.01$.

The parameters of what constitutes a detailed inspection are also presented. At this point all questions are addressed.

Present the Parameters of the Task to Evaluate

The next step involves discussing the parameters of the task at hand, that is, to discuss developing HEPs for the visual inspection for anomalies on composite aircraft panels. The aircraft manufacturer and repair person might start this discussion with a presentation on the types of anomalies that can be found. The inspectors might discuss the types of anomalies they have seen. It is up to the risk analyst and facilitator to ensure that only those items of interest are discussed. The parameters as to the size and shape of the anomalies that will be the focus of the Delphi session must also be presented. Of course, an aircraft panel that is destroyed will most likely be found. So, the panel members are focused on developing HEPs on anomalies that are 2 in diameter or smaller, since that is similar to the data presented in step 2.

Perform the Initial Round of Discussions on the Error Probability and Voting

The initial round of discussions on the HEP are initiated. Visuals, such as a scale showing relative probabilities, can be used to focus the panel. Figure 9.2 shows a scale used for this example.

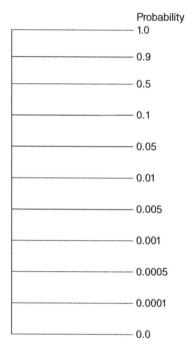

FIGURE 9.2   Relative probability scale.

The facilitator opens the discussion by telling the panel what size and type of anomaly the panel will be asked to develop an HEP for – in this case, a 0.5 in round dent on a composite panel that is 0.1 in depth. The panel is asked if they have any questions or need any additional information. For this first round, only a nominal value will be solicited. Meaning, the paint on the panel is in good shape, there is adequate lighting, and so forth. The panel will ask questions and discuss aspects of the task. At the end of the discussion, the panel is asked to mark on the scale, either in private or publicly, what they each think the probability of detecting that certain anomaly is. Each of the panel, except the risk analyst, marks the scale. Figure 9.3 shows how these marks might look. The panel takes a short break or is asked to consider the next item. In either case the risk analyst compiles the results.

Discuss the Results of the First-Round and Second-Round Voting

The results are presented to the panel after the votes are compiled by the risk analyst. In our example, the probability of detecting this anomaly ranges from a low of 0.0005 to a high of 0.9. This is a very broad range. However, the majority of the panel's votes were between 0.005 and 0.1. Discussion is elicited by the facilitator as to why panel members voted the way they did. It might be one panel has not had much experience with this type of anomaly or had a very bad experience missing a certain

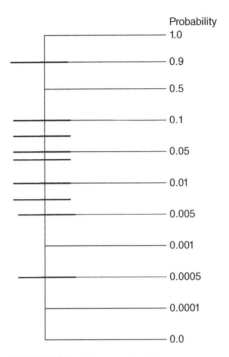

FIGURE 9.3 First-round voting.

size anomaly. These types of issues are discussed. In this example it was found that the very low probability of 0.0005 is due to the inspector's lack of experience with composite structure. The very optimistic 0.9 probability was also due to an inspector only working in a hangar with very bright lights. The facilitator and risk analyst discuss with the panel that these items might better be dealt with as PSFs and what they are trying to determine in a nominal value. At this point the second round of voting occurs. The results are again compiled.

Discuss the Results of the Second Round

Figure 9.4 shows the results of the second round of voting. As the figure shows, during this round the results are much tighter. The facilitator again would elicit discussion concerning the voting. The panel would then be asked to come to a consensus concerning the HEP. In this case, the natural value would be a probability of 0.05. This is the value the SMEs' votes clustered around. Discussion concerning this value would be elicited to see if it were a true consensus. If so, the voting ends, and if not, the voting continues until a consensus is reached.

Develop PSFs

The PSFs are developed in a similar fashion as the HEPs. Once the HEP is developed for a certain task, the panel is asked to consider how various parameters, such as

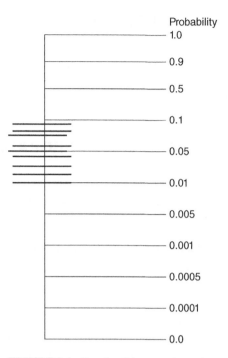

FIGURE 9.4   Results of the second round.

experience, stress, lighting, and noise, might affect the HEP. The panel first needs to come to consensus on which PSFs are important. For this example the panel decides that the following PSFs are deemed important for detecting composite panel anomalies:

- Training
- Experience
- Lighting
- Time

The panel comes to consensus that an inspector with one year or less supervisory experience lacks the experience to perform composite inspections with a high degree of reliability. A scenario is presented to the panel, such as what would be the probability of an inspector with one year of experience finding the same anomaly? The panel again votes on the topic and the process is repeated until a consensus is reached. A multiplication factor is then developed for that parameter for the HEP. In this case the panel feels that an inspector with only one year of experience has a probability of finding the anomaly of 0.5. Since the nominal HEP for this task was found to be 0.05, the multiplier for lack of experience is a multiplier of 10. This chapter discusses how these data are used.

9.1.13 Summary of the Delphi Process

The Delphi process is very useful for developing HEPs and corresponding PSFs for use in human reliability analyses. It is also useful for developing hardware failure rates. The keys to the process are having an appropriate panel and facilitator to ensure the panel keeps on track.

9.2 DEVELOPING HUMAN ERROR PROBABILITIES FROM OTHER SOURCES

As stated earlier, HEPs are not as readily available as hardware failure rates. However, the data can be developed from other sources, including

- Results from human factors and psychological studies.
- Operational and maintenance records.
- Specific research studies.

Though these HEPs, depending on the source, might not be as reliable as hardware failure rates, they are better than nothing. Take a complex maintenance task, for example, aircraft engine replacement. HEPs for specific aspects of the task can be developed from post-maintenance inspection records or even records from rework. Take a maintenance team replacing a control unit on an engine. In a given year, 100 of such control units are replaced. However, in five of these repairs, a functional check found the unit was not properly installed. A HEP can be developed for the task by dividing the five failures by the 100 total tasks:

$$\text{HEP} = \frac{5}{100} = 0.05.$$

Other factors surrounding this task can also be elucidated. Were all the failures at one facility, in certain weather conditions, on the ramp, or in the hangar?

These other factors are used to develop the PSFs for the HEP.

9.3 OVERALL SUMMARY

Hardware failure data and HEPs are available, but sometimes risk analysts have to dig for them. Hardware failure rate data is available from a variety of sources including handbooks, databases, research reports, and findings from accident investigations.

HEPs are available from standardized human reliability analysis processes like THERP, and they can be developed using tools like the Delphi process. The main point of this chapter is to use the best data available for risk assessments.

Self-Check Questions

1. Use the Delphi Technique to show how to develop failure probabilities for one or more of the following: 1. Failure to follow the normal protocol for a 4-way stop. 2. Failure to properly make a loaf of bread. 3. Success for entering an Ivy League College. 4. Failure to capture mice in a house. 5. Success for making a family road trip.

2. Identify a product or component and research the failure rates for that item. Discuss how you found the data and how you would manipulate it to develop a failure rate.

3. Identify an item in your home or your vehicle that frequently fails and discuss how you would develop failure rates for that item.

REFERENCES

1. Ebeling, C.E. (1997). *An Introduction to Reliability and Maintainability Engineering*. Boston: McGraw-Hill.
2. Reliability Information Analysis Center. Error rate data. https://www.quanterion.com/projects/reliability-information-analysis-center-riac (accessed 1 March 2019).
3. United States Military (1990). Mil-HDBK-217F, Military handbook: reliability prediction of electronic equipment (02-DEC-1991). http://everyspec.com/MIL-HDBK/MIL-HDBK-0200-0299/MIL-HDBK-217F_14591 (accessed 1 March 2019).
4. Cadwalleder, L. (1998). *Component Failure Rate Values from Fustion Safety Assessment*. Idaho Falls, ID: Idaho National Engineering and Environmental Laboratory.
5. Mannan, S. (2005). *Lees' Lost Prevention in the Process Industries: Hazard, Identification, Assessment and Control*, 3e. Boston: Butterworth-Heinemann.
6. National Transportation Safety Board (1989). Aloha Airlines. https://www.ntsb.gov/investigations/AccidentReports/Pages/AAR8903.aspx (accessed September 2018).
7. NASA. NASA Orbiter Fleet. https://www.nasa.gov/centers/kennedy/shuttleoperations/orbiters/orbiters_toc.html (accessed August 2018).
8. Wikipedia. Space shuttle program. https://en.wikipedia.org/wiki/Space_Shuttle_program (accessed August 2018).
9. Hammer, W. and Price, D. (2001). *Occupational Safety Management and Engineering*, 5e. Upper Saddle River, NJ: Prentice Hall.
10. Bureau of Labor Statistics (2018). Injuries, illnesses, and fatalities. http://www.bls.gov/iif (accessed 8 January 2011).
11. Last, J.M. (2001). *A Dictionary of Epidemiology*, 4e. New York: Oxford University Press.
12. Hubbard, D. (2007). *How to Measure Anything: Finding the Value of Intangibles in Business*. Hoboken, NJ: Wiley.
13. Vose, D. (2000). *Risk Analysis: A Quantitative Guide*, 2e. Chichester: Wiley.
14. Wolfe, T. (2008). *The Right Stuff*. New York: Picador.
15. Gertman, D.I. and Blackman, H.S. (1993). *Human Reliability and Safety Analysis Data Handbook*. New York: Wiley.
16. Ostrom, L. and Wilhelmsen, C.A. (2008). Developing risk models for aviation maintenance and inspection. *The International Journal of Aviation Psychology* 18: 30–42.

17. Brown, B.B. (1968). *Delphi Process: A Methodology Used for the Elicitation of Opinions of Experts*. Santa Monica, CA: RAND.

18. Delbecq, A.L., Van De Ven, A.H., Gustafson, D.H., and Van De Ven Delberg, A. (1986). *Group Techniques for Program Planning: A Guide to Nominal Group and Delphi Processes*. Middleton, WI: Green Briar Press.

19. Nelson, W., Haney, L., Richards, R. et al. (1997). Structured Human Error Analysis for Airplane Maintenance and Design. Idaho National Lab Report INEEL/EXT-97-01093, Idaho Falls.

20. Goranson, U. (1997). Fatigue issues in aircraft maintenance and repairs. *International Journal of Fatigue* 20: 413–431.

CHAPTER 10

Quantifying the Unquantifiable

10.1 INTRODUCTION

Other chapters discussed how to develop probabilistic data using techniques that include statistical analyses (Chapter 7), the Delphi technique (Chapter 9), and Bayesian analysis technique (Chapter 8). Chapter 19 also has information on qualitative research methods. In this chapter, the concept of thematic analysis will be discussed. Thematic analysis is a technique that can help quantify qualitative data.

10.2 THEMATIC ANALYSIS

Think of an analyst who needs to determine whether an aspect of a food product is preferable to another aspect. Of course, a study could be set up to count how many people like one aspect over another. However, if there are numerous possibilities, a quantitative study might be too expensive to run. Another possibility might be to conduct a thematic analysis. A thematic analysis is a very flexible and is relatively accessible tool that researchers use to quantify qualitative data (1, 2). In this case participants could taste the food in its various combinations of texture, taste combinations, colors, and sizes, and after each selection the participants note what they liked, disliked, or were neutral about. The data are then gathered and analyzed.

In such surveys the data are subjective in nature and are completely qualitative. In this sort of studies, thematic analysis can be applied to find important or frequent themes from the data. In the context of such a study, the theme is something important about the data in relation to the research question. This could be that a certain texture, taste, color, and portion size is the most desirable.

Risk Assessment: Tools, Techniques, and Their Applications, Second Edition. Lee T. Ostrom and Cheryl A. Wilhelmsen.
© 2019 John Wiley & Sons, Inc. Published 2019 by John Wiley & Sons, Inc.
Companion website: www.wiley.com/go/Ostrom/RiskAssessment_2e

Braun and Clarke (1, 2) develop a set of six steps that are commonly used to conduct a comprehensive thematic analysis. Some researchers or authors call these steps/phases. In either case, the steps/phases entail conducting the same activities. Table 10.1 contains these steps/phases.

10.2.1 Example Analysis

The best way to demonstrate how thematic analysis is conducted is through an example. The data that is usually collected during some sort of qualitative study is quite extensive. A limited data set will be provided here for demonstration purposes.

A real safety survey was conducted in a physics department at a large university that conducts both theoretical and applied physics research. The survey had multiple-choice questions, closed-ended questions, rank order questions, and open-ended questions. There were 75 potential respondents and 59 researchers responded. Therefore, the response rate was approximately 78%.

An example of a rank order question in the survey was "Select the five (5) potential hazards you feel are the most important in your laboratory." A list of potential hazards was provided. The results for that survey question are shown in Figure 10.1.

An example of a closed-ended question was "What learner style best describes you?" The results from this question are found in Figure 10.2.

The open-ended question that will be used in the example thematic analysis is "What was the most severe incident/accident you and or one of your coworkers experienced while at the university?"

The results from the survey are listed below:

1. Two colleagues in the mechanical workshop injured their hands in December 2017 just a couple of days apart: one due to an object falling on the hand and the other while swiping away splinters of metal.
2. I have not been here long enough to learn or experience such a situation.
3. None.
4. Someone tripped over a cable while giving a presentation.
5. The researcher was injured by the broken glass.
6. When there was a leakage of CO last summer, none of the faculty or staff members knew what to do. Essentially, faculty and staff have no safety training.
7. Although as such I have not witnessed any severe incident, however, the handling of chemicals and their waste needs serious attention, especially the perovskite materials based on lead metals and some unhealthy solvents DMF etc.
8. So far none that I am aware of.
9. Small injury with a needle bound to a syringe in the finger. Without chemicals in the needle.
10. "A chemical waste container made out of glass that exploded, many years ago. Glass bottles should not be used as waste containers. Luckily nobody was in the room.

TABLE 10.1

Steps/phases of a Thematic Analysis

Step/phase	Description of step/phase	Outcome
1	Familiarization with the data: Once the data are collected and aggregated, the researcher reads and rereads the data to become very familiar with it. Potential themes, patterns, and/or meanings are noted. It is important to be objective and not allow personal bias to taint this step/phase	During this step detailed notes and initial codes are developed
2	Generate initial codes: From the results of step/phase 1, initial codes are generated by documenting where and how patterns occur. This happens through data reduction where the researcher collapses data into labels to create categories for more efficient future analysis. Data complication is also completed here. This involves the researcher making inferences about what the codes mean	At the end of this step/phase, the researcher has comprehensive codes as to how data answers research questions
3	Combine codes: Codes are combined into overarching themes that accurately depict the data to the highest level possible. The researcher needs to accurately describe the themes' meaning(s). In some cases the theme might not seem to fit the data. At this point the researcher should also be able to describe what is missing from the data	At the end of this step/phase, the researcher has a set of themes for further study
4	Check step: The best way to describe this step/phase from an engineering perspective is as a quality control step. At this point the researcher looks at how the themes support the data and the overarching theoretical perspective. If the analysis seems incomplete, the researcher needs to go back and find what is missing	After this step/phase the themes should be able to represent the data and/or tell the story of the data
5	Interpreting the themes: During this step/phase the researcher defines what each theme is, the aspects of data that are being captured, and the interesting attributes of the themes	After this step/phase there should be a comprehensive understanding of the themes and how the themes contribute to the understanding of the data
6	Conclusions: During this final step/phase, the researchers write the report. At this point they must decide which themes make meaningful contributions to understanding the data. Researchers should also conduct "member checking." This is where the researchers go back to the sample at hand to see if their description is an accurate representation	At the conclusion of the analysis, the researcher will have produced a rich, thick description and understanding of the data

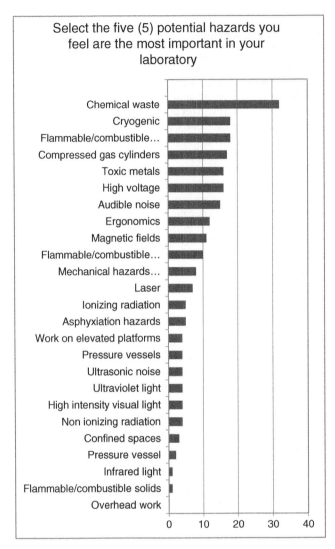

FIGURE 10.1 Results of rank order survey question.

11. There is a hazard missing. In the nano lab CO is used for making nanotubes, CO is toxic."

12. Accidentally (out of carelessness and ignorance) sublimating solid iodine in a vacuum oven into a substantial amount of iodine gas as a grad student at the chemistry department with chemist colleagues (also grad students).

13. I don't remember anything seriously dangerous.

14. None.

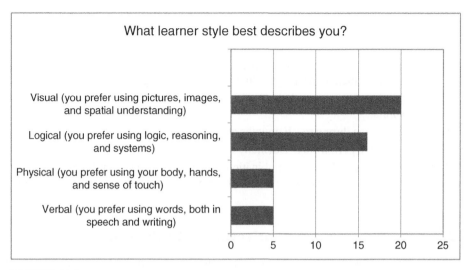

FIGURE 10.2 Results from a close-ended survey questions.

15. The cooling system of a weather chamber broke during a solar cell aging test, causing a rapid temperature rise since the solar simulator in the chamber produced heat. Temperature rose until the lamps of the solar simulator were broken.

16. I used Pasteur pipette as a funnel for propane (to make liquid propane by directing it into a cold container). I accidently shot the pipette when I applied too high flow. No one was injured as I did everything in a fume hood and used protective gear.

17. I'm not aware of any incidents in this department. In School of Chemical Engineering, breathing of harmful chemicals due to not using a fume hood.

18. No accidents during my time.

19. Dropping heavy item on finger./Sharp metal shears cutting wound./Hitting hand to machine part.

20. No accident has happened to people I work with since 2010.

21. Installation of a gas pressure regulator that was incorrectly assembled at the workshop so that the outlet was at bottle pressure (200 bar). As a result a gas pressure gauge in the gas line "exploded," but fortunately no personal injuries occurred.

22. Burning myself on items that came into contact with cryo liquids (liquid helium). They are only small burns, but it was those times that I didn't have easy access to Cryo-Gloves.

23. Continuous back pain from bad ergonomics.

24. Fire at the solar simulator. A cardboard sheet was used as a shade too close to the lamps and it caught fire.

25. Blister from opening a valve.

26. I don't remember anything too serious happening. I think there have been some cuts by broken glass or other sharp objects (knives, needles).
27. I'm a computer scientist, so the worst thing that is happening to me is that a cold air from the automatic ventilation is taking away a heat from my fingers and they are really cold then. Especially if I'm working for a long time.
28. During university years probably bruises in hands due to slipped wrench or something similar.
29. There was none during my 10 months at the university.
30. Second-degree burn from accidentally touching a hot heat gun.
31. Fortunately, none. This is a combination of luck and the fact that I have not been involved much with experimental work (mostly computational) despite almost 30 years at the university.

10.2.2 Step/Phase 1: Familiarization with the Data

The initial phase in thematic analysis is for researchers to familiarize themselves with the data. In this case the researcher should be familiar with the types of accidents that can happen in an applied physics laboratory. For this type of analysis, the researcher could already have a list of the types of accidents that can happen. This is called the "start list" of potential codes. These start codes should be documented. Analyzing data in an active way will assist researchers in searching for meanings and patterns in the data set. Reading and rereading the material until the researcher is comfortable is crucial to the initial phase of analysis. While becoming familiar with the material, note-taking is a crucial part of this step to begin developing potential codes.

When the list of accidents is reviewed, there are some attributes that are readily apparent. The first is that out of the 59 respondents to the survey, only 31 completed this question. This could be due to only 31 respondents have experienced an accident or that the 28 respondents who didn't answer the question got tired of taking the survey. This could be something for the researcher to follow up on.

Reading the responses shows that 12 indicate no injury accidents. However, the respondents wanted to provide that information. What are the major accidents? Are there any groupings?

It appears that hand injuries seem to be common. There appears to be broken glass, splintered metal, and other failed laboratory equipment.

10.2.3 Step/Phase 2: Generating Initial Codes

The second step in thematic analysis is generating an initial list of items from the data set that have a reoccurring pattern. This systematic way of organizing and gaining meaningful parts of data as it relates to the research question is called coding. The coding process evolves through an inductive analysis and is not considered to be linear process, but a cyclical process in which codes emerge throughout the research process. One way of doing this is to reorder the list and remove those responses where no injury occurred.

Hand Injuries

Two colleagues in the mechanical workshop injured their hands in December 2017 just a couple of days apart: one due to an object falling on the hand and the other while swiping away splinters of metal.

The researcher was injured by the broken glass.

Second-degree burn from accidentally touching a hot heat gun.

Dropping heavy item on finger./Sharp metal shears cutting wound./Hitting hand to machine part.

Small injury with a needle bound to a syringe in the finger. Without chemicals in the needle.

Blister from opening a valve.

I don't remember anything too serious happening. I think there have been some cuts by broken glass or other sharps objects (knives, needles).

During university years probably bruises in hands due to slipped wrench or something similar.

Chemical-Related Accidents/Incidents

When there was a leakage of CO last summer, none of the faculty or staff members knew what to do. Essentially, faculty and staff have no safety training.

Although as such I have not witnessed any severe incident, however, the handling of chemicals and their waste needs serious attention, especially the perovskite materials based on lead metals and some unhealthy solvents DMF etc.

A chemical waste container made out of glass that exploded, many years ago. Glass bottles should not be used as waste containers. Luckily nobody was in the room.

Accidentally (out of carelessness and ignorance) sublimating solid iodine in a vacuum oven into a substantial amount of iodine gas as a grad student at the chemistry department with chemist colleagues (also grad students).

I used Pasteur pipette as a funnel for propane (to make liquid propane by directing it into a cold container). I accidently shot the pipette when I applied too high flow. No one was injured as I did everything in a fume hood and used protective gear.

Temperature-Related Accidents

The cooling system of a weather chamber broke during a solar cell aging test, causing a rapid temperature to rise since the solar simulator in the chamber produced heat. Temperature rose until the lamps of the solar simulator were broken.

Burning myself on items that came into contact with cryo liquids (liquid helium). They are only small burns, but it was those times that I didn't have easy access to Cryo-Gloves.

Trips and Falls

Someone tripped over a cable while giving a presentation.

Pressure-Related Accidents

Installation of a gas pressure regulator that was incorrectly assembled at the workshop so that the outlet was at bottle pressure (200 bar). As a result a gas pressure gauge in the gas line "exploded," but fortunately no personal injuries occurred.

Ergonomic Related

Continuous back pain from bad ergonomics.

I'm a computer scientist, so the worst thing that is happening to me is that a cold air from the automatic ventilation is taking away a heat from my fingers, and they are really cold then. Especially if I'm working for a long time.

Fire

Fire at the solar simulator. A cardboard sheet was used as a shade too close to the lamps and it caught fire.

This cyclical process involves going back and forth between phases of data analysis as needed until you are satisfied with the final themes. Possible codes from this categorization and the number of injuries for each category/theme are:

- Hand injuries – 8
- Chemical-related accidents/incidents – 5
- Temperature related – 2
- Ergonomic related – 2
- Trips and falls – 1
- Pressure related – 1
- Fire – 1

The coding process is rarely completed the first time. Each time, researchers should seek to refine codes by adding, subtracting, combining, or splitting potential codes. Since this is a simple analysis, these themes/codes might be the finest detail that can be determined. Coding can be thought of as a means of reduction of data or data simplification. Using simple but broad analytic codes, it is possible to reduce the data. The process of creating codes can be described as both data reduction and data complication. Data complication can be described as going beyond the data and asking questions about the data to generate frameworks and theories.

10.2.4 Step/Phase 3: Combine Codes

In this step/phase the codes can be combined, and the process of developing themes and considering what works and what does not work within themes enables the researcher to begin the analysis of potential codes. In our analysis this process will be very minimal. It is important to begin by examining how codes combine to form overreaching themes in the data. The theme that is most evident in this analysis is that hand injuries are dominant. Though, accidents/incidents involving chemicals have the potential of being more severe. At this point in most thematic analysis studies, researchers have a list of themes and begin to focus on broader patterns in the data, combining coded data with proposed themes. Researchers also begin considering how relationships are formed between codes and themes. Themes differ from codes in that themes are phrases or sentences that identify what the data means. They describe an outcome of coding for analytic reflection. This is a slight stretch in this study, but there are two potential dominant themes:

1. Hand injuries are the leading result of accidents occurring to applied physics researchers.
2. Accidents/incidents involving chemicals have the most potential for causing serious injuries for applied physics researchers.

Themes of lesser importance include:

3. Accidents associated with temperature and pressure are less common but can cause multiple injuries and property damage.
4. Injuries associated with poor ergonomic conditions are less common and contained to one individual.

10.2.5 Phase 4: Check Step

This phase requires the researchers to search for data that supports or refutes the proposed theory. This allows for further expansion on and revision of themes as they develop. At this point, researchers should have a set of potential themes, as this phase is where the reworking of initial themes takes place. Since this data set is quite small, it is evident that the themes developed reflect the data. However, there are two levels of theme review:

Level 1
Reviewing coded data extracts allows researchers to identify if themes form coherent patterns. If this is the case, researchers should move onto Level 2.

Level 2
The researcher now considers the validity of individual themes and how they connect to the data set. This is crucial to completing this stage. It is very important to assess whether the potential thematic map accurately reflects the meanings in the data set. Once again, at this stage it is important to read and reread the data to determine if current themes relate back to the data set.

10.2.6 Phase 5: Interpreting the Themes

Defining and refining existing themes developed from the data that will be presented in the final analysis assists the researcher in analyzing the data within each theme. At this step/phase, identification of the themes' core meanings relates to how each specific theme affects the entire picture of the data. Analysis at this stage is characterized by identifying which aspects of data are being captured, what is interesting about the themes, and why themes are interesting.

10.2.7 Phase 6: Conclusions

Report writing occurs after the five previous steps/phases have been completed. The report contains the conclusions derived from the data and the process by which the researcher arrived to the conclusions. While writing the final report, researchers should decide on themes that make meaningful contributions to answering research questions that should be refined later as final themes. The report that is developed must contain a rich, thick description of the data and analysis. In many cases a qualitative report is much larger than a corresponding quantitative report. Every step must be well documented.

Developing Relative Quantitative Data

After a thematic analysis, researchers have a set of themes and coded data. A risk analyst can use these codes to provide some sort of binning to the data. The themes can be used as a sort of performance shaping factor. For instance, a risk ranking matrix is developed from the codes discussed above. An example matrix is shown in Table 10.2.

This table can be rearranged to reflect severity as shown in Table 10.3.

TABLE 10.2
Risk Ranking Matrix by Frequency

Accident/incident code	Likelihood	Potential severity
Hand injuries	Frequent	Minor to serious personnel injuries
Chemical-related accidents/Incidents	Frequent	Very severe and can cause severe multiple injuries and property damage
Temperature related	Infrequent	Moderately severe personnel injuries
Ergonomic in nature	Infrequent	Minor to serious personnel injuries
Pressure related	Very infrequent	Very severe and can cause severe multiple injuries and property damage
Fire	Very infrequent	Very severe and can cause severe multiple injuries and property damage
Trips and falls	Very infrequent	Minor to serious personnel injuries

TABLE 10.3
Risk Ranking Matrix by Severity

Accident/incident code	Likelihood	Potential severity
Chemical-related accidents/incidents	Frequent	Very severe and can cause severe multiple injuries and property damage
Pressure related	Very infrequent	Very severe and can cause severe multiple injuries and property damage
Fire	Very infrequent	Very severe and can cause severe multiple injuries and property damage
Temperature related	Infrequent	Moderately severe personnel injuries
Hand injuries	Frequent	Minor to serious personnel injuries
Trips and falls	Very infrequent	Minor to serious personnel injuries
Ergonomic in nature	Infrequent	Minor to serious personnel injuries

Self-Check Questions

1. Lidia wants to study how people feel after watching the newest animated movie. She asks each person coming out of the theater to fill out an open-ended question survey concerning their emotions while watching the movie. How could these data be analyzed?

2. Is coding best described as (i) measurement (i.e. creating variables for analysis), (ii) analysis, or (iii) a description of human information processing? Explain.

3. Suppose a consultant is brought in to see why a project went wrong at an organization. The objective is to learn the truth about what happened in an incident to guard against a similar failure in the future. The consultant visits the organization and

obtains from management a list of key players in the project. He then interviews these people and also interviews people that are named by them. He asks all of them what went wrong. He then puts together a composite picture (weighting some people's accounts more than others, based on his judgment of the accuracy of their reports) and presents the result to management. (i) Is this a good methodology given the research objective? (ii) Where would you place the attitude toward the interviewees that is implicit in this methodology along the continuum of subject–respondent–informant–expert?

4. Suppose someone wants to understand which attributes of entrepreneurs (e.g. personality, upbringing, resources, etc.) lead to long-term success. Should they use a survey to collect data or personal interviews? Why?

5. Coding is perhaps the most fundamental and universal element in text analysis. Discuss how coding is used in different kinds of studies, ranging from descriptive work to theory building.

6. You are interested in how musicians in the Austin area perceive the music world around them – the clubs, the bands, the public, the business. Design a survey to capture these perceptions.

7. Consider the following statement: "Very often, coding is not only how qualitative research is done, it is what we are studying when we do qualitative research." Explain what this refers to, and discuss.

REFERENCES

1. Braun, V. and Clarke, V. Using thematic analysis in psychology. *Qualitative Research in Psychology* 3: 77–101.
2. Guest, G., MacQueen, K.M., and Namey, E.E. (2011). *Applied Thematic Analysis*. New York: Sage.

CHAPTER 11

Failure Mode and Effects Analysis

When an analyst begins to perform a risk analysis, he/she first must determine what exactly they are analyzing. For this chapter we must first determine what we consider a failure. Is a failure the total loss of a spacecraft, aircraft, ship, or chemical plant? Or is it the failure to ensure there are enough funds in an account before using a debit card? On 2 November 2006 the NASA Mars Global Surveyor last communicated with Earth. Up to that point the spacecraft that had been launched in 1996 had operated four times as long as the design life and sent back huge amounts of geographical data on the Red Planet. Therefore, the mission was a great success. However, on 2 November 2006 after the spacecraft was directed to perform a routine adjustment of its solar panels, it sent back that it had experienced a series of alarms. The spacecraft then indicated that it had stabilized. However, that was its final transmission. Next, the spacecraft reoriented to an angle that exposed one of two batteries carried on the spacecraft to direct sunlight. This caused the battery to overheat and ultimately led to the loss of both batteries. The communication antenna was not oriented correctly and kept the orbiter from telling controllers its status. The system's programmed safety response did not include making sure the spacecraft orientation was thermally safe, and it failed (1).

However, since it had already outperformed its original mission, had it truly failed? We all would like things we buy to live longer than we expect. The B-52 is an example of an aircraft that has far outlived its design life. In 1952 when the first B-52 flew, no one would have expected it to still be a major player in the second decade of the 2000s. So, as originally stated, we have to have a firm understanding of what is a failure before we begin an analysis.

Risk Assessment: Tools, Techniques, and Their Applications, Second Edition. Lee T. Ostrom and Cheryl A. Wilhelmsen.
© 2019 John Wiley & Sons, Inc. Published 2019 by John Wiley & Sons, Inc.
Companion website: www.wiley.com/go/Ostrom/RiskAssessment_2e

11.1 INTRODUCTION

This section provides the basic instructions for performing a failure mode and effects analysis (FMEA) and a failure mode, effects, and criticality analysis (FMECA) for the purpose of analyzing procedures for risk. Also provided are examples of symbols and tables commonly used in the analysis process. An example of how these techniques are used for analyzing procedures is also provided.

11.1.1 Description

An FMEA is a detailed document that identifies ways in which a process or product can fail to meet critical requirements. It is a living document that lists all the possible causes of failure from which a list of items can be generated to determine types of controls or where changes in the procedures should be made to reduce or mitigate risk. The FMEA also allows procedure developers to prioritize and track procedure changes (2).

11.1.2 Why Is a Failure Mode and Effects Analysis Effective?

The process is effective because it provides a very systematic process for evaluating a system or a procedure, in this instance. It provides a means for identifying and documenting:

1. Potential areas of failure in process, system, component, or procedure.
2. Potential effects of the process, system, component, or procedure failing.
3. Potential failure causes.
4. Methods of reducing the probability of failure.
5. Methods of improving the means of detecting the causes of failure.
6. Risk ranking of failures, allowing risk informed decisions by those responsible.
7. A starting point from which the control plan can be created.

11.1.3 Types of Failure Mode and Effects Analyses

1. Procedure: Documents and addresses failure points and modes in procedures.
2. Process: Documents and addresses failure modes associated with the manufacturing and assembly process.
3. Software: Documents and addresses failure modes associated with software functions.
4. Design: Documents and addresses failure modes of products and components long before they are manufactured and should always be completed well in advance of prototype build.

5. System: Documents and addresses failure modes for system and subsystem level functions early in the product concept stage.

6. Project: Documents and addresses failures that could happen during a major program.

7. This document focuses on using the FMEA process for analyzing procedures.

11.1.4 Failure Mode and Effects Analysis Process

An FMEA is somewhat more detailed than a PHA and is conducted more on a step-by-step basis. Table 11.1 shows an example of an FMEA table. Note that a great deal of what is contained in a PHA is also contained in an FMEA. Therefore, this section will focus on the process of performing an FMEA.

The following constitutes the steps of an FMEA. These steps will be illustrated by the use of an example.

The first step is to create a flow diagram of the procedure. This is a relatively simple process in which a table or block diagram is constructed that shows the steps in the procedure. Table 11.2 shows the simple steps of starting a manual lawn mower. Note that this is a reasonable analysis and not an exhaustive analysis.

Table 11.3 shows the potential failure modes for each of the steps.

Table 11.4 shows the effect of the potential failures.

Table 11.5 lists the potential causes of the failures.

The basic process is complete once these four steps are completed. However, the next step in the FMEA process is very important for the procedure development process. This is providing a column listing the control measures for each of the potential failure causes. This step ensures that control measures are present and/or are adequate for each cause. It is very important to ensure that causes are not dismissed until there is an adequate control measure in place. Table 11.6 shows a listing of the control measures for each cause.

An additional technique used in FMEAs is to add the dimension of probability and criticality. This is known as an FMECA. An FMECA is an especially important technique for the assessment of risks in procedures because it can aid in:

1. The prioritization of steps/sections of procedures that need to be changed or the process changed to reduce risk.

2. Pointing out where warnings, cautions, or notes need to be added in procedures.

3. Pointing out where special precautions need to be taken or specialized teams/individuals need to perform tasks.

The criticality is mainly a qualitative measure of how critical the failure to the process really is. It is usually based on subject matter experts' opinion but can also be based on probability of occurrence and/or on the consequence or effect.

TABLE 11.1
Example FMEA Table

Item	Potential failure mode	Cause of failure	Possible effects	Probability	Criticality	Prevention

Step in procedure, part, or component	How it can fail:
	• Failures can be:
	  ○ Pump not working
	  ○ Stuck valve
	  ○ No money in a checking account
	  ○ Broken wire
	  ○ Software error
	  ○ System down
	  ○ Reactor melting down
		What caused the failure:
		Broken part
		Electrical failure
		Human error
		Explosion
		Bug in software
			Outcome of the failures:
			Nothing
			System crash
			Explosion
			Fire
			Accident
			Environmental release
				How possible is it:
				Can use numeric values:
				0.1, 0.01, or 1E-5
				Can use a qualitative measure:
				negligible, low probability, high probability
					How bad are the results:
					Can use dollar value:
					$10., $1 000., or $1 000 000
					Can use a qualitative measure:
					nil, minimal problems, major problems
						What can be done to prevent either failures or results of the failures?

TABLE 11.2
Process Steps For Starting a
Lawn Mower

FMEA, starting a lawn mower
Process steps
Check gas and oil
Fill as necessary
Set controls
Initiate starter

TABLE 11.3
Failure Modes Associated With Process Steps

FMEA, starting a lawn mower	
Process steps	Potential failure modes
Check gas and oil	Unable to remove gas cap
	Unable to remove oil plug
	Unable to determine depth of oil
	Oil or gas spill
Fill as necessary	No oil available
	Gas station closed
	No gas container
	Overfill gas
	Overfill oil
	Water in gas or oil
Set controls	Controls broken
	No instruction available
	Controls out of adjustment
Initiate starter	Starter malfunction
	Cord broken
	Engine flooded
	Ignition system malfunction

For the purposes of an FMECA, rough calculations can be developed using:

- Historical data
- A Delphi-like technique (3)
- Accident data
- Subject matter expert(s)
- Best estimate

Table 11.7 presents a way to calculate criticality based on probability.

TABLE 11.4
Effect of Potential Failures

	FMEA, starting a lawn mower	
Process steps	Potential failure modes	Potential failure effects
Check gas and oil	Unable to remove gas cap	Delay in process or personal injury
	Unable to remove oil plug	Delay in process
	Unable to determine depth of oil	Delay in process or the potential to overfill oil level
	Oil or gas spill	Environmental damage or potential for fire
Fill as necessary	No oil available	Delay in process
	Gas station closed	Delay in process
	No gas container	Delay in process
	Overfill gas	Potential for a fire or environmental damage
	Overfill oil	Environmental damage
	Water in gas or oil	Delay in process or engine damage
Set controls	Controls broken	Delay in process
	No instruction available	Delay in process
	Controls out of adjustment	Delay in process or engine damage
Initiate starter	Starter malfunction	Delay in process and/or repairs necessary
	Cord broken	Delay in process and/or repairs necessary
	Engine flooded	Delay in process
	Ignition system malfunction	Delay in process and/or repairs necessary

Note that the probability numbers in Table 11.5 provide an indication of the level of criticality and not an absolute failure probability.

Organizations have also developed risk matrices that can also be used to indicate criticality. Table 11.8 shows such a matrix. Note that these matrices provide a way to combine probability of occurrence with severity of consequence. Also note that these matrices are subjective in nature but do provide a way to systematically assess risk.

The following example (Table 11.9) shows all the elements of an FMECA developed for assessing the steps in the lawn mower-starting example. Note that probability can also be included. The first step in this process is to determine what does "criticality" mean in this context. Is it how bad might the consequences be or how critical the step is in the operation of the system? For this process we will make the assumption that criticality means how bad might the consequences be if we don't perform the step correctly.

11.2 SUMMARY

FMEA and FMECA are very effective tools. They can be applied to a broad range of applications and industries and are effective in elucidating the vulnerabilities of a system and its subsystems. Like PHA, FMEA and FMECA are applied early in the

TABLE 11.5
Failure Mode And Effects Analysis With Potential Causes Of Failures Listed

Process steps	Potential failure modes	FMEA, Starting a Lawnmower	
		Potential failure effects	Potential causes of failures
Check gas and oil	Unable to remove gas cap	Delay in process or personal injury	Cap rusted or broken
	Unable to remove oil plug	Delay in process	Operator error or plug cross threaded
	Unable to determine depth of oil	Delay in process or the potential to overfill oil level	Operator error or poor lighting
	Oil or gas spill	Environmental damage or potential for fire	Operator error
Fill as necessary	No oil available	Delay in process	Lack of planning
	Gas station closed	Delay in process	Lack of planning
	No gas container	Delay in process	Lack of planning
	Overfill gas	Potential for a fire or environmental damage	Lack of adequate equipment or operator error
	Overfill oil	Environmental damage	Lack of adequate equipment or operator error
	Water in gas or oil	Delay in process or engine damage	Poor practices
Set controls	Controls broken	Delay in process	Was not proper used on prior occasion
	No instruction available	Delay in process	Instructions not properly stored on prior occasion
	Controls out of adjustment	Delay in process or engine damage	Controls not properly maintained
Initiate starter	Starter malfunction	Delay in process and/or repairs necessary	Inadequate inspection or periodic maintenance
	Cord broken	Delay in process and/or repairs necessary	Inadequate inspection or periodic maintenance
	Engine flooded	Delay in process	Improper use of controls
	Ignition system malfunction	Delay in process and/or repairs necessary	Inadequate inspection or periodic maintenance

TABLE 11.6
Complete Table

FMEA, starting a lawn mower

Process steps	Potential failure modes	Potential failure effects	Potential causes of failures	Control measure
Check gas and oil	Unable to remove gas cap	Delay in process or personal injury	Cap rusted or broken	Cap maintenance program
	Unable to remove oil plug	Delay in process	Operator error or plug cross threaded	Operator training
	Unable to determine depth of oil	Delay in process or the potential to overfill oil level	Operator error or poor lighting	Operator training and provide additional lighting
	Oil or gas spill	Environmental damage or potential for fire	Operator error	Operator training
Fill as necessary	No oil available	Delay in process	Lack of planning	Ensure adequate oil is available
	Gas station closed	Delay in process	Lack of planning	Ensure fuel supply is available
	No gas container	Delay in process	Lack of planning	Provide equipment to minimize spill potential
	Overfill gas	Potential for a fire or environmental damage	Lack of adequate equipment or operator error	Provide equipment to minimize spill potential
	Overfill oil	Environmental damage	Lack of adequate equipment or operator error	Ensure fuel and oil containers are not exposed to sources of water
	Water in gas or oil	Delay in process or engine damage	Poor practices	
Set controls	Controls broken	Delay in process	Inspection and periodic maintenance	Institute inspection and periodic maintenance program
	Lack of labeling on the controls	Delay in process	Instructions not properly stored on prior occasion	Ensure controls are adequately labeled
	Controls out of adjustment	Delay in process or engine damage	Controls not properly maintained	Institute inspection and periodic maintenance program
Initiate starter	Starter malfunction	Delay in process and/or repairs necessary	Inadequate inspection or periodic maintenance	
	Cord broken	Delay in process and/or repairs necessary	Inadequate inspection or periodic maintenance	
	Engine flooded	Delay in process	Improper use of controls	
	Ignition system malfunction	Delay in process and/or repairs necessary	Inadequate inspection or periodic maintenance	

TABLE 11.7
Criticality Based on Probability

FMECA criticality		
Criticality	Relative probability rates	Probability rates
Very high: Failure is almost inevitable	1 in 3 to 1 in 2	0.33 to >0.50
High: Generally associated with processes similar to previous processes that have failed	1 in 20 to 1 in 8	0.05 to 0.125
Moderate: Generally associated with processes that have experienced occasional failures	1 in 2 000 to 1 in 80	0.005 to 0.0125
Low: Isolated failures associated with similar processes	1 in 15 000	0.000 067
Very low: Only isolated failures associated with almost identical processes	1 in 150 000	0.000 006 7
Remote: Failure unlikely. No failure ever associated with an almost identical processes	1 in 1 500 000	0.000 000 67

TABLE 11.8
Example Risk Matrix

Risk matrix

Consequence	Probability of failure				
	Very low probability <1 in 1 000 000	Low probability 1 in 1 000 000 to 1 in 100 000	Moderate probability 1 in 100 000 to 1 in 10 000	High probability 1 in 10 000 to 1 in 100	Very high probability >1 in 100
No effect					
Minor consequence (repair costs less than $100 or down time <1 h)	Low risk	Low risk	Low risk	Minor risk	Minor risk
Moderate consequence (repair costs from $100 to 10 000 or down time from 1 to 24 h)	Low risk	Low risk	Minor risk	Moderate risk	High risk
High consequence (repair costs from $10 000 to 100 000 or down time from 24 to 120 h or minor environmental spill or minor personal injury)	Low risk	Minor risk	Moderate risk	High risk	Very high risk
Severe consequence (repair costs >100 000 or down time >120 h or major environmental spill or severe injury or fatality)	Minor risk	Moderate risk	High risk	Very high risk	Severe risk

TABLE 11.9
Criticality Analysis

	FMECA, starting a lawn mower				
Process steps	Potential failure modes	Potential failure effects	Potential causes of failures	Control measure	Criticality of step
Check gas and oil	Unable to remove gas cap	Delay in process or personal injury	Cap rusted or broken	Cap maintenance program	Low criticality
	Unable to remove oil plug	Delay in process	Operator error or plug cross threaded	Operator training	
	Unable to determine depth of oil	Delay in process or the potential to overfill oil level	Operator error or poor lighting	Operator training and provide additional lighting	
	Oil or gas spill	Environmental damage or potential for fire	Operator error	Operator training	
Fill as necessary	No oil available	Delay in process	Lack of planning	Ensure adequate oil is available	High criticality if filling process is done incorrectly
	Gas station closed	Delay in process	Lack of planning	Ensure fuel supply is available	
	No gas container	Delay in process	Lack of planning		
	Overfill gas	Potential for a fire or environmental damage	Lack of adequate equipment or operator error	Provide equipment to minimize spill potential	
	Overfill oil	Environmental damage	Lack of adequate equipment or operator error	Provide equipment to minimize spill potential	
	Water in gas or oil	Delay in process or engine damage	Poor practices	Ensure fuel and oil containers are not exposed to sources of water	

155

TABLE 11.9
(*Continued*)

| | | FMECA, starting a lawn mower | | |
Process steps	Potential failure modes	Potential failure effects	Potential causes of failures	Control measure	Criticality of step
Set controls	Controls broken	Delay in process	Inspection and periodic maintenance	Institute inspection and periodic maintenance program	Low criticality
	Lack of labeling on the controls	Delay in process	Instructions not properly stored on prior occasion	Ensure controls are adequately labeled	
	Controls out of adjustment	Delay in process or engine damage	Controls not properly maintained	Institute inspection and periodic maintenance program	
Initiate starter	Starter malfunction	Delay in process and/or repairs necessary	Inadequate inspection or periodic maintenance		Low criticality
	Cord broken	Delay in process and/or repairs necessary	Inadequate inspection or periodic maintenance		
	Engine flooded	Delay in process	Improper use of controls		
	Ignition system malfunction	Delay in process and/or repairs necessary	Inadequate inspection or periodic maintenance		

The highly critical step in this process concerns adding oil or fuel. In these cases, then, warnings/cautions should be included in the procedure, or the system should be modified to include controls to prevent adding fuel to a hot engine.

design life of a system and used to ensure the system has no unidentified failure points. As with the Mars Global Surveyor, we have to determine up front what is a failure and what is success.

Self-Check Questions

1. Perform an FMEA on a small appliance.

2. In some cases an FMEA might be as detailed of a risk assessment that is needed. Why or why not?

3. Perform an FMEA on your house, apartment, or dorm room. Discuss what you found.

4. Perform an FMEA on your car, SUV, pickup, or public transportation system. Discuss what you found.

5. Do you think an FMEA should be performed on social media platform? Discuss what such an analysis might reveal.

REFERENCES

1. NASA (2007). Report Reveals Likely Causes of Mars Spacecraft Loss|NASA. http://www.nasa.gov/mission_pages/mgs/mgs (accessed August 2018).
2. Department of Defense (2012). *Standard Practice System Safety, Mil std 882E*. US Military.
3. Gertman, D. and Blackman, H.S. (1994). *Human Reliability and Safety Analysis Handbook*. New York: Wiley.

CHAPTER 12

Human Reliability Analyses

12.1 INTRODUCTION

12.1.1 Purpose

The purpose of this chapter is to provide guidance in "how to" perform a human reliability analysis (HRA). The methodology provided is not intended to be used as a cookbook. Each situation requiring an HRA is different. The references provide more detailed instructions, alternative methods, and estimations of human error probability (HEP) values (1).

12.1.2 Background

The need for a methodical approach to performing HRAs originated with the need to perform probabilistic risk assessments (PRAs) and probabilistic safety assessments (PSAs). For this chapter, PRA and PSA are considered interchangeable, and only PRAs will be referred to. While nuclear power plants may be the first processes to come to mind when discussing PRAs and HRAs, PRAs are performed on many processes or activities. Examples are the assembly and disassembly of nuclear weapons, petroleum refinery operations, chemical processing plants, etc.

PRAs typically focus on equipment failures. HRAs may be used to analyze the human response to an equipment failure. The PRA may also include a section that discusses the probability of human failure being the initiating event. Even if a full-scale PRA is not performed, an HRA may be beneficial. Any process or activity that involves humans is susceptible to human error. HRAs are used to quantify the probability of human errors. HRAs can also be used to identify steps or activities in the process that can be targeted for changes that could reduce the probability of human error (2).

Risk Assessment: Tools, Techniques, and Their Applications, Second Edition. Lee T. Ostrom and Cheryl A. Wilhelmsen.
© 2019 John Wiley & Sons, Inc. Published 2019 by John Wiley & Sons, Inc.
Companion website: www.wiley.com/go/Ostrom/RiskAssessment_2e

The basic steps to performing an HRA include:

1. Bounding the system
2. Task analysis
3. HRA modeling
4. Quantifying the HEP
5. Documentation
6. Methodology

12.1.3 Bounding the System

Bounding of the system to be analyzed is probably one of the most important steps in the performance of an HRA. The system being discussed is not a physical system but is a series of steps or actions that involve the potential for human failure.

As stated above, HRAs are typically performed in support of a PRA. The PRA may identify an equipment or human failure that requires a human to respond to mitigate the failure. For example, the failure may be the trip of a pump, and human is required to start the other pump. While the PRA would analyze the entire failure sequence, the HRA would focus only on the human involvement (i.e. the probability of human failure, in the sequence of events).

The same holds true for the failure sequences in the PRA that are initiated by a human error. There may be automatic system responses that occur, as a result of the initiating human error, but again, the HRA would focus on the probability of human failure.

For example, one accident usually analyzed in a nuclear power plant PRA is a rupture in the reactor piping. This accident is commonly called a loss-of-coolant accident or LOCA. A small portion of the sequence of events in a LOCA would be:

1. Reactor coolant system ruptures.
2. Drywell pressure reaches 2 psig.
3. Emergency core cooling system (ECCS) initiates.
4. Reactor SCRAM signal initiated.
5. Containment building isolates.
6. Emergency ventilation systems start.

To help keep track of the different HRAs, each should be given a descriptive title. The sequence of events for a LOCA is referred to as the LOCA sequence. The titles in the different sequences are identified in parentheses and italics.

The sequences and their relationship should be laid out using logic diagrams.

The operators must take numerous actions as part of the LOCA sequence. For example, the operators must:

1. Verify the ECCS initiates (ECCS Initiate).
2. Take the actions for the reactor SCRAM (SCRAM Actions).
3. Verify that the containment building isolates (Cnmt Isolation).
4. Verify that the emergency ventilation systems start (Emerg Vent Start).

In addition to verifying each of the automatic actions, if an automatic action does not occur, the operators must manually perform the action or take alternative action. For example, if the ECCS pumps do not automatically initiate, the operators must manually start the pumps. There are also numerous manual actions that must be taken such as shutting down the generator and starting the equipment to mitigate hydrogen production. Each of the sequence of steps would have a probability of failure.

As you can see this can easily get very complex, very fast. As a rule of thumb, each individual HRA should be bounded to a series of steps that have a measurable beginning and end (i.e. the start point and successful completion of the steps can be distinctly identified). These will be broken down to the lowest level possible.

As an example, look at the SCRAM Actions sequence. This sequence breaks down into nine separate steps:

1. All control rods insert into the reactor core.
2. The operator places the reactor mode switch in the shutdown position (MS Shutdown).
3. The operator verifies all control rods fully inserted (Verify Rods Inserted).
4. The operator verifies reactor power is decreasing (Verify Rx Power).
5. The operator selects and inserts the source range and intermediate range monitors (they monitor reactor power at low power levels) (Insert SRM/IRM).
6. The operator verifies reactor vessel level to be within the correct band (Verify Rx Level).
7. The operator verifies reactor pressure to be within the correct band (Verify Rx Pressure).
8. The operator verifies the reactor coolant pumps shift to slow speed (Verify RCP Downshift).
9. The operator shifts the feed water control system to single element (FW to Single Element).

Step 1 does not require human involvement. Therefore, they are excluded from the HRA but are included for continuity. Steps 2 through 9 require the involvement of a human, the operator, to be completed. The HRA for the SCRAM Actions sequence of steps would consist of several separate HRAs (i.e. an HRA for each step). A total probability of failure for the entire SCRAM Actions sequence can then be determined. The total probability for human failure of the SCRAM Actions sequence can then be used along with the probabilities determined for the other steps to determine the probability of failure for the LOCA sequence of events.

The steps will be broken down using task analysis into the separate actions the operator must perform. The HEPs will be determined for each of these steps.

12.1.4 Summary Points

- The system will typically be defined as an accident sequence by the PRA.
- The accident sequence should be broken down into its separate sequence of actions and steps.
- These sequences should be further broken down into a series of steps that have a measurable beginning and an end.
- Task analysis will break these steps down into the separate actions the operator must perform.

12.2 TASK ANALYSIS

The task analysis breaks down the steps to be analyzed into its smallest set of actions. This breakdown must be very detailed. The task analysis may include steps that do not require human involvement. These additional steps are for continuity and will be eliminated during the HRA modeling stage of the HRA. (See Section 12.3 for more information on task analysis.)

To perform the task analysis, information must be gathered on what actions are needed to complete the system to be analyzed. This information can come from procedures or other documentation that describes the system and the sequence to be analyzed (3).

Walk-throughs of the sequence with personnel familiar with the sequence are almost mandatory to ensure accuracy of the end result. Walk-throughs with more than one person will improve the credibility of the end result. These may also identify hidden problems (i.e. missing procedural steps, poorly designed instrumentation, poorly designed control layout, etc.), which may increase the probability of failure. Copious notes should be taken. Photographs or sketches of the instrumentation and control layout should also be obtained.

As part of the task analysis, the recovery from failure steps needs to be identified. During the walk-through, the operations personnel should be asked what steps they would take to recover from a failure to perform a step or failure to perform a step properly.

The task analysis does not have a specified format; you can use whatever is comfortable and fits the situation. An example sequence and method for the task analysis is:

1. Make a text list of the steps based on the procedure. This is typically a good place to start. Information from the walk-through can be used to flesh out the sequence.

2. Validate the list by walking through the sequence with the personnel that operate the equipment. Add notes concerning factors that may influence the completion of the actions, for example, switch positions, panel layout, lighting, color coding, etc.

3. Lay out the steps in sequence showing the path to be followed, branches to other steps, iteration of steps, etc. This can be done using flowcharting software, text lists, sketching the symbols by hand, etc. Another method is to list each step on a sticky note and then stick them in sequence on a wall or board. The advantage of this method is that as you gain information about the sequence, it is easy to rearrange the sticky notes.

4. Have the operating personnel validate the sequence.

This sequence should be done as many times as needed in order to thoroughly understand the actions and the factors affecting their performance.

We will use step 3, Verify Rods Inserted, from the SCRAM Actions sequence as an example task analysis as shown in Figure 12.1.

The operations personnel should validate this sequence of steps and the sequence changed or notes added as needed.

12.2.1 Summary Points

- Task analysis breaks down the steps to be analyzed into its smallest set of actions.
- The task analysis needs to be very detailed and include recovery actions.
- The operators must validate the sequence.

12.3 HRA MODELING

The purpose of the HRA model is to give a visual representation of the sequence of steps that:

- Shows the potential human errors and their mechanisms.
- Shows recovery paths.
- Enables error quantification.

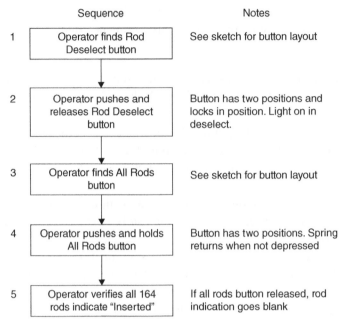

FIGURE 12.1   SCRAM actions for verify rods inserted.

There are numerous methods for modeling the HRA. There are first-, second-, and third-generation models. The first-generation models were developed to help predict risk and quantify the likelihood of human error. These methods encourage the risk assessor to break the task into subparts and then consider the potential impact of modifying factors such as time pressure, equipment design, and stress. By doing so they can determine a nominal HEP.

The second-generation models began in the 1990s and are still being developed. This generation considers context and errors of commission in human error prediction. These methods are not all validated. The third-generation methods are using additional tools based on the first-generation methods.

Let's look at some examples in Table 12.1 (4).

More examples are HRA event trees, fault trees, generic error modeling (GEM) causation diagrams, human error modeling/investigation tool (HERMIT), and the human error rate assessment and optimizing system (HEROS) (5).

Regardless of the methods used, the end result needs to be each step from the task analysis visually depicted as a failure with the applicable recovery actions. For example, we will use the HRA event tree method to model the task analysis done above on step 3, Verify Rods Inserted. The first step is to identify and number the actions in the task analysis that involves a potential for human error as shown in Figure 12.2.

TABLE 12.1

HRA Methods

Tool/generation	Description	Application
ASEP – Accident Sequence Evaluation Program – first gen	A shortened version of THERP developed for the USNRC Nuclear	Generic
ATHEANA – A Technique for Human Error Analysis – second gen	Resource intensive and would benefit from further development. Developed by the USNRC	Nuclear with wider application
CAHR – Connectionism Assessment of Human Reliability – second gen	A database method that is potentially useful. Available by contacting the authors (CAHR website)	Generic
CREAM – Cognitive Reliability and Error Analysis Method – second gen	Fully bidirectional, i.e. the same principles can be applied for retrospective analysis as well as performance prediction. The model is based on a fundamental distinction between competence and control	Nuclear with wider application
HEART – Human Error Assessment and Reduction Technique – first gen	Relatively quick to apply and understood by engineers and human factors specialists	Generic
NARA – Nuclear Action Reliability Assessment – third gen	A nuclear specific version of HEART (different author to the original). A proprietary tool	Nuclear
SPAR-H – Simplified Plant Analysis Risk Human Reliability Assessment – second gen	Useful approach for situations where a detailed assessment is not necessary. Developed for the USNRC. Based on HEART	Nuclear with wider application
THERP – Technique for Human Error Rate Prediction – first gen	A comprehensive HRA approach developed for the USNRC	Nuclear with wider application

Note that the step involving pushing the All Rods button is not numbered. This is because the button only has two positions pushed or not pushed. The Rod Deselect button is included because it must be pushed until it locks in position.

In an HRA event tree, the steps are worded as negatives (i.e. a failure to perform the action). The recovery actions are worded as positives (i.e. successfully performing the recovery action). Written as text the example steps would look like Table 12.2.

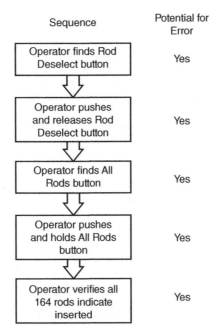

FIGURE 12.2 HRA event tree method.

TABLE 12.2
Recovery Action

Step	Failure action	Recovery action
1	Operator fails to find Rod Deselect button	Operator finds Rod Deselect button
2	Operator fails to lock Rod Deselect push button in Deselect position	Operator locks Rod Deselect push button in Deselect position
3	Operator fails to find All Rods button	Operator finds All Rods button
4	Operator fails to verify all 164 rods indicate – – (double dash)	Operator verifies all 164 rods indicate – – (double dash)

The HRA event tree for step 3, Verify Rods Inserted, is depicted in Figure 12.3 with the failures on the right side of the tree and successes on the left side. Figure 12.4 shows another depiction of the same task.

Note that dashed lines are used to indicate the recovery actions.

If the sequence were laid out in a logic tree, it would look like this.

12.3.1 Summary Points

- The purpose of the HRA model is to give a visual representation of the sequence of steps.

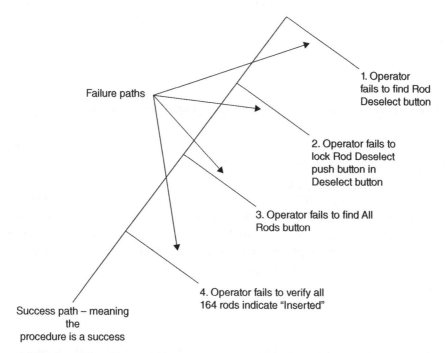

Failure paths

1. Operator
fails to find Rod
Deselect button

2. Operator fails to
lock Rod Deselect
push button in
Deselect button

3. Operator fails to find All
Rods button

4. Operator fails to verify all
164 rods indicate "Inserted"

Success path – meaning
the
procedure is a success

FIGURE 12.3   Verify rods inserted.

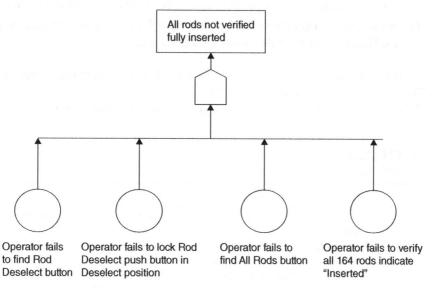

All rods not verified
fully inserted

| Operator fails
to find Rod
Deselect button | Operator fails to lock Rod
Deselect push button in
Deselect position | Operator fails to
find All Rods button | Operator fails to verify
all 164 rods indicate
"Inserted" |

FIGURE 12.4   Another view of verified rods.

- Regardless of the methods used, the end result needs to be each step from the task analysis visually depicted as a failure with the applicable recovery actions.

- The steps are worded as negatives. The recovery actions are worded as positives.

12.4 QUANTIFYING HUMAN ERROR PROBABILITY (HEP)

HEP is defined as the probability that when a given task is performed, an error will occur. This should not be confused with "human reliability," that is, the probability that the task will be correctly performed.

Like other parts of the HRA, there are several methods for quantifying the HEP. One of the most common is through the use of the technique for human error rate prediction (THERP) tables provided in NUREG/CR-1278, "Handbook of Human Reliability Analysis with Emphasis on Nuclear Power Plant Applications" (6). These tables provide HEPs and the other necessary information to quantify HEPs in many activities.

If HEP values are not in NUREG/CR-1278, then other sources that may be used are:

- Historical data – This could be for the specific activity or similar activities.

- Expert advice – Personnel with experience in performing HRAs can be consulted to obtain their estimation of the HEP.

- Testing – Simulations of the activity can be performed, and data collected.

- Other literature – There have been numerous PRAs performed on various processes. Data for same/similar activities can be used.

The HEP value used for the step must be recorded, including the source of the HEP. This can be done on the HRA event tree or in table format.

THERP tables (Table 12.3) will be used as a source to continue the example.

TABLE 12.3
THERP Tables

Step	Failure action	HEP	Source
1	Operator fails to find Rod Deselect button	0.003	Table 20-12, item 2
2	Operator fails to lock Rod Deselect push button in Deselect position	0.003	Table 20-12, item 10
3	Operator fails to find All Rods button	0.003	Table 20-12, item 2
4	Operator fails to verify all 164 rods indicate – – (double dash)	0.001	Table 20-11, item 1

TABLE 12.4
Control Switches

All rods	CRDM fault	Over travel in	Over travel out	Group
Individual drive	Gang drive	—	Rod Select	Rod Deselect

It must be noted that item 2 of THERP table 20-12 states, "Select wrong control on a panel of similar appearing controls identified by labels only." The other choices on the table are controls that are arranged in well-delineated functional groups and controls that are part of a well-defined mimic layout. To choose the most correct item, the analyst must refer back to the photographs, drawings, and sketches that were made of the panel layout. In this case, the switches are laid out as shown in Table 12.4.

The switches are color coded but the color is only visible when they are lit. The All Rods push button only lights up when depressed. The Rod Deselect push button lights up in the Deselect position. There are no specific instructions as to position of the switches during normal operations.

For the verification of all rods, the operator must look at a well-laid out mimic of the control rods relative positions to each other. The background is flat black and the indication lights (LEDs) are orange. Normally the LEDs show a number that is representative of the rod position for any rods that are selected. If no rods are selected, the display is blank. During an SCRAM, an emergency shutdown of a nuclear reactor, the indication shows the double dashed line as the position for each rod.

The shift technical advisor (STA) using the computer also checks the position of the control rods.

Other factors, commonly called performance shaping factors (PSFs), need to be considered. These are such things as use of a check-off procedure, lighting, switch layout, experience, time pressure, and dependence.

The example considers the following PSFs:

Is there a procedure that is followed? The steps are described in a procedure; however, operators are required to commit the procedure steps to memory. They are expected to complete the full SCRAM Actions sequence and then refer to the procedure to verify if they have completed all steps. Because of the importance of the step used in the example, as soon as the SCRAM is identified, the senior reactor operator is required to ask the reactor operator if all rods are fully inserted. The individual steps are considered second nature. Therefore, a value of 0.01 from Table 9.4 will be used.

Is there time pressure to complete the sequence? There is a high time pressure to complete the sequence. All of the actions in the SCRAM Actions sequence must be completed as soon as practical. A rough estimate of the time it normally should take to complete the SCRAM Actions sequence is three minutes. Therefore, the HEP must be modified by the values listed in Table 12.5. In our example, there is a heavy task load and a dynamic situation. The operators have

TABLE 12.5
Total HEP Calculations

Step	Original HEP	Procedure	Time pressure	Final HEP
1	0.003	×0.01	×5	0.00015
2	0.003	×0.01	×5	0.00015
3	0.003	×0.01	×5	0.00015
4	0.001	×0.01	×5	0.00005

experienced numerous SCRAMs during their training, so they should be considered skilled. The value from Table 12.5 is ×5.

To calculate the total HEP for the task, the HEPs originally determined for the steps must be modified by the factors described above.

The total HEP for a sequence is either the sum or the product of the individual HEPs. Looking back at the sequence laid out in a logic tree, it can be seen that the steps in the Verify Rods Inserted sequence are connected as an AND statement. Therefore, the product of the individual HEPs is the HEP for the Verify Rods Inserted. This value is

$$(0.000\,15) \times (0.000\,15) \times (0.000\,15) \times (0.000\,05) = 1.7 \times E - 16$$

It is clear that this is a very small probability of failure.

If the actions are connected by OR statements, then the total HEP is the sum of the HEPs for the individual steps.

A factor that may need to be considered in some sequences is dependence. Dependence is the relationship between the tasks. NUREG/CR-1278, Chapter 10, provides a thorough discussion of how to determine dependence. In our example, though there is an independent check performed, dependence does not need to be considered. That is because the STA's check does not verify that the operator did the Verify Rods Inserted task. It only verifies that the rods are fully inserted. Dependence would exist if:

- A second person verified any of the steps that the operator did, as he did them.
- The STA's check in some way verified that the operator had done his check.

If the check is performed on each step, then the dependence PSF is applied to each step. If the check is performed at the end (i.e. on the entire sequence), then the dependence PSF is applied to the HEP for the sequence (7).

12.4.1 Summary Points

- HEP is defined as the probability that when a given task is performed, an error will occur.

- HEP values may come from NUREG/CR-1278, historical data, expert advice, testing, or other literature.
- The HEP value used for the step must be recorded, including the source of the HEP value.
- PSFs need to be considered. They may apply to the HEP values for the individual steps or the HEP value for the sequence.
- For sequences logically connected using AND gates, the HEP values are the product of the individual HEP values.
- For sequences logically connected using OR gates, the HEP values are the sum of the individual HEP values.

12.5 DOCUMENTATION

As a rule of thumb, the documentation must be adequate so that the results of the HRA can be reproduced. This means that it must include the HRA models and the task analyses. All assumptions and references must be identified. Sketches and photographs that support decisions or assumptions also should be included. This does not mean that all field notes, operator interviews, etc. need to be in the final report. They should be retained as supporting documents.

The methodology described above shows several tables and lists. These can be consolidated in the final product.

12.5.1 Summary Points

- The documentation must be adequate so that the results of the HRA can be reproduced.
- Lists, tables, etc. can be combined in the final report.

12.6 USE OF HUMAN RELIABILITY ANALYSIS TECHNIQUES FOR ANALYZING PROCEDURES

HRA techniques should be reserved for the most critical procedures. For instance, HRA should not be used for analyzing simple maintenance procedures because of the cost of performing the analysis. However, for critical operations it is appropriate. Determining where recovery actions, such as inspection steps, should be in procedures is the best use of HRA techniques. Tools like THEA [1] can be used for such analyses. The following analysis shows how HRA can be used to determine where recovery actions should be placed.

A modified coolant flush procedure will be used as the basis of the example. Procedure is listed below.

12.6.1 Procedure

Warning – Cooling system must be below 100 °F prior to draining.

1. Begin with the engine cold and ignition off. Remove the radiator pressure cap.

Warning – Ethylene glycol coolant is toxic and must be disposed of in an appropriate manner.

2. Open the petcock at the bottom of the radiator and drain the coolant into a bucket.
3. Close the petcock and fill the radiator with water.
4. Start the engine and turn the heater control to hot. Add cooling system cleaner and idle the engine for 30 minutes (or as per the instructions on container).

Warning – Cooling system must be below 100 °F prior to draining.

5. Stop the engine and allow it to cool for five minutes. Drain the system.
6. Close the petcock, fill the radiator with water and let the engine idle for five minutes.
7. Repeat step No.5. Close the petcock.
8. Install new 50/50 mixture of water and nontoxic antifreeze/coolant.

Figure 12.5 is an HRA event tree representation of the procedure. Eight failure paths were found for this procedure. Some of these failure paths are more critical than others. Adding a recovery step will greatly reduce the probability of a failure (see Swain and Guttman (6) for a complete discussion of recovery and dependency). However, an inspection or other recovery step should be included if the PHA or FMEA/FMECA has shown these steps as being critical. For instance, if there is a 0.01 probability of a failure and an inspection step is added that has a probability of failure of 0.1, then the probability of failure is reduced by a factor of 10:

$$\text{(Procedure step)}\, 0.01 \times \text{(inspection step)}\, 0.1 = 0.001$$

In this procedure the critical steps are those dealing with the temperature of the cooling system and the toxicity of the coolant. Therefore, inspection steps are added after steps 1, 2, 5, and 8. Steps 1 and 2 are listed below with the inspection steps.

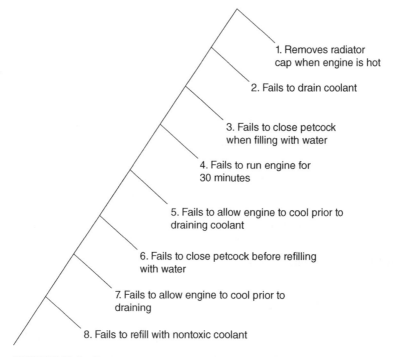

1. Removes radiator cap when engine is hot

2. Fails to drain coolant

3. Fails to close petcock when filling with water

4. Fails to run engine for 30 minutes

5. Fails to allow engine to cool prior to draining coolant

6. Fails to close petcock before refilling with water

7. Fails to allow engine to cool prior to draining

8. Fails to refill with nontoxic coolant

FIGURE 12.5   HRA event structure of coolant flush procedure.

12.6.2 Procedure with Inspection Steps

Warning – Cooling system must be below 100 °F prior to draining.

1. Begin with the engine cold and ignition off.
2. Second mechanic verifies engine is cool.
3. Remove the radiator pressure cap.

Warning – Ethylene glycol coolant is toxic and must be disposed of in an appropriate manner.

4. Open the petcock at the bottom of the radiator and drain the coolant into a bucket.
5. Coolant disposal technician is contacted for disposing of coolant.

Figure 12.6 shows how the HRA event tree would be modified for steps 1 and 2.

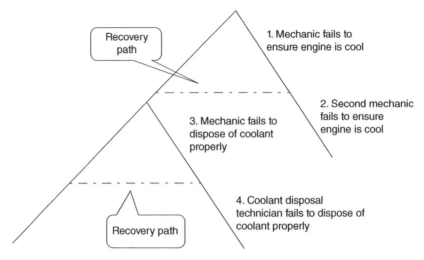

FIGURE 12.6   Modified event tree.

Self-Check Questions

1. What is one of the most important steps in the performance of an HRA?

2. What are some of the methods used for modeling an HRA?

3. What is an HEP?

4. Should HRA techniques be used for analyzing all procedures?

REFERENCES

1. Ostrom, L.T. and Wilhelmsen, C.A. (1998). THEA – tool for human error analysis. Presented at the System Safety Society Conference, Seattle, WA (September 1998).
2. Galliers, J., Sutcliffe, S., and Minocha, S. (1999). An impact analysis method for safety: critical user interface design. *ACM Transactions on Computer-Human Interaction* 6 (4): 341–369.
3. Battelle Memorial Institute. Columbus Laboratories; American Institute of Chemical Engineers. Center for Chemical Process Safety (1985). *Guidelines for Hazard Evaluation Procedures*. New York, NY: American Institute of Chemical Engineers.
4. Bell, J. and Holroyd, J. (2009). *Review of human reliability assessment methods RR679*. Buxton, UK: Health and Safety Executive https://humanfactors101.files.wordpress.com/2016/04/review-of-human-reliability-assessment-methods.pdf (accessed 17 July 2018).
5. Richei, A., Hauptmanns, U., and Unger, H. (2001). The error rate assessment and optimizing system (HEROS). *Reliability Engineering and System Safety* 72: 153–164.
6. Swain, A.D. and Guttman, H.E. (1983). *Handbook of Human Reliability Analysis with Emphasis on Nuclear Power Plant Applications*. Washington, DC: US Nuclear Regulatory NUREG/CR-1278.
7. Gertman, D.I. and Blackman, H.S. (1994). *Human Reliability and Safety Analysis Handbook*. New York: Wiley.

CHAPTER 13

Critical Incident Technique

13.1 INTRODUCTION

A technique that has applicability to a wide range of risk assessments is the critical incident technique (CIT). Forms of the CIT have been in existence since the 1930s and were further developed by Colonel John C. Flanagan during World War II (1, 2). CIT is a set of procedures used for collecting direct observations of human behavior that have critical significance and meet methodically defined criteria. CIT aids in the development of specific failure pathways that can later be further developed into event, fault, or human reliability event trees. The following discusses CIT and presents examples of its use.

13.2 WHY CONDUCT A CIT?

Critical incidents need to be written down in order to communicate the information clearly. Enough information needs to be written in order for others to understand the background and the behavior of the person involved in the incident. Table 13.1 shows a few examples of some critical incident information.

It is easy for us to provide a simple answer to questions, which does not provide useful information about the incident or any behavioral examples. It is helpful to ask the person to describe details about the behavior of a person or give examples of that behavior. Latham and Wexley (3) suggested several practical tips in collecting critical incidents. The person being questioned should not be asked about their own performance because they tend to recall effective incidents more than ineffective incidents. The incident should have occurred no longer than 6–12 months because the memory

Risk Assessment: Tools, Techniques, and Their Applications, Second Edition. Lee T. Ostrom and Cheryl A. Wilhelmsen.
© 2019 John Wiley & Sons, Inc. Published 2019 by John Wiley & Sons, Inc.
Companion website: www.wiley.com/go/Ostrom/RiskAssessment_2e

TABLE 13.1
Sample Critical Incidents

Poor	Better
The student quickly extinguished the fire in the chemical lab. It was lucky that no one was hurt, and the damage was minimal	A fire started in the chemical lab at the university. The student was walking by the lab and happened to smell smoke as he passed. He entered the lab and quickly extinguished the fire with the wall fire extinguisher. His quick action prevented a major fire, an explosion from the stored chemicals, and injuries to anyone in the building
The janitor ignored the water on the floor in the hall and made the cost of damages rise drastically	The janitor spotted some water in the hallway of the office building he was cleaning but went about performing his nightly duties and ignored the small water puddle. He completed his duties and left the building. When the employees arrived for work the next morning, the building had flooded on two floors, making it impossible to work and causing major damage to equipment, furniture, and important information contained in several file cabinets
The girl was very cognizant of a strange car following her as she journeyed to her home after the school day. She called for help and was safe	A young girl left the school on her daily route home when she noticed a red car following her. She stayed on the sidewalk and did not attempt to cross the road. She knew her route well as she traveled the same one every day, and when she came to a good neighbor's home, she noticed they were just leaving in the car. She ran to the car and asked for help. The police were called, and the girl was able to give a good description of the car and the driver as well as the license plate number. The police were able to find the guy and found out he was wanted for child kidnapping in another state. The young girl's awareness to her surroundings saved her from being kidnapped

tends to distort after a prolonged timeframe from the actual incident. Positive recall should come before the negative information to eliminate the fear of a witch hunt.

1. Get the key context or circumstances of the behavior.
2. Get the behavior for one incident. Make it specific to that incident. Decide whether the behavior is effective or ineffective.
3. Get the outcome. The outcome usually tells you whether the behavior was effective or ineffective.
4. Try using the term behavioral examples, rather than critical incidents, and start with the good examples.

13.3 METHOD

As with the Delphi process, CIT can be conducted in many ways. Kanki and Hobbs (4) used this form of CIT to develop process maps of aviation maintenance and inspection tasks.

Process mapping is a technique that produces visual representations of the steps involved in industrial or other processes. Therefore, it can be such a useful technique for developing initial failure paths. It was originally developed for use in manufacturing industries but has subsequently become a common tool in many industries worldwide. Process mapping has been standardized by the American National Standards Institute (ANSI); however, the system included with the Microsoft Visio software has also become widely used. Process maps have been found to be invaluable in revealing areas for improvement in processes, including obstacles, bottlenecks, areas where personnel lack formal guidance, coordination problems, and problems with equipment or other hardware. Because of this, process maps can also be used as the bases of simulations of industrial processes. Process mapping is used in holistic operational assessments of efficiency that develops or analyzes solutions that are sustainable.

Even in businesses where established systems have been in place for many years, process maps can help to clarify the steps that are carried out, the people who perform the tasks, and the points at which communication occurs. Although manuals and procedure documents are usually available, in many cases, only when information is gathered from the people on the "shop floor" can the actual operational steps in the process be established.

Process mapping involves describing the steps involved in a process and placing them in time order. Each step is described and identified as involving either decisions, reference to documents, communication, inspection activities, or other specialized steps. Process mapping is ideally suited to processes such as the maintenance of aircraft (5).

The following discusses how Kanki and Hobbs (4) are using CIT to develop process maps of aviation maintenance and inspection tasks related to aircraft structural composite material. Their objective was to describe the activities involved in the maintenance and inspection of composite structures at different maintenance, repair, and overhaul (MRO) operators, ultimately producing a single generic process map for this process that will have applicability across operators. The process-based modeling method could then be used to identify areas of operational risk where future tools, techniques, or other aids could prove useful.

The CIT process used began with structured interviews conducted with representatives of each of the following aviation professional and trade groups:

- Aviation maintenance technicians (AMTs)
- Ramp personnel
- Engineering staff
- Inspectors
- Composite shop personnel
- Maintenance control

Structured interviews were held with a cross section of aviation personnel drawn from the groups outlined above. The interviews followed a standard format.

Participants received an introduction to the project, including an explanation of the aims of the project and how their expertise would assist in achieving the goals of the project. Participants were then asked to describe their job and their work experience. During this introductory segment of the interview, participants were free to mention any issues they wished to raise relevant to composite structures.

A general flow of the process as it was understood at that point was explained; participants were then asked to describe from their own experience a specific incident in which potential damage to a structure was discovered, either via a scheduled inspection, an unrelated task, or the detection of a hazardous event. Their experience with previous interviews found it most beneficial to focus the participant's attention on a single case, rather than asking them to consider the entire range of potential airworthiness issues during the interview.

As participants described an incident, the sequence of events was laid out in front of them using paper notes or labels. This enabled participants to comment on the flow of events, rearrange the order of events if necessary, and add detail to ensure that a complete picture of the process was obtained. As the events were displayed in front of the participant, open-ended probe questions were used to ensure that the trade group of the people involved in each stage were identified (e.g. AMTs, ramp personnel, pilots), and the location of the step was specified (e.g. gate, hangar, composite shop). Green-colored notes were used to identify the tools, documents, and other resources necessary to perform the step. Red notes were used to identify problem areas that could interfere with the step.

The nature of the step was also identified as belonging to one of the following categories:

- Decision
- Inspection
- Creation of document, entry of information, signature, or stamp
- Communication (face to face, phone, or email)
- Movement of aircraft or component
- Repair
- Obtain approval
- Other

The goal was to have a process flow or map developed at the end of each participant interview. Figure 13.1 shows the type of process map that would be developed from each participant interview.

The desktop process maps created during the interviews were converted into initial process maps using the methods and symbols of the Microsoft Visio process mapping system (Mic11). Figure 13.2 shows the symbols used to populate the process charts.

FIGURE 13.1   Process map from interview. *Source*: Courtesy of Kanki and Hobbs.

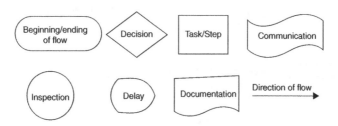

FIGURE 13.2   Process map symbols.

Initial process maps for each participant were then combined into a generic process map for each trade group at each MRO organization as shown in Figure 13.3. In many cases, this involved removing case-specific details in order to express the general sequence of events. Process maps for each trade group at each MRO organization were then combined into a generic overall process map for each operator.

To date Kanki and Hobbs have developed generic process maps for four MRO operations sites.

The beauty of the technique is it provides insights into a process from the performer's perspective. The technique is simple to apply, but a large number of participants might be required to provide a broad enough perspective of the process. Qualitative researchers use the term "saturation" when enough data is collected from interviews for the purpose of developing an understanding of a phenomenon (6). It is the judgment of the researcher as to when saturation is reached. Therefore, an exact number of participants is difficult to calculate before the interview process begins. Once data saturation occurs no further data need be collected.

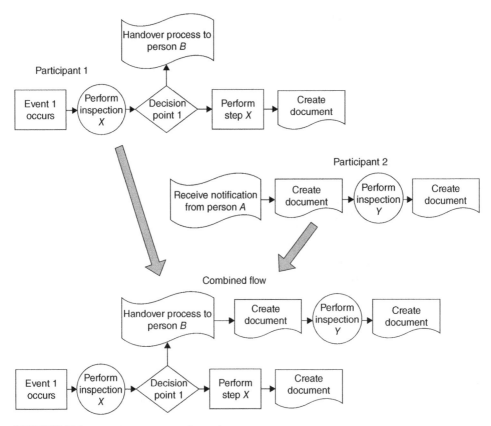

FIGURE 13.3   Generic process map for trade group.

13.4 BUILDING ON THE RESULTS OF A CRITICAL INCIDENT TECHNIQUE SESSION

The process flow or map is just the first step in the development of a risk assessment, using the CIT. The process map can be used to guide future interviews on the process with the intent of eliciting other pertinent facts about a process of a failure path. Figure 13.4 is a simple process map of an incident.

FIGURE 13.4   Simple incident process map.

The risk analyst can then use this process map as a guide for interviewing participants on the attributes of the process. There are five steps in this process. The following are questions a risk analyst might ask for each step:

Step 1: Ramp Agent Observes Anomaly on Aircraft Cargo Door
   a. Approximately, how big was the anomaly?
   b. Could you describe its appearance?
   c. Do you normally perform a visual inspection of the cargo door, and/or is this a part of your normal job requirements?
   d. Were you following a procedure when you performed the inspection?
   e. What time of day was it?
   f. Was it overcast?
   g. Was it bright sunshine?
   h. If at night, did you use a flashlight?
   i. How did you perceive the lighting to be?
   j. Do you wear corrective lenses?
   k. Was it very noisy?
   l. Were you wearing hearing protection?
   m. Was anything obscuring your vision?
   n. Was it hot or cold at that time?
   o. Was the panel generally clean or dirty when you observed the damage?
   p. How often do you find anomalies?
   q. In your rough estimation, what is the probability that you could find such anomaly in the future?

Step 2: Ramp Agent Discusses Damage with Supervisor
   a. How is your relationship with your supervisor?
   b. Did you show the damage to your supervisor?
   c. What was your supervisors response?
   d. Did the supervisor provide any guidance on reporting the damage without your input?

Step 3: Ramp Agent Discusses Decision to Report Damage to Maintenance Control with Supervisor
   a. How receptive was the supervisor in your desire to report the damage?
   b. Did the supervisor provide any insight into reporting the damage?
   c. Where did you discuss this?

Step 4: Ramp Agent Decides to Report Damage to Maintenance Control
   a. Why did you decide to report the damage to maintenance control?
   b. Were you following a procedure or protocol when you reported the damage?

    c. In the past were you rewarded for reporting damage?

    d. What do you feel the consequences would be if you did not report the damage?

    e. How was your report received?

Step 5: Ramp Agent Discusses the Maintenance Control Damage

    a. Had conditions changed from your initial observations?

    b. How did maintenance control respond to your finding?

    c. How did you feel after this interaction with maintenance control?

    d. Did maintenance control provide you any feedback from your finding?

The answers from these questions would then be used to further flesh out the parameters of the risk assessment.

The same type of process could be used for gaining insights into hardware failures as well. In this case the process map would be created with the hardware system as the focus. The questions would then relate to the hardware components' successes and failures. There is nothing that would prevent a combined human and hardware analysis either. In fact, an integrated approach is always of benefit.

13.5 SUMMARY

CIT is a very useful tool in the risk analyst's kit. The results from these types of analyses can provide great insights into a process. It is an easy technique to use but does require enough participants to attain data saturation prior to the development of process maps or failure paths. Though the focus of this chapter was on CIT, other accident analysis techniques can be used as well with good results in the development of risk assessments. Root cause analysis is, of course, a great technique to use for aiding in a risk assessment (7). Whatever the technique, the goal is always the same – develop the best risk assessment model as possible.

Self-Check Questions

1. What is CIT? Can this method be used for industrial processes only? If it can be used for other processes, name a few.

2. How would you use process mapping within the CIT method?

3. Do you need to follow any standard format within the CIT process?

4. How many participants does this process require?

5. Once the process map is developed, how many steps are involved in the process that can be used by the risk analyst?

REFERENCES

1. Flanagan, J. (1954). The critical incident technique. *Psychological Bulletin* 51 (4): 327–358.
2. Carlilsle, K. (1986). *Analyzing Jobs and Tasks*. Englewood Cliffs, NJ: Educational Technology Publications.
3. Latham, G. and Wexley, K. (1993). *Increasing Productivity Through Performance Appraisal*, 2e. Los Angeles: Sage.
4. Kanki, B. and Hobbs, A. (2010). *Development of An Approach for the Identification of Operational Risks Associated with Inspection, Maintenance, and Repair of Damage: An Application to Composite Structures*. Moffet Field, CA: NASA Ames Research Center.
5. Eiff, G. and Suckow, M. (2008). Reducing accidents and incidents through control of process. Special Issue on Aircraft Maintenance Human Factors. *International Journal of Aviation Psychology* 18 (1): 43–50.
6. Denzin, N.K. and Lincoln, Y.S. (2005). *The Sage Handbook of Qualitative Research*, 3e. Thousand Oaks, CA: Sage.
7. Eicher, R. and Knox, N. (1992). *Mort User's Manual*. Idaho Falls, ID: Idaho National Engineering Laboratory. https://www.osti.gov/servlets/purl/5254810 (accessed 5 March 2019) https://doi.org/10.2172/5254810.

CHAPTER 14

Basic Fault Tree Analysis Technique

The fault tree analysis (FTA) technique is proven to be an effective tool for analyzing and identifying areas for hazard mitigation and prevention while in the planning phase or anytime a systematic approach to risk assessment is needed. FTA is used as an integral part of a probabilistic risk assessment. In this chapter we will cover the very basics of FTA. The NASA Fault Tree Handbook with Aerospace Applications (1) is a complete guide to FTA.

14.1 HISTORY

Knowledge of the history of the need for FTA is useful for understanding the simple yet powerful potential of the tool. This history begins with the inception of mechanical vehicles. One common problem that plagued vehicles was malfunction and failures caused by "little things."

Steam engines blew up when pressure relief valves stuck closed. Early autos scattered parts across the countryside as nuts and bolts separated. Airplanes fell to earth because poorly designed fittings tore apart. Always it was the little things that failed and set up potentially deadly chain reactions.

Despite major advances in design and manufacturing techniques, significant numbers of accidents and failures continued to occur. Airplane accidents, attributable to training, accounted for over one-third of the losses during the WWII years 1941–1945. Over 14 000 major accidents were recorded in the United States alone.

Often, the airplane accidents were attributed to "pilot error." However, the majority of crashes should have been linked to a malfunction of little things … a failed hydraulic pump … a broken feathering stop … a missing lock nut.

Risk Assessment: Tools, Techniques, and Their Applications, Second Edition. Lee T. Ostrom and Cheryl A. Wilhelmsen.
© 2019 John Wiley & Sons, Inc. Published 2019 by John Wiley & Sons, Inc.
Companion website: www.wiley.com/go/Ostrom/RiskAssessment_2e

As technology became more exotic, technological advances exceeded the average skill level for operation and maintenance of advanced air vehicles. Because of the complexity of systems, nuts and bolts errors became even more frequent. An improvement in safety analysis was needed.

This technique had to be capable of handling systems of enormous complexity and allow detailed analysis at the nuts and bolts level. The basic premise behind the development of the tool was LITTLE THINGS CAUSE ACCIDENTS. The first FTA was developed and applied by Bell Telephone Laboratories in 1962, with the requirements in mind. The tool was initially applied to the Minuteman ICBM. As a result of the FTA of that extremely complex system and taking corrective measures, the missile was rated as one of the safest in the USAF inventory (2).

14.2 APPLICATION

Fault trees show graphically the interaction of failures and other events in a system. Basic events are depicted at the bottom of the fault tree and are linked via logic symbols (known as gates) to one or more of the top (TOP) events. These TOP events represent hazards or system failure modes for which predicted reliability or availability data is required. Typical TOP events might be:

- Total loss of production
- Explosion
- Toxic emission
- Safety system unavailable

As indicated, the fault tree begins at the end, so to speak. This top-down approach starts by supposing that an accident takes place. It then considers the possible direct causes that could lead to this accident. Next it looks for the origins of these causes. Finally it looks for ways to avoid these origins and causes. The resulting diagram resembles a tree, thus the name.

Fault trees can also be used to model success paths as well. In this regard they are modeled with the success at the top, and the basic events are the entry-level success that put the system on the path to success.

14.3 FAULT TREE CONSTRUCTION

The goal of fault tree construction is to model the system conditions that can result in the undesired event. The analyst must acquire a thorough understanding of the system before beginning the analysis. A system description should be part of the analysis documentation. The analysis must be bounded, both spatially and temporally, in order to define a beginning and endpoint for the analysis.

The fault tree is a model that depicts graphically and logically represents the various combinations of possible events, both fault and normal, occurring in a system, leading to the TOP event. The term "event" denotes a dynamic change of state that occurs to a system element. System elements include hardware, software, human, and environmental factors.

14.4 EVENT SYMBOLS

The symbols shown in Table 14.1 show the most common fault tree symbols. These symbols represent specific types of fault and normal events in FTA. In many simple trees only the basic event, undeveloped event, and output event are used.

Events representing failures of equipment or humans (components) can be divided into failures and faults. A component failure is a malfunction that requires the component to be repaired before it can successfully function again. For example, when a pump shaft breaks, it is classified as a component failure. A component fault is a malfunction that will "heal" itself once the condition causing the malfunction is corrected. An example of a component fault is a switch whose contacts fail to operate because they are wet. Once they are dried, they will operate properly.

Output events include the top event, or ultimate outcome, and intermediate events, usually groupings of events. Basic events are used at the ends of branches since they are events that cannot be further analyzed. A basic event cannot be broken down without losing its identity. The undeveloped event is also used only at the ends of event branches. The undeveloped event represents an event that is not further analyzed either because there is insufficient data to analyze or because it has no importance to the analysis.

14.5 LOGIC GATES

Logic gates are used to connect events. The two fundamental gates are the "AND" and "OR" gates. Table 14.2 describes the gate functions and also provides insight to their applicability.

Electrical circuits are used to illustrate the use of AND and OR gates. Figure 14.1 is a picture of switches in series and the corresponding fault tree. In order for the bulb to be lit, all of the switches must be in the closed position. The logic gate is an "AND" gate.

Figure 14.2 represents the OR gate logic. The bulb will be lit if any of the switches are closed.

TABLE 14.1

Common Fault Tree Symbols

Symbol name	Symbol	Description
Basic event		A basic initiating fault (or failure event)
Undeveloped event		An event that is not further developed. It is a basic event that does not need further resolution
Output event		An event that is dependent on the logic of the input events
External event (house event)		An event that is normally expected to occur
		In general, these events can be set to occur or not occur, i.e. they have a fixed probability of 0 or 1
Conditioning event		A specific condition or restriction that can apply to any gate
Transfer		Indicates a transfer to a subtree or continuation to another location

TABLE 14.2
Logic Gates

Description	Symbol	Truth table		
AND gate. The AND gate indicates that the output occurs if and only if all of the input events occur		Input A	Input B	Output
		T	T	T
		T	F	F
		F	T	F
		F	F	F
OR gate. The OR gate indicates that the output occurs if and only if at least one of the input events occur		Input A	Input B	Output
		T	T	T
		T	F	T
		F	T	T
		F	F	F

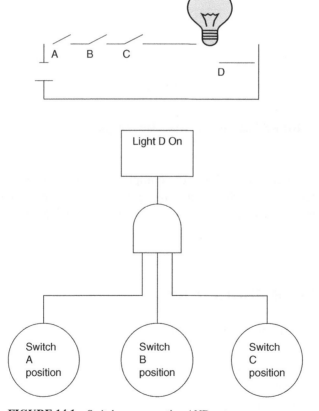

FIGURE 14.1   Switches representing AND gate.

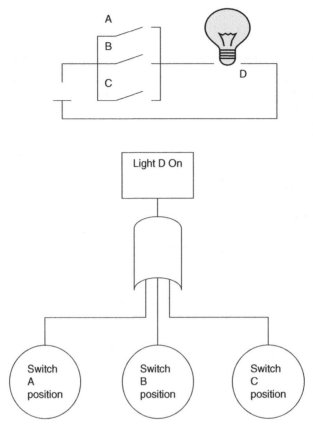

FIGURE 14.2   Switches representing OR gate.

Other gates that can be used in more complicated trees are shown in Table 14.3. These logic gates are used when representing complex systems. Other gates can be used as well. These are usually very specialized in nature and do not have widespread application.

14.6 ANALYSIS PROCEDURE

There are four steps to performing an FTA:

1. Defining the problem.
2. Constructing the fault tree.
3. Analyzing the fault tree qualitatively.
4. Documenting the results.

TABLE 14.3
Fault Tree Symbols

Symbol name	Symbol	Description
Voting OR (*k* out of *n*)	K	The output event occurs if *k* or more of the input events occur
Inhibit		The input event occurs if all input events occur and an additional conditional event occurs
Priority AND		The output event occurs if all input events occur in a specific sequence

14.6.1 Defining the Problem

A top event and boundary conditions must be determined when defining the problem. Boundary conditions include:

- System physical boundaries
- Level of resolution
- Initial conditions
- Not allowed events
- Existing conditions
- Other assumptions

Top events should be precisely defined for the system being evaluated. A poorly defined top event can lead to an inefficient analysis.

14.6.2 Constructing the Fault Tree

Construction begins at the top event and continues, level by level, until all fault events have been broken into their basic events. Several basic rules have been developed to promote consistency and completeness in the fault tree construction process. These rules, as listed in Table 14.4, are used to ensure systematic fault tree construction (excerpted from *Guidelines for Hazard Evaluation Procedures*, Center for Chemical Process Safety of the American Institute of Chemical Engineers (3)).

TABLE 14.4
Rules for Constructing Fault Trees

Fault tree statements	Write the statements that are entered in the event boxes and circles as malfunctions. State precisely a description of the component and the failure mode of the component. The "where" and "what" portions specify the equipment and its relevant failed state. The "why" condition describes the state of the system with respect to the equipment, thus telling why the equipment state is considered a fault. Resist the temptation to abbreviate during construction
Fault event evaluation	When evaluating a fault event, ask the question "Can this fault consist of an equipment failure?" If the answer is yes, classify the fault event as a "state-of-equipment" fault. If the answer is no, classify the fault event as a "state-of-system" fault. This classification aids in the continued development of the fault event
No miracles	If the normal functioning of equipment propagates a fault sequence, assume that the equipment functions normally. Never assume that the miraculous and totally unexpected failure of some equipment interrupts or prevents an accident from occurring
Complete each gate	All inputs to a particular gate should be completely defined before further analysis of any other gate. For simple models, the fault tree should be completed in levels, and each level should be completed before beginning the next level. This rule may be unwieldy when constructing a large fault tree
No gate to gate	Gate inputs should be properly defined fault events; that is, gates should never be directly connected to other gates. Shortcutting the fault tree development leads to confusion because the outputs of the gate are not specified.

14.6.3 Analyzing the Fault Tree

Many times it is difficult to identify all of the possible combinations of failures that may lead to an accident by directly looking at the fault tree. One method for determining these failure paths is the development of "minimal cut sets." Minimal cut sets are all of the combinations of failures that can result in the top event. The cut sets are useful for ranking the ways the accident may occur and are useful for quantifying the events, if the data is available. Large fault trees require computer analysis to derive the minimal cut sets, but some basic steps can be applied for simpler fault trees:

Uniquely identify all gates and events in the fault tree. If a basic event appears more than once, it must be labeled with the same identifier each time.

Resolve all gates into basic events. Gates are resolved by placing them in a matrix with their events.

Remove duplicate events within each set of basic events identified.

Delete all supersets that appear in the sets of basic events.

By evaluating the minimal cut sets, an analyst may efficiently evaluate areas for improved system safety.

14.6.4 Documenting the Results

The analyst should provide a description of the system. There should be a discussion of the problem definition, a list of the assumptions, the fault tree model(s), lists of minimal cut sets, and an evaluation of the significance of the minimal cut sets. Any recommendations should also be presented.

14.6.5 Examples of Fault Tree Analysis

Simple Example

The following examples will show the fundamentals of FTA. We will start with analyzing a simple cooling system flushing procedure. This procedure can also be analyzed using human reliability analysis (HRA) techniques, but we will use FTA at this point. The procedure reads as follows:

Warning – Cooling system must be below 100 °F prior to draining

1. Begin with the engine cold and ignition off.
2. Remove the radiator pressure cap.

Warning – Ethylene glycol coolant is toxic and must be disposed of in an appropriate manner.

3. Open the petcock at the bottom of the radiator and drain the coolant into a bucket.
4. Close the petcock and fill the radiator with water.
5. Start the engine and turn the heater control to hot. Add cooling system cleaner and idle the engine for 30 minutes (or as per the instructions on container).

Warning – Cooling system must be below 100 °F prior to draining.

6. Stop the engine and allow it to cool for five minutes. Drain the system.
7. Close the petcock, fill the radiator with water, and let the engine idle for five minutes.
8. Open petcock and drain the water.
9. Repeat Step 6–8.
10. Close the petcock.
11. Fill cooling system with 50/50 mixture of water and *nontoxic* antifreeze/coolant.

The first step will be to determine the credible top events. In this case it will be:

Mechanic is Burned
Cooling Flushing Failed

In fact, the "Mechanic is Burned" top event can be grouped under the "Cooling Flushing Failed" top event because if the mechanic were burned, then the task would fail in a sense.

From the procedure we can identify several basic events. These are shown in Table 14.5, along with their credibility.

TABLE 14.5

Car radiator Failures

Failure	Description	Credible failure
Engine not below 100 °F before beginning flushing procedure	In this failure the mechanic begins the coolant draining process without ensuring the engine is cool enough	This is a credible error. It happens all the time to professional as well as amateur mechanics. Since the system is under pressure, a severe burn can occur
Failure to remove radiator cap	As it says, the mechanic fails to remove the radiator cap	This is not a credible error, unless we are modeling the fact that the mechanic does not do the task at all
Failure to drain radiator	The mechanic fails to drain the radiator	This is not a credible error, unless, again, we are modeling the fact that the mechanic does not do the task at all
Failure to close petcock valve	This failure involves the mechanic not closing or incorrectly closing the petcock valve. This error can occur at least four times in the procedure	This is a credible error and can lead to an environmental spill
Failure to add flushing agent	The mechanic fails to add the flushing agent	This is a credible error because the mechanic can get busy and forget where they are in the process
Failure to remove flushing agent	The mechanic fails to remove the flushing agent	This is a credible error because, once again, the mechanic can get busy and forget where they are in the process. OR a shift change occurs or job change over and the second mechanic does not know where in process they are. This happens in the airline industry every day
Failure to rinse engine	The mechanic fails to rinse the remaining flushing agent from the engine	This might not be a catastrophic error, once the engine is drained of the flushing agent. It depends on how corrosive the flushing agent is
Failure to fill engine with 50/50 nontoxic antifreeze mix	The mechanic fails to fill the cooling system with the proper mixture, the right amount of coolant, or coolant at all	This again is a credible error

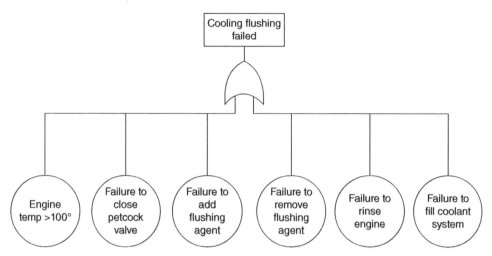

FIGURE 14.3   Fault tree analysis of coolant flushing task.

Once done assessing the credibility of the failures, we next construct the fault tree. We will not include errors that were deemed to be credible. Our top event will be the Cooling Flushing Failed. All the basic events will be entry points into the tree. The tree is shown in Figure 14.3.

Notice that we were able to build the fault tree only using an OR gate and the basic events. That is because only one of the credible failures can lead to the failure of the task.

Next we will model a simple hardware system failure, one that most homeowners have experienced. That is of the sprinkler system failure. The top event is Sprinkler System Failure. Table 14.6 contains the credible failures that can lead to the top event.

In Figure 14.4 we have constructed a partial fault tree from these failures. A full tree was not constructed to save space.

14.6.6 Modeling Success Using Fault Tree Analysis

One of the useful attributes of FTA is that it can also be used to model success paths as well as its more traditional use of modeling failure paths. For instance, say that a hiker wants to climb Mt. Everest. What must happen in order to do such a climb? Or say that someone wants to pass a certification examination or even complete a project on time. What does that person need to do to succeed in those goals? Obviously, there are many project management techniques that are great at modeling success paths for projects. Network diagrams are one example of a tool that can be used. However, FTA can also be used to model this type of process.

So, let's say our goal is to write a technical book that has 12 chapters and meet the contractual requirements of the publisher. As with modeling in the failure space,

TABLE 14.6

Sprinkler Head Failures

Failure	Description	Credible failure
Sprinkler head failure 1	Sprinkler head fails because it wore out	Yes – it is a credible failure
Sprinkler head failure 2	Sprinkler head fails because neighbor hits it with their lawn mower	Yes – it is a credible failure
Sprinkler valve failure 1	Sprinkler valve wears out	Yes – it is a credible failure
Sprinkler valve failure 2	Sprinkler valve breaks due to freezing	Yes – this failure though is contingent on the system not being properly drained the fall before. So, we will model it in this manner
Sprinkler controller failure 1	The battery that backs up the memory fails, and after a power failure the system has lost its mind	Yes – it is a credible failure
Sprinkler controller failure 2	The sprinkler controller fails	Yes – it is a credible failure
Sprinkler pipe failure 1	The sprinkler pipe breaks due to freezing	Yes – this failure though is contingent on the system not being properly drained the fall before. So, again, we will model it in this manner
Sprinkler pipe failure 2	The sprinkler pipe breaks due to digging in the yard	Yes – it is a credible failure

we need to develop the list of credible events that must occur to succeed. Table 14.7 lists these.

Though in real life an event tree is probably a better tool to model this with, we will develop the model using a fault tree. Chapter 16, Event Trees and Decision Analysis Trees, will show how this process can be modeled using an event tree. Figure 14.5 will show a fault tree for this process.

14.6.7 Fault Tree Analysis for Use in Accident Investigation

The following provides a description of an actual accident involving TAM Linhas Aéreas Flight 3054. An FTA will be constructed from the information provided (4–7).

On 17 July 2007 TAM Linhas Aéreas Flight 3054, an Airbus A320-233 aircraft, left Salgado Filho International Airport in Porto Alegre, only to land in wet conditions and crash at Congonhas-São Paulo International Airport in São Paulo, Brazil. When the flight first touched down, it was raining, causing the plane to overrun the runway, cross a highly busy main road during rush hour traffic, and crash into the TAM Express warehouse, which happened to be next to a gas station that exploded with the force of

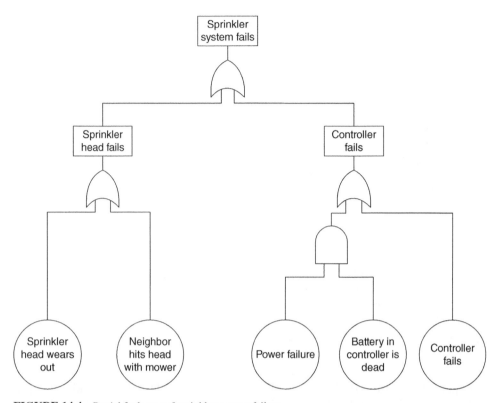

FIGURE 14.4 Partial fault tree of sprinkler system failure.

TABLE 14.7
Author Failures

Success	Description	Credible
Author 1 completes six chapters	Author 1 completes the chapters assigned to him/her	Yes – this has to occur to succeed
Author 2 completes six chapters	Author 2 completes the chapters assigned to him/her	Yes – this has to occur to succeed
Editor's changes are appropriate	The editor's changes must not change the technical content of the book and must be grammatically appropriate	Yes – this has to occur to succeed
Artwork meets requirements	The artwork has to meet the publisher's requirements to be included	Yes – this has to occur to succeed
Manuscript is formatted correctly	Besides the book needing to meet technical and grammatical requirements, it also has to be formatted correctly	Yes – this has to occur to succeed
Manuscript is submitted on time	The manuscript has to be submitted on time to be accepted	Yes – this has to occur to succeed

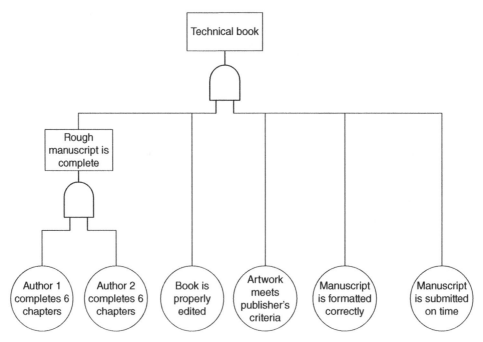

FIGURE 14.5 Fault tree for success modeling.

the impact of the Airbus A320-233, #789 (4). With 187 people on board, and 12 people on the ground, there totaled 199 fatalities (5), causing this crash to be the highest in deaths of any Latin American aviation accident. Not only was it the most devastating in Latin America, but it was the world's worst Airbus A320 crash involving fatalities anywhere in the world (6).

Airbus A320-233 was registered as PR-MBK and had the manufacturer's serial number of 789. The A320-233 was powered by two International Aero V2527E-A5 engines. The A320-233, #789 was built in February 1998 and took its first flight in March of 1998 and had its last flight in July of 2007 (7). TAM Linhas Aéreas was the last of four companies to operate the A320-233, #789, in less than a decade. TAM Linhas Aéreas did not come into position of the A320-233, #789, until December of 2006. Data collected from *Flight International* shows that as of April 30, the A320-233, #789 had mounted up to 20 379 flying hours and 9313 cycles (7).

The aircraft was dispatched for the Flight 3054 with a jammed thrust reverser, a braking device on the aircraft. According to TAM, the fault in the thrust reverser did not make the landing any more dangerous, and the mechanical problem was not known of at the time. It was later reported that the plane had trouble braking on the São Paulo runway on 16 July, the day before the crash, indicating that they had prior knowledge that something was wrong with the braking system (7).

Once the aircraft touched down in São Paulo, the pilots were unable to slow the aircraft down at a normal rate. The aircraft was still traveling at approximately 90 knots toward the end of the runway. The aircraft took a hard left and overshot the runway where it cleared the major roadway since the runway was elevated but eventually collided with the TAM Express building. Surveillance videos showed that the aircraft touched down at a normal speed and at a normal spot on the runway but the aircraft failed to properly slow down (www.News.com).

Authorities uncovered the flight data recorder, which contained information about what happened in the plane during flight. The data showed the following information. The thrusters had been in the climb position just prior to touchdown as the engines were being controlled by the computer system (4). An audio warning was given by the computer two seconds before touchdown warning the pilots that they should manually take control of the throttle. When the aircraft touched down, it was found that one thruster was in the idle position while the other was stuck in the climb position. In order for the spoilers to deploy and assist in slowing the aircraft down, both thrusters must be in the idle position. With different forces being applied to each side, it created a force that caused the plane to veer off to the left uncontrollably (4, 7).

Prior to the accident, the airport became under increased scrutiny due to a mid-air collision in September of 2006. The airport was known to have safety issues regarding operations in the rain as well as runway characteristics for the traffic going through it. One of these characteristics involved the length of the runway (7). There are so many variables that can affect the landing distance of an airplane that the airport had failed to consider.

For example, if the aircraft's approach speed is 20 knots higher than normal, it will take the aircraft 25% longer to slow down. The runway had been seen as a problem prior to the incident, and in February 2007, a judge had actually banned flights using Fokker 100, Boeing 737-700s, and Boeing 737-800s, stating that the runway needed an additional 1275 ft in order to operate safely. The A320 was not banned because the manufacture stated a shorter braking distance than the banned aircraft. However, the ban was quickly lifted as the airline industry stated that they would be inconveniencing thousands of passengers (6).

The root causes for the crash were that one of the reverse thrusters was known to be out prior to the flight, the runway was wet, and the runway should have been longer. While TAM claims that the thrusters should not have caused the crash, it is obvious that had the reverse thruster been functioning, the aircraft would have most likely been able to stop. Having grooves cut into the pavement to help reduce the risk of hydroplaning could have prevented the moisture on the runway, and had the runway been longer, the aircraft would have had more time to stop (4).

Pilot training might have also contributed to the accident even though both pilots were very well trained and had plenty of experience. Commander Kleyber Aguiar Lima, from Porto Velho, was born on 22 March 1953, and worked for TAM from November 1987 to July 2007 and had over 14 000 flight hours in his career, and Commander Henrique Stephanini Di Sacco, from São Paulo, who was born on 29 October 1954, joined TAM in 2006 and also had over 14 000 flight hours in his career (5). They knew that one thruster was not functional and should have planned the landing as if neither thruster would work. They knew that the landing strip was short; they also knew that the strip was wet. The combination of the short landing strip and the wet landing strip along with the malfunctioning thrusters should have alerted the pilots to take a different course of action.

Precautions should be taken to improve traction during wet weather. This includes cutting grooves in the pavement to allow the water to flow off of the runway increasing the traction when an aircraft lands. This airport just finished with major renovations on the landing strip; it should have been mandatory that the strip be 100% finished before being allowed for use.

Also, warnings from the government stating that the runway was much too short for larger airliners to land on were passed off way too quickly. Governmental rulings should be respected, and with that, the changes must be made no matter how necessary to ensure the safety of all the passengers on all the planes. The airport officials knew that the runways were too short to handle such large planes, and yet they continued to allow those planes to fly to not disrupt the economy.

As with most aircraft crashes, there are several factors that lead to the crash. The following is a list of items that contributed to the crash. This list is then used to construct the fault tree that is depicted in Figure 14.6:

- Runway was wet.
- Rain in area.

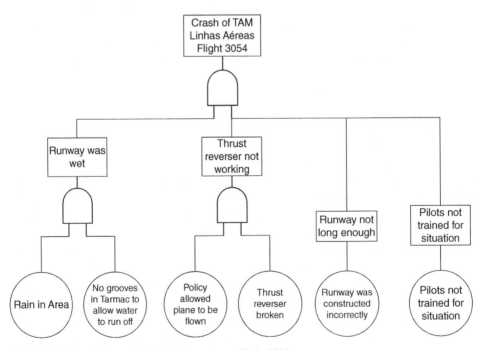

FIGURE 14.6 Fault tree for TAM Linhas Aereas Flight 3054.

- No grooves cut in runway.
- Thrust reverser broken.
- Airline policy allowed aircraft to be flown with broken thrust reverser.
- Runway too short.
- Airport policy allowed larger planes to land on runway.
- Experienced pilots had not had training in this situation.

The fault tree is very useful to showing how all the individual factors come together to cause the flight to crash. Is this all of them? No, there are some decision processes that the pilots went through that are not shown. These can be better modeled using HRA techniques. These are discussed in Chapter 12.

14.7 SUMMARY

The FTA technique has proven to be a very rigorous and valuable tool for analyzing complex systems. Strengths include an ability to analyze down to a great level of detail, a simple presentation, and a systematic method for analysis. Fault trees are used in a variety of disciplines and can model many types of systems.

The Probabilistic Risk Assessment (PRA) chapter will discuss how FTA is used in conjunction with other techniques to analyze complex systems.

Self-Check Questions

1. What advantage does FTA have over other risk assessment techniques?

2. What disadvantages does FTA have?

3. Pick a recent accident or event and develop an FTA for it.

4. FTAs can be used for financial events as well. How could an FTA be used to determine how a business might overspend their resources.

5. How can FTA be used in enterprise risk management?

6. Describe how an FTA can be used in conjunction with an FMEA or PHA.

REFERENCES

1. Vesely, W., Dugan, J., Fragola, J. et al. (2002). *Fault Tree Handbook with Aerospace Applications*. Washington, DC: National Aeronautics and Space Agency https://elibrary.gsfc.nasa.gov/_assets/doclibBidder/tech_docs/25.%20NASA_Fault_Tree_Handbook_with_Aerospace_Applications%20-%20Copy.pdf (accessed 5 March 2019).

2. Fuetz, R.J. and Tracy, J.P. (1965). *Introduction to Fault Tree Analysis*. Seattle: Boeing Product Development Airplane Group.

3. Center for Chemical Process Safety (CCPS) (1992). *Guidelines for Hazard Evaluation Procedures*. New York: American Institute of Chemical Engineers.

4. CNN (2007). Brazil Plane 'Flew with Mechanical Fault (27 July 2007, 2 December 2007). http://www .news.com.au/story/0,23599,22104315-23109,00.html (accessed 2011). Pilot in Brazilian crash tried to abort landing, official says.

5. TAM (2007). Press release (22 July 2007). http://www.tam.com.br/b2c/jsp/default.jhtml? adPagina=3&adArtigo=10743 (accessed 2011). Commanders had more than 14 thousand flight hours. *TAM*.

6. Accident Description (2007). Aviation Safety Network (2007-07-06). https://aviation-safety.net/ database/operator/airline.php?var=5196 (accessed 2011).

7. Ionides, N. (2007). BREAKING NEWS: TAM A320 crashes in Sao Paolo. *Flight International*, 4 (18 July 2007). http://www.flightglobal.com/articles/2007/07/18/215565/breaking-news-tam-a320-crashes-in-sao-paolo.html (accessed 2011).

CHAPTER 15

Critical Function Analysis

15.1 INTRODUCTION

When one types in "Functional Analysis" on any search engine, a wide range of tools pop up, some of which involve analyzing behavior in psychological studies (1), computer networks (2), and involve mathematical analyses (3). Functional analysis techniques can also be used to analyze complex systems to ensure that needed components of a system are in place and available for use (4).

When a web search is conducted for the word "function," a wide range of definitions come up, along with where a person can buy "Function," whatever that means. www.Dictionary.com defines function as (5):

noun

1. the kind of action or activity proper to a person, thing, or institution; the purpose for which something is designed or exists; role.

This is the definition that will be used in this chapter. Take a screwdriver, for instance. A screwdriver is designed to tighten screws. There are different types of screwdrivers: flat, Phillips, star, hex head, electric, battery operated, and plain old manual ones. However, sometimes screw drivers are used for other purposes, such as a hammer, a pry bar, and a weapon to name a few. However, the function a screw driver was designed to perform is to tighten or loosen screws. In a full tool kit, a screw driver has only those functions.

The Space Shuttle Columbia made its first flight on 12 April 1981 (6). There had been several flights by the Space Shuttle Enterprise before that date. However, these were test flights and the Orbiter had been carried on or released from a Boeing 747 (6). Space Shuttle Atlantis made the last flight on 8 July 2011 (6). The major functions of the Space Shuttle fleet were to carry crews and cargo to the Earth's orbit. The Shuttle was composed of several major components.

Risk Assessment: Tools, Techniques, and Their Applications, Second Edition. Lee T. Ostrom and Cheryl A. Wilhelmsen.
© 2019 John Wiley & Sons, Inc. Published 2019 by John Wiley & Sons, Inc.
Companion website: www.wiley.com/go/Ostrom/RiskAssessment_2e

TABLE 15.1

Functions of the Major Components of a Space Shuttle

Component	Major function(s)
Orbiter	Crew life support, propulsion, return vehicle, cargo transport, docking capability
Boosters	Propulsion
External tank	Fuel storage and supply for main engine in orbiter

TABLE 15.2

Lower Level Crew Life Support Functions of Orbiter

Component	Function
Crew cabin	Ensures breathable gases are contained in vessel, provides living space, and contains other life support systems
Atmospheric control	Supply oxygen to the crew, remove carbon dioxide and other trace gases, ensure that the gas components were in the proper ratio, and ensure the humidity was at the correct level
Water system	Produce water in the fuel cells, store water in one of four tanks, remove bacteria, warm or chill the water, and route excess water to waste disposal system
Food storage	Supply food to the crew
Radiation shielding	Protect crew from radiation
Heat shield	Protect crew from excessive heat during re-entry
Temperature control	Ensure crew areas are within a livable temperature

Table 15.1 contains the major components of any of the Space Shuttles of the now retired fleet and their corresponding functions (7). For instance, it is apparent that the Orbiter has many more functions than the boosters. Each component of the Orbiter system there are many if not hundreds of lower level functions. These functions are associated with the numerous subcomponents of the system. Table 15.2 lists some of the lower level functions that comprise the Crew Life Support function of the Orbiter.

Each of the functions listed in Table 15.2 is important. Atmospheric control, temperature control, radiation shielding, and heat shield are some of the more important functions. The Columbia accident was due to the failure of the heat shield component. The heat shield was damaged during liftoff when a piece of foam insulation from the external tank broke off and hit the leading edge of the wing of the Orbiter (8). If that function were maintained, the Orbiter would not have failed.

Nelson and Bagian (9) discussed the concept of critical function approach or analysis (CFA) and its application to space systems. CFA was developed as an analysis tool after the Three Mile Island nuclear accident (10). The basic concept behind this approach is that within each system, there are a set of critical functions that must be maintained so that the system does not fail. The heat shielding on the Space Shuttle Columbia is an example of a critical component that maintained a critical function. Once the shielding failed, the Space Shuttle also failed. There was no recovery at that point because there was no backup for that critical function. The crew's space suits were backup for the atmospheric system on the Orbiter. The space suits had a supply of oxygen that could support the crew until the shuttle landed.

Two very dramatic commercial airline accidents demonstrate the importance of functions to safe operation and airworthiness of airplanes:

- United Flight 232
- Air Canada Flight 143

15.1.1 United Flight 232

United Airlines Flight 232 was a scheduled flight from Stapleton International Airport in Denver, Colorado, to O'Hare International Airport in Chicago (11–13). The flight was then scheduled into Philadelphia International Airport. On 19 July 1989, the DC-10, with tail registration number N1819U, took off normally at 2:09 p.m. (14 : 09). The plane was in a shallow right turn at 3 : 16 p.m. (15 : 16) at 37 000 ft when the fan disk of its tail-mounted General Electric CF6-6 engine failed. The disk disintegrated and the debris was not contained by the engine's nacelle, a housing that protects the engine. The disintegrated disk, along with pieces of nacelle, penetrated the aircraft tail section in numerous places. This included the horizontal stabilizer. Airplanes are designed with redundant systems. One of the problems with the DC-10 design is that components of all three redundant hydraulic systems are positioned closely together in the horizontal stabilizer. The shrapnel from the failed engine punctured the lines of all three hydraulic systems and the hydraulic fluid drained away rapidly (11).

Despite the loss of all three hydraulic systems, the crew was able to attain and then maintain limited control by using the two remaining engines. The crew steered by applying power to one engine over the other, the gained altitude by applying power to both engines, and they decreased altitude by reducing power on both engines. The crew flew the crippled jet to the Sioux Gateway Airport. They lined the airplane up for landing on one of the runways. Without flight controls, they were unable to slow down the airplane for landing. The crew was forced to attempt landing at much too high a speed and rate of descent. On touchdown, the aircraft broke apart, caught fire, and rolled over. The largest section came to rest in a cornfield next to the runway. The crash was very intense, killing 111 people, but two-thirds of the occupants survived. The cause of the engine failure was traced back to a manufacturing defect in the fan disk. Microscopic cracks were found and determined to be impurities in the castings. The cracks were present during maintenance inspections and could have possibly been detected by maintenance personnel (11).

The accident is considered a prime example of successful crew resource management because of the manner that the flight crew handled the emergency. The flight crew, Captain Alfred C. Haynes, a 30 000 hour pilot, First Officer William Records, and Flight Engineer Dudley Dvorak, became well known as a result of their actions that day, in particular the captain, Alfred C. Haynes, and a DC-10 instructor, Dennis E. Fitch, who happened to be onboard and who offered his assistance (13).

15.1.2 Air Canada Flight 143

On 23 July 1983, Air Canada Flight 143, a Boeing 767-200 jet, registration C-GAUN, c/n 22520/47, ran out of fuel at 41 000 ft (12 500 m) mean sea level (MSL) altitude (13–15). The flight was approximately halfway through its flight from Montreal, Quebec, to Edmonton, Alberta, via a stop in Ottawa, Ontario. The airplane was flown from Toronto, Ontario, to Edmonton on 22 July 1983 where it underwent routine checks. The next day it flew Montreal and then departed following a crew change as Flight 143 for the return trip to Edmonton via Ottawa. The captain for this flight was Robert (Bob) Pearson and First Officer Maurice Quintal at the controls.

At 41 000 ft (12 500 m), over Red Lake, Ontario, the aircraft's cockpit warning system sounded, indicating a fuel pressure problem on the aircraft's left side. The pilots turned it off assuming it was a fuel pump failure (14). The crew knew that gravity would still feed fuel to the aircraft's two engines even though the pump failed. At this point the aircraft's fuel gauges were inoperative. However, the flight management computer (FMC) indicated that there was still sufficient fuel for the flight. The pilots subsequently realized the fuel entry calculation into the FMC was incorrect. A few moments later, a second fuel pressure alarm sounded. At this point the pilots decided to divert to Winnipeg, Manitoba. Within seconds, the left engine failed and the crew began preparing for a single-engine landing (14).

They tried to restart the left engine and began communicating their intentions to controllers in Winnipeg. The cockpit warning system sounded again, this time with a long "bong" that no one in the cockpit could recall having heard before (14). This was the "all engines out" sound, an event that had never been simulated during training. Within seconds most of the instrument panels in the cockpit went dark. In addition, the right-side engine stopped and Boeing 767 lost all power.

This Boeing 767-200 was one of the first commercial airliners to include an electronic flight instrument system (EFIS). This system required the electricity generated by the aircraft's jet engines to operate. The system went dead without electrical power. The auxiliary power unit (APU) is a small jet engine in the tail section of large jet aircraft and its purpose is to supply electricity, hydraulic power, and pneumatic air for starting the other jet engines (16). Since there was no fuel, this engine failed as well. The airplane was left with only a few basic battery-powered emergency flight instruments. These provided basic information with which to land the aircraft. However, there was no working vertical speed indicator that would be needed for landing.

The main engines and APU supply power for the hydraulic systems without which the aircraft cannot be controlled. Commercial aircrafts require redundant systems for such power failures. Boeing aircraft, like the 767-200, usually achieve this through the automated deployment of a ram air turbine (RAT) (17). This is a small generator driven by a small propeller that is driven by the forward motion of the aircraft. The higher the airspeed, the more power the RAT generates (17).

As the pilots were descending through 35 000 ft (11 000 m), the second engine shut down. The crew immediately searched their emergency checklist for the section on flying the aircraft with both engines out. There was no such section to be found. Fortunately, Captain Pearson was an experienced glider pilot. He knew some flying techniques almost never used by commercial pilots. To have the maximum range and therefore the largest choice of possible landing sites, he needed to fly the 767 at the "best glide ratio speed." Making his best educated guess as to this airspeed for the 767, he flew the aircraft at 220 knots (410 km h^{-1}; 250 mph). First Officer Maurice Quintal began making calculations to see if they could reach Winnipeg. He used the altitude from one of the mechanical backup instruments. The distance traveled was supplied by the air traffic controllers in Winnipeg. From this he calculated that the aircraft lost 5000 ft (1500 m) in 10 nautical miles (19 km; 12 miles), giving a glide ratio of approximately 12 : 1 for the airplane. The controllers and Quintal all calculated that Flight 143 would not make it to Winnipeg (14).

First Officer Quintal proposed landing at the former RCAF Station Gimli. This was a closed air force base where he once served as a Canadian Air Force pilot. However, to complicate matters, the airstrip had been converted to a race track complex that was in use that day (14). Without power, the pilots had to try lowering the aircraft's main landing gear via a gravity drop, but, due to the airflow, the nose wheel failed to lock into position. The decreasing forward motion of the aircraft also reduced the effectiveness of the RAT, making the aircraft increasingly difficult to control because of the reduced power being generated (14).

It became apparent as the flight approached the runway that the aircraft was too high and too fast. This raised the danger of running off the runway before the aircraft could be stopped safely. The lack of adequate hydraulic pressure prevented flap/slat extension. These devices are used under normal landing conditions to reduce the stall speed of the aircraft for a safe landing. The pilots considered executing a 360° turn to reduce speed and altitude. However, they decided that they did not have enough altitude for this maneuver. Pearson decided to execute a forward slip to increase drag and lose altitude. This maneuver is commonly used with gliders and light aircrafts to descend more quickly without gaining forward speed.

When the wheels touched the runway, Pearson "stood on the brakes," blowing out two of the aircraft's tires (14). The unlocked nose wheel collapsed and was forced back into its well, causing the aircraft's nose to scrape along the ground. The plane also slammed into the guard rail, separating the strip, which helped slow it down (14). Because there was no fuel onboard, there was little chance of a major fire. A minor fire in the nose area was extinguished by racers and course workers armed with fire extinguishers. None of the 61 passengers were seriously hurt, but there were some minor injuries when passengers exited the aircraft via the rear slides. The accident has been nicknamed the "Gimli Glider" in recognition of the flight crew's handling of the situation (13, 14).

15.2 CRITICAL FUNCTIONS

Obviously in both of these accidents, critical functions of the airplanes failed. In both cases they were hardware components. The human component in both of these accidents succeeded in performing their tasks. The components of an airplane are listed in Table 15.3, along with the functions that failed our succeeded in each event.

It is apparent from Table 15.3 that critical functions are dependent on each other in modern commercial airliners. Figure 15.1 shows how these functions are interconnected. Figure 15.2 shows how the diagram would change if the engines failed, for instance.

CFA then can be used to determine which critical functions are needed to support the desired mission. A recent web article on www.globalsecurity.org listed the flight critical systems for the F-22 Raptor (17). See Table 15.4.

An F-22 crashed in Alaska on 16 November 2011, killing the pilot (18). The cause of the crash was found to be the bleed air system, part of the air cycle system, which is listed above as one of the critical functions required by the F-22 to fly (19).

15.3 CONDUCTING A CRITICAL FUNCTION ANALYSIS

Conducting a CFA is relatively straightforward. If anything, the problem with conducting a CFA is getting everyone involved to agree on what constitutes a critical function. This approach can be used for a wide variety of analyses. It can be used for a complex system design, as shown above, such as emergency planning (20), financial activities (21), and military operations (22). Different organizations might use various terms for aspects of a CFA. Nelson and Bagian (9) use the terms:

- Mission
- Critical functions
- Tasks
- Resources
- Support systems

It really depends on the depth of the analysis and how your organization defines the layers. For the purposes of this book, the following terms and their definitions are:

- Mission – Goal of process, organization, or task.
- Critical function – What has to be in place to achieve or maintain the mission.
- Component – System, job/person, part, tool, or other thing that performs the activities that make up the critical function.
- Support – Utilities, materials, activities, or other items that support the components.

TABLE 15.3

Critical Functions of a Commercial Airplane

Component	Critical function	United Flight 232	Air Canada Flight 143
Fuselage	Pressure control. Pressure vessel for crew and passenger cabin	Succeeded	Succeeded
Wings	Lift	Succeeded	Succeeded
Engines	Propulsion. Also motive force for hydraulic system electrical system, pneumatic system for control systems, and others	Two engines succeeded. One engine in the tail failed due to catastrophic failure of fan disk. Critical function was maintained	Both engines failed due to lack of fuel
Fuel system	Supports propulsion	Succeeded	Failed due to incorrect calculation during refueling
Auxiliary power unit (APU)	Instruments. Supplies auxiliary electrical power, hydraulic power, pneumatic power	Not needed	Failed due to lack of fuel
Electrical system	Instruments. Supplies electrical power to avionics, control systems, cabin comfort systems	Succeeded	Failed due to loss of engines
Hydraulic system	Control systems. Supplies hydraulic power for aircraft control systems	Failed due to catastrophic failure of engine in tail and subsequent destruction of the three redundant hydraulic systems	Succeeded, but to a limited degree. Was powered by the RAT
Landing gear system	Landing safely. Provides a controlled means to land the aircraft	Succeeded	Succeeded, but to a limited degree
Control surfaces	Control systems. Provides aircraft the ability to turn, to control lift, to descend at a controlled rate, and to control aircraft attitude	Failed due to failure of hydraulic system	Succeeded, but to a limited degree

209

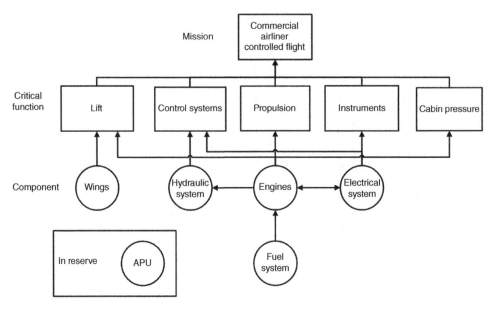

FIGURE 15.1 Interconnection of critical functions.

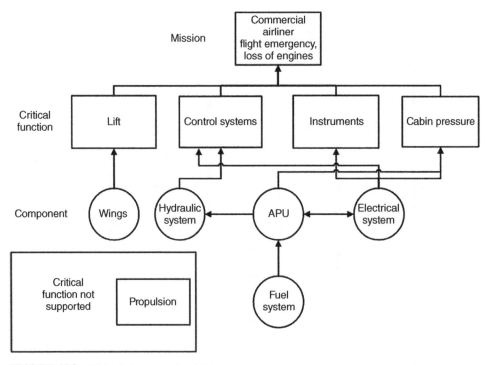

FIGURE 15.2 Flight during an engine failure event.

TABLE 15.4

Flight Critical Systems for F-22 Raptor

Flight critical system	Function
Vehicle Management System (VMS)	The Vehicle Management System (VMS) provides integrated flight and propulsion control. The VMS enables the pilot to aggressively and safely maneuver the F-22 to its maximum capabilities
Utilities and subsystems (U&S)	• Integrated vehicle subsystem controller • Environmental control system • Fire protection • Auxiliary power generation system (APGS) • Landing gear • Fuel system • Electrical system • Hydraulics • Arresting system
Integrated vehicle subsystem controller (IVSC)	The integrated vehicle subsystem controller (IVSC) is the system responsible for aircraft integration, control and diagnostics The F-22 uses a totally integrated environmental control system (ECS) that provides thermal conditioning throughout the flight envelope for the pilot and the avionics
Environmental control system	The five basic safety critical functions the ECS must take care of include: avionics cooling; adequate air to the pilot; canopy defog; cockpit pressurization; and fire protection
Air cycle system	The air cycle conditions air from the engines for various uses
Liquid cooling system	Unlike other fighter aircraft, the F-22 uses liquid cooling, rather than air cooling for the mission avionics
Thermal management system (TMS)	The thermal management system (TMS) is used to keep the fuel cool
Fire protection	Fire protection is provided for the aircraft's engine bays, the auxiliary power unit (APU), and for dry bays The aircraft uses infrared and ultraviolet sensor for fire detection and Halon 1301 for fire suppression
Auxiliary power generation system (APGS)	The APGS consists of an APU, and a self-contained stored energy system (SES)
Landing gear	The F-22 utilizes tricycle landing gear, with the standard two main gears (each with a single tire) and a single-wheel, steerable nose landing gear assembly
Fuel system	There are eight fuel tanks on the F-22, including one (designated F-1) in the forward fuselage behind the pilot's ejection seat. The others are located in the fuselage and the wings. The F-22 will run on JP-8, a naphthalene-based fuel with a relatively high flash point
Electrical, hydraulic, and arresting systems	The F-22 uses a Smiths Industries 270 V, direct current (DC) electrical system. It uses two 65 kW generators. The hydraulic system includes four 72 gal min^{-1} pumps and two independent 4000 psi systems

The following presents several examples of analyses and how the layers of a CFA are linked. These sample analyses are:

- Small business
- Chemical reactor
- Emergency planning

15.4 CRITICAL FUNCTION OF A SMALL BUSINESS

Take any small business and there are several critical functions that must be maintained for the business to operate. A convenience store (C Store) will be used for this example. Most everyone reading this book has been in a C Store. This store has the following attributes:

- Four (4) gas pumps
- 1 ea. 10 000 gal 87 octane gasoline tank
- 1 ea. 10 000 diesel tank
- 30 ft of beer and soda cooler
- 100 ft of room temperature storage
- Two soda fountains
- One microwave food station

The first step for conducting a CFA is to define the mission and critical functions of the store. The mission of the business is to make money. The critical functions that are needed to allow the store to make money are:

- A location/building
- A shopper friendly environment
- Products to sell
- Means to perform transactions (cashier or self-service station)
- Appropriate operating licenses

In this example the goal is not to optimize the cash flow, but only to stay open. Figure 15.3 shows the relationship of the critical functions to the mission and the support functions.

It goes without saying that in the absence of any of the critical functions the store could not make money or not for very long. The Appropriate Licenses critical function could be eliminated, but within short order, in most well run cities, this would cause the business to close soon.

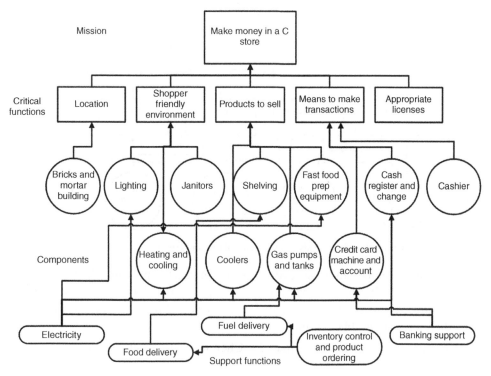

FIGURE 15.3   Critical functions analysis for a convenience store.

15.5 CHEMICAL REACTOR CRITICAL FUNCTION ANALYSIS

In this example a CFA will be performed on a chemical reactor system. The mission of the reactor is to produce a 5000 gal batch of Chemical D within the quality limits. The attributes of the reactor are the following:

- It is a 5500 gal capacity batch reactor.
- Three chemicals are combined in the reactor to produce Chemical D: Chemicals A, B, and C.
- The ratio of the three (3) chemicals are:
  - ○ 10% Chemical A
  - ○ 30% Chemical B
  - ○ 60% Chemical C
- Chemicals have to be mixed in the proper ratio for 30 minutes to ensure a successful batch.
- The reaction is exothermic. For each degree over 300 °F the reactor reaches, the quality of the product is reduced. Chemical E is the contaminant produced. The batch becomes 1% Chemical E for degree over the 300° level.

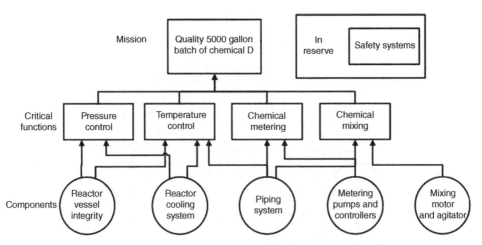

FIGURE 15.4 Critical functional analysis of a chemical reactor system.

- Increased temperature can cause a spike in reactor pressure. If the pressure reaches 310 °F, the reactor pressure will near the safety factor limits of the reactor. At this point a rupture disk will break and the gases produced will be directed to a scrubber column.
- The product must have less than 2% Chemical E to be successful.

In this regard, the critical functions of the system are:

- Pressure control
- Temperature control
- Chemical metering
- Chemical mixing
- Safety systems

Figure 15.4 shows the CFA for the reactor system. In the event of a process upset, the mission is to prevent a catastrophic failure of the pressure vessel. Figure 15.5 shows the relationship of the critical functions needed to succeed in this mission.

15.6 EMERGENCY MANAGEMENT PLANNING

One of the attributes of a modern emergency management plan is a continuity of operations (COOP) plan. COOP is a collection of resources, actions, procedures, and information that is developed, tested, and held in readiness for use in the event of a major disruption of operations. The COOP Multi-Year Strategy and Program Management Plan (MYSPMP) should provide the guidance, objectives, performance measures, enabling tasks, and resources necessary for the organization to accomplish its overall mission, and its priority and secondary mission essential functions.

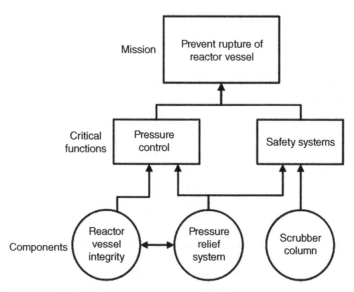

FIGURE 15.5 Critical functions during process upset.

The Federal Emergency Management Agency (FEMA) provides guidance for ensuring critical functions of an organization are maintained via a COOP plan (23).

FPC 65 lists the objectives of a viable COOP program (X):

(1) Ensuring the performance of essential functions/operations.

(2) Reducing loss of life and minimizing damage and losses.

(3) Executing, as required, successful succession to office with accompanying authorities in the event a disruption renders agency leadership unable, unavailable, or incapable of assuming and performing their authorities and responsibilities.

(4) Reducing or mitigating disruptions to operations.

(5) Ensuring that alternate facilities are available from which to continue to perform their essential functions.

(6) Protecting essential facilities, equipment, vital records, and other assets.

(7) Achieving a timely and orderly recovery from a COOP situation and maintenance of essential functions to both internal and external clients.

(8) Achieving a timely and orderly reconstitution from an emergency and resumption of full service to both internal and external clients.

(9) Ensuring and validating COOP readiness through a dynamic, integrated test, training, and exercise program to support the implementation of COOP plans and programs.

FEMA states that there are critical essential functions that government organizations must be able to perform, either continuously or without significant disruption, during and following a crisis, if required, in the assurance of COOP (20). The National Essential Functions (NEF) are listed in Table 15.5. Secondary functions are listed in Table 15.6.

TABLE 15.5

Critical Functions for Continuity of Operations.

Critical function category	Critical functions
National Essential Functions Functions that represent the overarching responsibilities of the Executive Branch to lead and sustain the country and will generally be the primary focus of the President	Preserve our constitutional form of government. Ensure the continued functioning of our duly elected representative form of government and, in particular, the functioning of the three independent branches of government. This NEF includes department and agency functions that respect and implement the check and balance relationship among the three branches of the Federal government Provide visible leadership to the nation; maintain the trust and confidence of the American people. This NEF includes department and agency functions to demonstrate that the Federal government is viable, functioning, and effectively addressing the emergency Defend the country against all enemies, foreign or domestic, and prevent and interdict future attacks. This NEF includes department and agency functions to protect and defend the worldwide interests of the United States against foreign or domestic enemies, to honor security agreements and treaties with allies, and to maintain military readiness and preparedness in furtherance of national interests and objectives Maintain and foster effective relationships with foreign nations. This NEF includes department and agency functions to maintain and strengthen American foreign policy Protect against threats to the homeland and bring to justice perpetrators of crimes or attacks against the nation, its citizens or interests. This NEF includes department and agency functions to protect against, prevent, or interdict attacks on the people or interests of the nation and to identify, incarcerate and punish those who have committed violations of the law Provide rapid and effective response to and recovery from the domestic consequences of an attack or other incident. This NEF includes department and agency functions to implement response and recovery plans, including, but not limited to, the National Response Plan Protect and stabilize the nation's economy; ensure confidence in financial systems. This NEF includes department and agency functions to minimize the economic consequences of an attack or other major impact on national or international economic functions or activities Protect and stabilize the nation's economy; ensure confidence in financial systems. This NEF includes department and agency functions to minimize the economic consequences of an attack or other major impact on national or international economic functions or activities Provide for critical Federal government services that address the national health, safety and welfare needs of the Nation. This NEF includes department and agency functions that ensure that the critical national level needs of the nation are met during an emergency with regard to Federal government activity

TABLE 15.6
Secondary Critical Continuity of Operations Functions

Critical function category	Critical functions
Priority mission essential functions (PMEFs)	Those department-specific mission essential functions that support the NEFs and flow directly up from supporting activities or capabilities within department or agency COOP plans
Secondary mission essential functions (SMEFs)	Those mission essential functions that the department must perform in order to bring about full resumption of its normal functions, but which are not PMEFs. Resumption of SMEFs may need to occur within a very short period of time or only after several days depending on the nature of the department's mission and the nature of the disruption to normal department functions
Supporting activities	Those specific activities that the department must conduct in order to perform its mission essential functions
Capabilities	Communications, facilities, information, trained personnel, and other assets necessary to conduct the department's mission essential functions and supporting activities

COOP is also important for corporations, state and county governments, and small businesses. Within a rural county in Idaho, the critical functions might not be as extensive as a major metropolitan county like New York County. However, they are critical for the residents who reside there. Rural counties in the western states can be the size of entire states in the east. Idaho County in Idaho comprises approximately 8 500 mile^2 and only has approximately 17 000 residents (24). Rhode Island is approximately 1 200 mile^2 in size and has over 1 000 000 residents (25). In fact, Idaho county is half the size of Switzerland ($\sim$15 000 mile^2) (26).

The issues with such a big county, with so few residents from a critical function perspective are:

- Tax base
- Distances between municipalities
- Emergency services to serve remote areas
- Schools and medical services
- Social services

From a COOP perspective, ensuring that such a county can operate under disaster conditions is very important for ensuring the well-being of its residents. Figure 15.6 shows how the critical functions for a rural county might look.

15.7 CRITICAL SAFETY FUNCTION OF A WATER HEATER

If you want hot water in your home, you need to have a water heater, or some type of heating device. Most water heaters nowadays are relatively safe to use. You can have

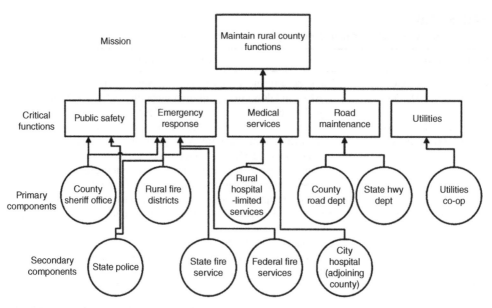

FIGURE 15.6   Rural county critical functions.

an electric or a gas-operated one. Most people go about their business without giving their water heaters a second thought. Some homeowners will not even inspect them, until they are required to do so when they rent or sell a house, or until something goes really bad, and they need to be replaced.

Water heaters are not to be confused with water boilers. The latter are much more complex and are also used to heat the house. Water heater design has come a long way over the years. Water heater safety can be addressed in three major categories: design, installation, and end usage. Critical safety function of a water heater will be analyzed in this chapter. Figure 15.7 shows a diagram of a hot water tank.

Design: Over the years the design of the water heater has become safer. For example, if the gas valve, pilot, thermostat, hi-limit temperature switch, or vent damper fails to signal that they are ready for combustion, the ignition system is designed not to ignite the main burner. Heaters are certified to ANSI Z21.10.1-2002. Also, new water heaters are designed so the unit cannot ignite flammable vapors caused by spilled gasoline outside the water heater unit.

Installation: The safe installation of a water heater is very important as well. One must take into account the local and federal codes and regulations.

End User: When you hear on the TV or news about water heater explosions, the most common cause is end user error. This is because proper maintenance checks need to be conducted regularity to ensure that the water heater does not fail. For example, it is recommended that the pressure relief valve is activated monthly to ensure proper function. It is highly unlikely that people go around their water heaters monthly and do these safety checks. Fortunately, there are other safety functions in place that can prevent an explosion. Below are the most critical systems of a gas water heater.

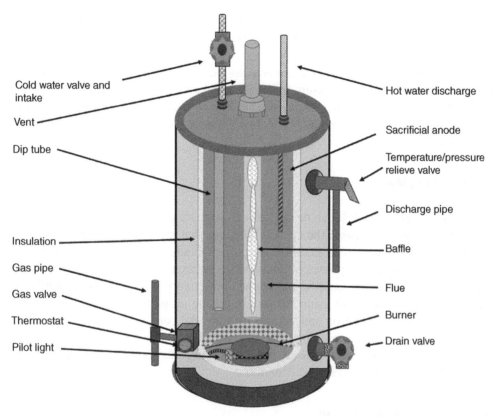

Cold water valve and intake

Vent

Dip tube

Insulation

Gas pipe

Gas valve

Thermostat

Pilot light

Hot water discharge

Sacrificial anode

Temperature/pressure relieve valve

Discharge pipe

Baffle

Flue

Burner

Drain valve

FIGURE 15.7   Depiction of gas hot water heater.

Critical system	Function
Gas valve	This valve is used to shut of the gas to the water heater
Thermocouple	A device that monitors the flame of a pilot on a water heater
Thermostat	Keeps water at safe temperature to reduce risk of scalding
Vent damper/flue pipe	To evacuate the deadly gases that contain carbon monoxide created upon combustion
Temperature/pressure relief valve	This valve allows excess heat or pressure to be released from a hot water tank so it does not explode
Shutoff valve	Stops the water from going to the system
Anticorrosion anode rod (sacrificial anode)	Simply put through electrolysis the anode rods will corrode before the exposed metal in the tank

Although water heater explosions are rare, other heater failures are very common. The most common reason water heaters fail is because of corrosion and scaling according to the Institute for Business and Home Safety (IBHS), about 44% of the failure are as result to corrosion and to scaling. Corrosion happens when the sacrificial anode is consumed, and not replaced. Scaling occurs when hard water deposits are not regularly removed.

Explosions may occur when there is a gas leak and the ignition flame can ignite the gas, or when the water pressure builds up and the safety relief valve fails.

Water heaters are designed to be safe, but as with many things, human error, such as not maintaining the water heater per manufactures recommendations, is likely to result in failures.

15.8 SUMMARY

CFA is a very useful tool. It can be applied to a wide range of activities, systems, and organizations to determine if the critical functions that are required for a mission to be successfully achieved or maintained are properly and adequately supported. It can also be applied to various phases of an activity, system, or organization to determine if the critical functions would be maintained under varying or adverse conditions. One of the difficulties with using the tool is getting all parties to agree as to what constitutes a critical function and what constitutes the components that support the critical functions. Also, terminology does vary between analysts who use the tool. As with all the tools and techniques discussed in this book, once the analysis is performed, the results are used to improve the situation.

Self-Check Questions

1. Critical function analysis (CFA) is a good tool for what purposes?

2. In recent years, high-tech companies have discussed commercializing a flying car, based on the helicopter drone design. What would be the critical functions of that sort of flying car?

3. What are the critical functions of a
   (a) Microwave
   (b) Motorcycle
   (c) Firearm
   (d) Fishing rod
   (e) House

4. Contrast a CFA with a failure mode and effect analysis (FMEA).

REFERENCES

1. Wikipedia. Functional analysis (psychology). http://en.wikipedia.org/wiki/Functional_analysis_(psychology) (accessed August 2011).

2. lists.r-forge.r-project.org Mailing Lists. http://netpro.r-forge.r-project.org (accessed August 2011).
3. Wikipedia. Functional analysis. http://en.wikipedia.org/wiki/Functional_analysis (accessed August 2011).
4. Thomson, F., Mathias, D., Go, S., and Nejad, H. (2010). Functional risk modeling for lunar surface systems. NASA Technical Reports Server (NTRS).
5. Function. http://dictionary.reference.com/browse/function (accessed August 2011).
6. NASA. Space Shuttle. http://www.nasa.gov/mission_pages/shuttle/main (accessed August 2011).
7. NASA. Space Shuttle Components. http://spaceflight.nasa.gov/history/shuttle-mir/spacecraft/s-orb-sscomponents-main.htm (accessed August 2011).
8. NASA. Columbia. http://www.nasa.gov/columbia/home/index.html (accessed August 2011).
9. Nelson, W.R. and Bagian, T.M. (2000). Critical Function Models for Operation of the International Space Station. INEEL Report INEEL/CON-00-01017.
10. Corcoran, W.R., Finnicum, D.J., Hubbard, F.R. III et al. (1981). Nuclear power plant safety functions. *Nuclear Safety* 22: March/April 1981.
11. United States. National Transportation Safety Board (1990). *United Airlines Flight 232 Mcdonnell Douglas Dc-1040 Sioux Gateway Airport Sioux City, Iowa, 19 July 1989*, United Flight 232, PBSO-910406 NTSB/AAR-SO/06. Washington DC: National Transportation Safety Board.
12. Surhone, L.M., Tennoe, M.T., and Henssonow, S.F. (2011). *United Airlines Flight 232* [Paperback]. W.W. Norton & Company, (1 March 2011).
13. Reason, J. (2008). *The Human Contribution* [Paperback], 1e. Arena (1 December 2008).
14. Nelson, W.H. (2013). The Gimli Glider. *Soaring: The Journal of the Soaring Society of America* (October): http://www.wadenelson.com/gimli.html.
15. Williams, Merran (2003). The 156-tonne Gimli Glider. Flight Safety Australia: 27. https://students.cs.byu.edu/~cs345ta/reference/Gimli%20Glider.pdf (accessed 2 February 2019).
16. Wikipedia. Auxiliary power unit. http://en.wikipedia.org/wiki/Auxiliary_power_unit (accessed August 2011).
17. Globalsecurity.org. F-22 Raptor Flight Critical Systems. https://www.globalsecurity.org/military/systems/aircraft/f-22-fcas.htm (accessed 2 February 2019).
18. AF.mil. http://www.af.mil/news/story.asp?id=123232976 (accessed August 2018).
19. ABC News. Air Force Blames Oxygen-Deprived Pilot in Deadly F-22 Crash. https://abcnews.go.com/Blotter/air-force-blames-oxygen-deprived-pilot-22-crash/story?id=15162509 (accessed 2 February 2019).
20. FEMA (2014). Continuity of Operations (COOP) Multi-Year Strategy and Program Management Plan Template Federal Emergency Management Agency. Federal Government Report.
21. Elliott, D. (2010). *Business Continuity Management: A Critical Management Routledge*, 1e.
22. Wade, N.M. (2005). *Battle Staff Smartbook: Doctrinal Guide to Military Decision Making and Tactical Operations*. Lightning Press.
23. Federal Emergency Management Agency (2004). Federal Preparedness Circular 65, 15 June 2004.
24. Official Idaho County Site. www.idahocounty.org (accessed 2 February 2019).
25. Wikipedia. Rhode Island. http://en.wikipedia.org/wiki/Rhode_Island (accessed August 2011).
26. Wikipedia. Switzerland. http://en.wikipedia.org/wiki/Switzerland (accessed August 2011).

CHAPTER 16

Event Tree and Decision Tree Analysis

Two of the largest nuclear accidents, Three Mile Island and Chernobyl, partially resulted from decisions made on the part of the power plant operating or engineering staff. In the case of Chernobyl, the accident was the result of a very bad decision on the part of the engineering staff who decided to run a test while the reactor was at low power. Three Mile Island accident, on the other hand, was initiated by a hardware failure but escalated when the operating crew decided to switch off the automatic safety system. The case study at the end of the chapter will analyze a portion of the Chernobyl event using an event tree.

16.1 EVENT TREES

An event tree is a graphical representation of a series of possible events in an accident sequence. Using this approach assumes that as each event occurs, there are only two outcomes, failure or success. A success ends the accident sequence and the postulated outcome is either the accident sequence terminated successfully or was mitigated successfully. For instance, a fire starts in a plant. This is the initiating event. Then the fire suppression system is challenged. If the system actuates, the fire is extinguished or suppressed and the event sequence ends. If the fire suppression system fails, then the fire is not extinguished or suppressed and the accident sequence progresses. Table 16.1 shows this postulated accident sequence. Figure 16.1 shows this accident sequence via an event tree.

As in most of the risk assessment techniques, probabilities can be assigned to the events and combined using the appropriate Boolean logic to develop an overall probability for the various paths in the event. Using our example from above, we will

Risk Assessment: Tools, Techniques, and Their Applications, Second Edition. Lee T. Ostrom and Cheryl A. Wilhelmsen.
© 2019 John Wiley & Sons, Inc. Published 2019 by John Wiley & Sons, Inc.
Companion website: www.wiley.com/go/Ostrom/RiskAssessment_2e

TABLE 16.1

Accident Sequence

Event	Description	Possible outcomes
Fire	This is the initiating event	
Fire suppression system actuates	The fire suppression system detects the fire and it actuates	Success – system actuates and controls the fire
		Failure – system fails to control the fire
Fire alarm system actuates	Fire alarm system detects the fire and sends a signal to the appropriate fire department. Fire department arrives in time to extinguish the fire before major damage occurs	Success – fire is extinguished prior to major damage occurring
		Failure – fire causes major damage

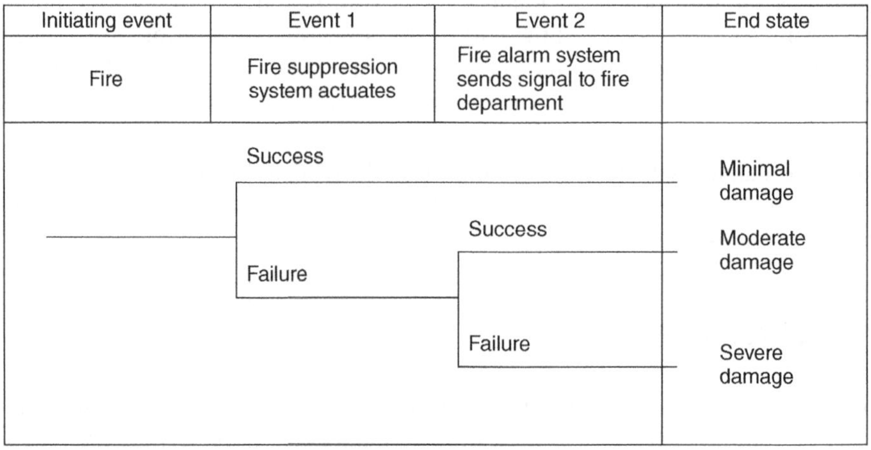

Initiating event	Event 1	Event 2	End state
Fire	Fire suppression system actuates	Fire alarm system sends signal to fire department	

FIGURE 16.1   Event tree.

now add probabilities to the events and show how the probabilities combine for each path. Figure 16.2 shows the addition of path probability to the event tree. Table 16.2 summarizes the probabilities for this event sequence.

The result of this analysis tells us that the probability derived for a fire in which the fire suppression system actuates and the consequence is minimal damage is approximately 1/1000 or 1×10^{-3}. The probability derived for a fire in which the fire suppression system fails to actuate but a fire alarm signal is successfully transmitted to the local fire department and there is only moderate damage is 1/100 000 or 1×10^{-5}. Finally, the probability that a fire occurs and both the fire suppression system and the fire alarm system fail and severe damage occurs is 5/10 000 000 or 5×10^{-8}.

This approach is considered inductive in nature. Meaning the system uses forward logic. A fault tree, discussed in this chapter, is considered deductive because

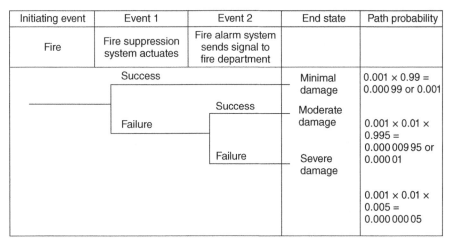

Initiating event	Event 1	Event 2	End state	Path probability
Fire	Fire suppression system actuates	Fire alarm system sends signal to fire department		
	Success		Minimal damage	0.001 × 0.99 = 0.000 99 or 0.001
	Failure	Success	Moderate damage	0.001 × 0.01 × 0.995 = 0.000 009 95 or 0.000 01
		Failure	Severe damage	0.001 × 0.01 × 0.005 = 0.000 000 05

FIGURE 16.2   Event tree with path probabilities.

TABLE 16.2
Event Sequence with Probabilities

Event	Description	Possible outcomes	Probability
Fire	This is the initiating event		0.001
Fire suppression system actuates	The fire suppression system detects the fire and it actuates	Success – system actuates and controls the fire	0.99
		Failure – system fails to control the fire	0.01
Fire alarm system actuates	Fire alarm system detects the fire and sends a signal to the appropriate fire department. Fire department arrives in time to extinguish the fire before major damage occurs	Success – fire is extinguished prior to major damage occurring	0.995
		Failure – fire causes major damage	0.005

usually the analyst starts at the top event and works down to the initiating event. In complex risk analyses, event trees are used to describe the major events in the accident sequence and each event can then be further analyzed using a technique most likely being a fault tree. Figure 16.3 shows a much more complicated event tree. It is part of the events from an analysis of a small-break loss-of-coolant accident (LOCA) (1). Again, note this is not the complete event tree, only a portion. Each of the events then comprises a fault tree of its own. The entire analysis then feeds into a probabilistic risk assessment.

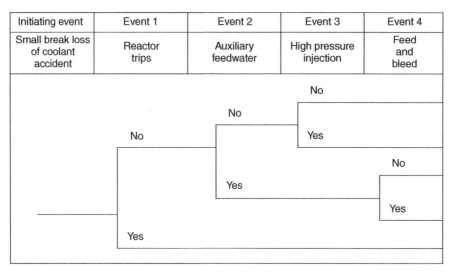

Initiating event	Event 1	Event 2	Event 3	Event 4
Small break loss of coolant accident	Reactor trips	Auxiliary feedwater	High pressure injection	Feed and bleed

FIGURE 16.3   Event tree for a portion of a small-break LOCA.

16.1.1 Case Study

Transportation accidents involving hazardous materials (HazMat) occur every day in the United States and around the world. On 1 May 2001, in Ramona, Oklahoma, an accident occurred that resulted in the release and explosion of hydrogen gas. A semi-trailer that contained horizontally mounted gas-filled cylinders and a pickup truck were involved in the accident. Both of these vehicles were traveling northbound on Interstate 75 south of the town of Ramona, OK. According to witnesses, the pickup truck veered into the path of the semitrailer, causing it to go out of control, flipping on its side, and traveling 300 ft before it came to rest, the pickup truck also ran off the road causing a rupture in the vehicles gas line, igniting a fire. During the initial stages of the accident, relief valves were broken off, causing a leak of the hydrogen gas. According to the National Transportation Safety Board (NTSB) report, the semitrailer driver was killed and the pickup truck driver was seriously injured.

At 14 : 15, the 911 dispatch center was notified of the incident, and the local fire department (Ramona Volunteer Fire Department) was notified. At 14 : 16, the Washington County Emergency Management Agency (EMA) started responding to the scene. Between 14 : 16 and 14 : 21, an Air Gas employee driving southbound saw the vehicle and notified the Air Gas Corporation. The driver of the pickup truck was extricated from the vehicle by local bystanders. At 14 : 21, the Washington County Emergency Medical Service (EMS) requested assistance from a private HazMat response team. At 14 : 22, the chief of the local fire department arrived and directed the firefighters to "cool" the burning cylinders. The Washington County EMA requested additional manpower for mutual aid companies. At 14 : 22, other firefighters not involved in active firefighting attempted to extricate the truck driver. At 14 : 25, the Emergency Operations Center (EOC) and emergency plan were activated by the Washington County EMA. At 14 : 30, Washington County EMA and Oklahoma State Patrol officer trained in HazMat arrived on scene. At 14 : 40, extrication attempts were halted because of the lack of water to suppress the fire coming from the cylinders. The Washington County EMA and Phillips Petroleum DART commander advised the firefighters to stay away from the ends of the burning tanks in the event of an explosion. At 15 : 00, the Air Gas executive team arrived on the scene and offered assistance to the incident command team. At 15 : 11, the Tulsa OK HazMat team was requested to aid. At 15 : 15, the truck driver was extricated from the vehicle. By 15 : 30–15 : 35, the Tulsa HazMat team arrived on the scene and assumed command. At 16 : 00, Air Gas executive team member and safety director offered assistance to the Tulsa OK HazMat team with their own response team called *AERO*. At 17 : 15, the AERO team members arrive on scene. At 18 : 30–18 : 40, the scene was declared controlled and the Tulsa OK HazMat team went into service. At 00 : 20, the AERO team commanded that the cylinders be vented and properly cooled. At 00 : 55, command was turned over to the Oklahoma State Patrol, active firefighting, HazMat response is terminated. At 06 : 00, the highway then reopened (2).

An event tree can be used to represent the major events in this accident sequence and can be used in the future by emergency response personnel to determine what the critical events in the sequence were (Figure 16.4).

16.2 DECISION TREES

Decision analysis is a very large topic, and there are risk professionals who specialize only in decision analysis. In this book, we will not attempt to present even a fraction of

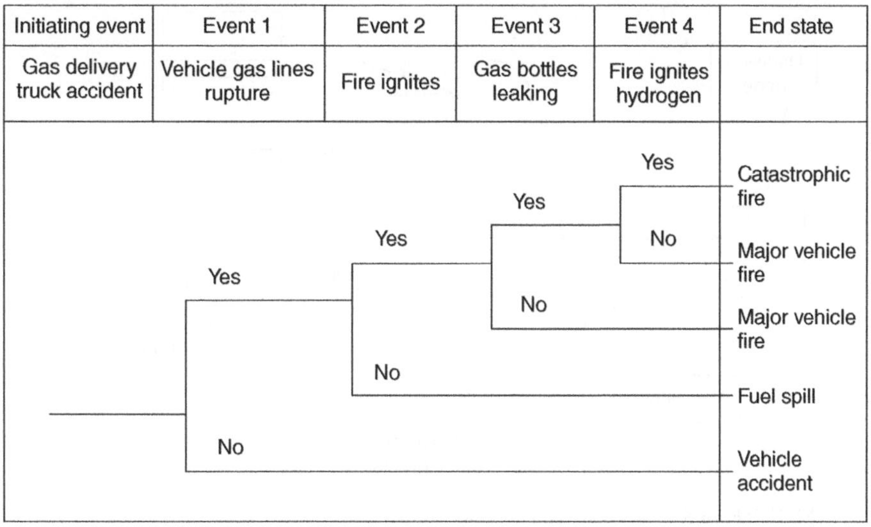

Initiating event	Event 1	Event 2	Event 3	Event 4	End state
Gas delivery truck accident	Vehicle gas lines rupture	Fire ignites	Gas bottles leaking	Fire ignites hydrogen	

FIGURE 16.4   Example event tree for this accident sequence.

all the decision analysis tools available. In fact, we will focus on two types of decision trees. Decision analysis and their associated tree analysis techniques assume that a decision is being made under risk. All decisions have some risk associated with them. The launch of the Challenger in 1986 was made under risk. The obvious consequence of this decision was the loss of the orbiter (3). We all make decisions everyday that affect our family's safety, financial security, and health. Some of our decisions also have the potential of affecting many other people as well. Elected officials, military leaders, heads of large corporations, and even university presidents make decisions that can have far-reaching effects.

The decisions concerning driving are some of the most important ones we make. Driving is still one of the most hazardous things we do. In an average year, approximately 35 000 people die from car accidents in the United States. Decision drivers contribute to a large number of these accidents. Drinking alcohol while driving is directly associated with approximately 16 000 persons dying in alcohol-related traffic accidents in 2008 (4).

There are several styles of decision trees. The style presented here is one that is commonly used in industry. Influence diagrams are a close cousin to this style of decision tree (5). This style of decision tree is commonly used in production settings to help determine the best alternative among many. Figure 16.5 shows the general configuration of this decision tree. For this tree, it is best to provide an example and then work through the process.

In this scenario, you are evaluating four systems. Table 16.3 contains the information about the systems.

Figure 16.6 shows the decision tree for this scenario.

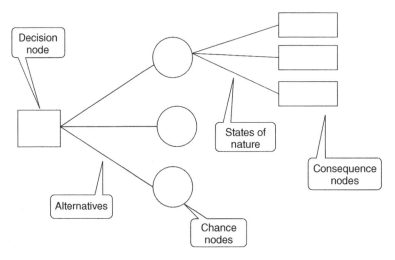

FIGURE 16.5 Decision tree 1 format.

TABLE 16.3
Decision Tree 1 Analysis

| System | Cost | Probability of success and postulated benefits | | |
		Good, 0.30	Moderate, 0.50	Poor, 0.20
A	1 000 000	100 000 000	25 000 000	−4 000 000
B	750 000	80 000 000	35 000, 000	−1 000 000
C	1 500 000	125 000 000	45 000 000	0
D	450 000	95 000 000	65 000 000	−80 000 000

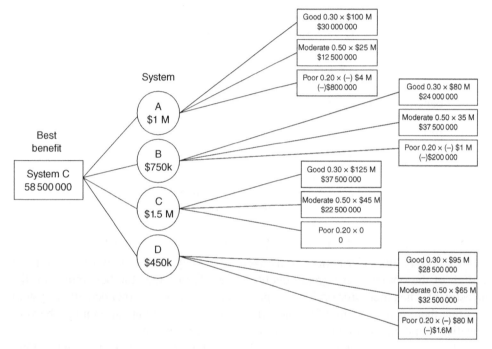

FIGURE 16.6   Decision tree 2 format.

The following shows how the final outcomes are derived from the data.

System A

$$0.3 \times 100\,000\,000 = 30\,000\,000$$
$$0.5 \times 25\,000\,00 = 12\,500\,000$$
$$0.2 \times (-)4\,000\,000 = (-)800\,000$$
$$\text{Average benefit} = \$41\,700\,000$$
$$\text{Cost is } 1\,000\,000$$
$$\text{Net benefit} = \text{Benefit} - \text{Cost} = 41\,700\,000 - 1\,000\,000 = \$40\,700\,000$$

System B

$$0.3 \times 80\,000\,000 = 24\,000\,000$$
$$0.5 \times 35\,000\,000 = 17\,500\,000$$
$$0.2 \times (-)1\,000\,000 = (-)200\,000$$
$$\text{Average benefit} = \$41\,300\,000$$
$$\text{Cost is } 750\,000$$
$$\text{Net benefit} = \text{Benefit} - \text{Cost} = 41\,300\,000 - 750\,000 = \$40\,550\,000$$

System C

$$0.3 \times 125\,000\,000 = 37\,500\,000$$
$$0.5 \times 45\,000\,000 = 22\,500\,000$$
$$0.2 \times 0 = 0$$
$$\text{Average benefit} = \$60\,000\,000$$
$$\text{Cost is } 1\,500\,000$$
$$\text{Net benefit} = \text{Benefit} - \text{Cost} = 60\,000\,000 - 1\,500\,000 = \$58\,500\,000$$

System D

$$0.3 \times 95\,000\,000 = 28\,500\,000$$
$$0.5 \times 65\,000\,000 = 32\,500\,000$$
$$0.2 \times (-)80\,000\,000 = (-)1\,600\,000$$
$$\text{Average benefit} = \$45\,000\,000$$
$$\text{Cost is } 450\,000$$
$$\text{Net benefit} = \text{Benefit} - \text{Cost} = 45\,000\,000 - 450\,000 = \$44\,550\,000$$

So, how is this information used? Well, obviously, this is a contrived example and all of the options have great benefit. System C provides the best return on the investment. The other three options are pretty much a wash. Another benefit of System C is that there is no potential for a negative outcome. An organization might be very sensitive to not having any potential for a negative consequence.

This style of decision tree can easily be modified to provide insights into safety attributes as well. Organizations place monetary value on accident risks and consequences. For this example, we will use the relationships in Table 16.4 to help develop this example.

In this scenario, instead of subtracting out the cost of the system, we add it to the calculated cost. Performing the math generates the following decision criteria.

System A

$$0.10 \times \$1\,000\,000 = \$100\,000$$
$$0.30 \times \$300\,000 = \$90\,000$$
$$0.20 \times \$1\,030\,000 = \$206\,000$$
$$0.40 \times \$5\,000 = \$2\,000$$
$$\text{Total cost} = \$100\,000 + \$90\,000 + \$206\,000 + \$2\,000 + \$2\,000\,000$$
$$\text{(system cost)} = \$2\,398\,000$$

TABLE 16.4
Accident Classification and Associated Cost

Accident type	Description	Relative cost ($)
Class A	Fatality or loss of property in excess of $1 000 000	1 000 000
Class B	Disabling Injury or property loss between $100 000 and $1 000 000	100 000
Class C	Lost work day injury or property loss between $10 000 and $100 000	10 000
Class D	OSHA recordable injury or minor property damage	5 000

System B
$$0.05 \times \$2\,000\,000 = \$100\,000$$
$$0.40 \times \$1\,400\,000 = \$560\,000$$
$$0.30 \times \$540\,000 = \$162\,000$$
$$0.25 \times \$50\,000 = \$12\,500$$
Total cost $= \$100\,000 + \$560\,000 + \$162\,000 + \$12\,500 + \$1\,000\,000$
(system cost) $= \$1\,824\,500$

System C
$$0.20 \times \$3\,000\,000 = \$600\,000$$
$$0.40 \times \$2\,500\,000 = \$1\,000\,000$$
$$0.30 \times \$1\,400\,000 = \$420\,000$$
$$0.10 \times \$20\,000 = \$2000$$
Total cost $= \$600\,000 + \$1\,000\,000 + \$420\,000 + \$2000 + \$500\,000$
(system cost) $= \$2\,522\,000$

In this example, System B is the clear winner. However, this example does not take into account the cost of lost business or reputation in the event of a Class A accident. These items can be factored in as well.

For instance, besides the direct cost of a Class A accident estimated at $1 000 000, there might be a loss of business estimated at a cost of $1 000 000 per occurrence (Tables 16.5 and 16.6). Therefore, this type of decision tree is very utilitarian and can be adapted to a wide range of uses.

16.3 CASE STUDY: CHERNOBYL

On 25 April 1986, prior to a routine shutdown, the reactor crew at Chernobyl 4 began preparing for an experiment to determine how long turbines would spin and supply power to the main circulating pumps following a loss of main electrical power supply. This test had been carried out at Chernobyl the previous year, but the power from the turbine ran down too rapidly, so new voltage regulator designs were to be tested (6).

Multiple operator actions, including the disabling of automatic shutdown mechanisms, preceded the attempted experiment on the morning of 26 April. By the time that the operator began to shut down the reactor, it was in an extremely unstable condition.

TABLE 16.5
Accident Comparison

System	Description	Cost ($)	Probability and associated cost
A	This system will provide a high degree of safety but has a high cost	2 000 000	0.10:1 Class A accident 0.30:3 Class B accidents 0.20:1 Class A and 3 Class C accidents 0.40:1 Class D accident
B	The system will provide a moderate degree of safety at a moderate cost	1 000 000	0.05:2 Class A accidents 0.40:1 Class A accident and 4 Class B accidents 0.30:5 Class B accidents and 4 Class C accidents 0.25:2 Class C accidents and 6 Class D accidents
C	This system will provide a marginal degree of safety at a low cost	500 000	0.20:3 Class A accidents 0.40:2 Class A accidents and 5 Class B accidents 0.30:1 Class A accidents and 4 Class B accidents 0.10:4 Class D accidents

TABLE 16.6
Accident Analysis Table

System	Description	Cost ($)	Probability of accidents
A	This system will provide a high degree of safety, but has a high cost	2 000 000	0.10: $1 000 000 0.30: $300 000 0.20: $1 030 000 0.40: $5000
B	The system will provide a moderate degree of safety at a moderate cost	1 000 000	0.05: $2 000 000 0.40: $1 400 000 0.30: $540 000 0.25: $50 000
C	This system will provide a marginal degree of safety at a low cost	500 000	0.20: $3 000 000 0.40: $2 500 000 0.30: $1 400 000 0.10: $20 000

The design of the control rods caused a dramatic power surge as they were inserted into the reactor.

The interaction of extremely hot fuel with the cooling water led to fuel disintegration, along with rapid steam production and an increase in reactor pressure. The design characteristics of the reactor were such that substantial damage to even three or four fuel assemblies could – and did – result in the failure of the reactor vessel. Extreme pressure in the reactor vessel caused the 1000 ton cover plate of the reactor to become partially detached. The fuel channels were damaged and the control rods jammed, which by that time were only halfway down. Intense steam generation then spread throughout the entire core. The steam resulted from water being dumped into the core because of the rupture of the emergency cooling circuit. A steam explosion resulted and released fission products to the atmosphere. A second explosion occurred a few seconds later that threw out fuel fragments and blocks of hot graphite. The cause of the second explosion has been disputed by experts, but it is likely to have been caused by the production of hydrogen from zirconium–steam reactions.

Two workers died as a result of these explosions. The graphite (about a quarter of the 1200 tons of it was estimated to have been ejected) and fuel became incandescent and started a number of fires, causing the main release of radioactivity into the environment. A total of about 14 EBq (14×10^{18} Bq) of radioactivity was released, over half of it being from biologically inert noble gases.

About 200–300 tons of water per hour was injected into the intact half of the reactor using the auxiliary feed water pumps. However, this was stopped after half a day because of the danger of it flowing into and flooding units 1 and 2. From the second to tenth day after the accident, some 5000 tons of boron, dolomite, sand, clay, and lead were dropped on to the burning core by helicopter in an effort to extinguish the blaze and limit the release of radioactive particles.

It is estimated that all of the xenon gas, about half of the iodine and cesium, and at least 5% of the remaining radioactive material in the Chernobyl 4 reactor core (which had 192 tons of fuel) was released in the accident. Most of the released material was deposited close to the reactor complex as dust and debris. Lighter material was carried by wind over the Ukraine, Belarus, Russia, and to some extent over Scandinavia and Europe.

The casualties included firefighters who attended the initial fires on the roof of the turbine building. All these were put out in a few hours, but radiation doses on the first day were estimated to be up to 20 000 mSv, causing 28 deaths – 6 of which were firemen – by the end of July 1986.

The Soviet Government made the decision to restart the remaining three reactors. To do so, the radioactivity at the site would have to be reduced. Approximately 200 000 people ("liquidators") from all over the Soviet Union were involved in the recovery and cleanup during the years 1986 and 1987. Those individuals received high doses of radiation, averaging around 100 mSv. Approximately 20 000 of the liquidators received about 250 mSv and a few received 500 mSv. Later, their numbers swelled to over 600 000 but most of them received only relatively low radiation doses.

Causing the main exposure hazard were short-lived iodine-131 and cesium-137 isotopes. Both of these are fission products dispersed from the reactor core, with half-lives of 8 days and 30 years, respectively (1.8 EBq of I-131 and 0.085 EBq of Cs-137 were released). About 5 million people lived in areas contaminated (above 37 kbq m^{-2} Cs-137), and about 400 000 lived in more contaminated areas of strict control by authorities (above 555 kbq m^{-2} Cs-137).

Approximately 45 000 residents were evacuated from within a 10 km radius of the plant, notably from the plant operators' town of Pripyat on 2 and 3 May. On 4 May, all those living within a 30 km radius – a further 116 000 people from the more contaminated area – were evacuated and later relocated. Approximately 1000 of those evacuated have since returned unofficially to live within the contaminated zone. Most of those evacuated received radiation doses of less than 50 mSv, although a few received 100 mSv or more.

Reliable information about the accident and the resulting contamination was not made available to affected people for about two years following the accident. This led the populace to be distrustful of the Soviet Government and led to much confusion about the potential health effects. In the years following the accident, a further 210 000 people were moved into less contaminated areas, and the initial 30 km radius exclusion zone (2800 km^2) was modified and extended to cover an area of 4300 km^2. This resettlement was owing to the application of a criterion of 350 mSv projected lifetime radiation dose, although, in fact, radiation in most of the affected area (apart from half a square kilometer) fell rapidly after the accident so that average doses were less than 50% above normal background of 2.5 mSv per year.

Recent studies have found that the area surrounding the reactors is recovering, although background radiation levels are approximately 35 times normal background level (7). In fact, in 2010, the area surrounding the reactor site was opened for tourism (8).

The consequences of this event are still felt today. Nuclear power is viewed as being very dangerous by a large percentage of Americans. Those exposed during the event have developed various forms of cancer. For instance, in Belarus, thyroid cancer rates have risen 2400% (9). The aftereffects of the event will take many generations and possibly hundreds of years to resolve.

16.3.1 Analysis of the Event

This event has been analyzed and reanalyzed thousands of times over the past two decades. In fact, in our teaching careers, we have received hundreds of papers on the subject. Our goal here is to show how an event can be used to aid in the analysis process. We will also show how a decision tree could have helped to prevent the accident.

16.3.2 Event Tree Analysis

As shown above, in the event tree analysis process, the entry point into the tree is the initiating event. In the Chernobyl event, what was the initiating event? Was it the reactor shutdown procedure or was it the experiment the engineers were running or something else? Table 16.7 lists the important events we will model in this demonstration analysis. Note that some of the events one might want to model occurred well before the accident, like the design of the reactor itself. Those events that occurred after rod insertion were the result of the accident and the reactor design.

Event and decision trees are highly effective tools that have broad application in providing visual and analytical means to better understand both simple as well as

TABLE 16.7
Events to be Analyzed in Chernobyl Event

Event	Description	Consequence	Alternative
Initiating event	Decision to conduct experiment during reactor shutdown	The decision to conduct the experiment directly led to the accident	If the decision to not conduct this experiment on an operating reactor, the event would not have occurred
Event 1	Reactor in shutdown mode	This design of reactor is unstable at low power	Maintaining reactor power and not doing the shutdown would have prevented the accident
Event 2	Automatic shutdown mechanism disabled	Once the automatic shutdown mechanisms were disabled, the reactor during the shutdown phase lost the ability to be controlled adequately	Not disabling the mechanism would have allowed the reactor to shut down normally
Event 3	Control rod insertion	In this particular reactor design, when the control rods are inserted, there can be and was a rapid increase in power before the rods are fully inserted	It is pure speculation, but there is a possibility that if the rods had not been inserted or inserted at a different pace, the event would not have occurred as it did

complex events. One can use the tools to model events in failure, as well as success space to help better understand them.

Self-Check Questions

1. A property owner has to choose: (i) A large-scale investment (A) to improve her investments. This could produce a substantial pay-off in terms of increased revenue net of costs but will require an investment of $1 400 000. After extensive market research, it is considered that there is a 40% chance that a pay-off of $2 500 000 will be obtained, but there is a 60% chance that it will be only $800 000. (ii) A smaller scale project (B) to re-decorate her premises. At $500 000 this is less costly but will produce a lower pay-off. Research data suggests a 30% chance of a gain of $1 000 000, but a 70% chance of it being only $500 000. (iii) Continuing the present operation without change (C). It will cost nothing, but neither will it produce any pay-off. Clients will be unhappy and it will become harder and harder to rent the apartments out when they become free. How will a decision tree help the taking of the decision?

2. The Mining Group Limited (MGL) is a company set up to conduct geological explorations of parcels of land in order to ascertain whether significant metal deposits (worthy of further commercial exploitation) are present or not. Current MGL has an option to purchase outright a parcel of land for $3 million. If MGL purchases this parcel of land, then it will conduct a geological exploration of the land. Past experience indicates that for the type of parcel of land under consideration geological explorations cost approximately $1 million and yield significant metal deposits as follows:

   ◦ Manganese: 1% chance
   ◦ Gold: 0.05% chance
   ◦ Silver: 0.2% chance

   Only one of these three metals is ever found (if at all), i.e. there is no chance of finding two or more of these metals and no chance of finding any other metal.
   If manganese is found, then the parcel of land can be sold for $30 million. If gold is found, then the parcel of land can be sold for $250 million. If silver is found, the parcel of land can be sold for $150 million.
   MGL can, if they wish, pay $750 000 for the right to conduct a three-day test exploration before deciding whether to purchase the parcel of land or not. Such three-day test explorations can only give a preliminary indication of whether significant metal deposits are present or not, and past experience indicates that three-day test explorations cost $250 000 and that significant metal deposits are present 50% of the time.
   If the three-day test exploration indicates significant metal deposits, then the chances of finding manganese, gold, and silver increase to 3, 2, and 1%, respectively. If the three-day test exploration fails to indicate significant metal deposits, then the chances of finding manganese, gold, and silver decrease to 0.75, 0.04, and 0.175%, respectively.
   What would you recommend MGL should do and why?
   A company working in a related field to MGL is prepared to pay half of all costs associated with this parcel of land in return for half of all revenues. Under these circumstances, what would you recommend MGL should do and why?

3. A company is trying to decide whether to bid for a certain contract or not. They estimate that merely preparing the bid will cost $10 000. If their company bids, then they estimate that there is a 50% chance that their bid will be put on the "short-list;" otherwise their bid will be rejected.
   Once "short-listed" the company will have to supply further detailed information (entailing costs estimated at $5000). After this stage their bid will either be accepted or rejected.
   The company estimate that the labor and material costs associated with the contract are $127 000. They are considering three possible bid prices, namely, $155 000, $170 000, and $190 000. They estimate that the probability of these bids being accepted (once they have been short-listed) is 0.90, 0.75, and 0.35, respectively.

What should the company do and what is the expected monetary value of your suggested course of action?

4. Develop an event tree for the following accident.
National Transportation Safety Board
Marine Accident Brief
Capsizing and sinking of towing vessel *Gracie Claire*

Accident no.	DCA17FM025
Vessel name	*Gracie Claire*
Accident type	Capsizing
Location	Tiger Pass, 1.3 miles southwest of Lower Mississippi River in Venice, Louisiana 29°15.06′ N, 089°21.49′ W
Date	23 August 2017
Time	0756 Central Daylight Time (coordinated universal time − 5 h)
Injuries	None
Property damage	$565 000 est.
Environmental damage	Narrow oil sheen, 1 mile long, observed; 1100 gal of diesel fuel discharged; approximately 655 gal recovered
Weather	Cloudy, winds northwest at 3 mph, air temperature 80 °F
Waterway information	Tiger Pass is an outlet to the Gulf of Mexico near mile marker 10 on the Lower Mississippi River. The channel is 14 ft deep and 150 ft wide

On 23 August 2017, at 0756 local time, the towing vessel *Gracie Claire* was moored in Tiger Pass near mile marker 10 on the Lower Mississippi River in Venice, Louisiana. While taking on fuel and water, the towboat began to slowly list to starboard. After the wake of a passing crew boat washed onto the *Gracie Claire*'s stern, the list increased. In a short period of time, water entered an open door to the engine room and flooded the space. The towboat sank partially, its bow being held above the water by the lines connected to the dock. All three crew members escaped to the dock without injury. Approximately 1100 gal of diesel fuel were discharged into the waterway. The damage to the Gracie Claire shown in Figures 16.7 and 16.8 was estimated at $565 000.

Accident Events
At 2135, on 20 August 2017, the *Gracie Claire*, a 45-ft-long towboat, departed Morgan City, Louisiana, pushing two empty hopper barges. On board were a

FIGURE 16.7 *Gracie Claire* moored for repair following sinking. *Source*: Adapted from the NTSB Report.

captain, a mate, and a deckhand. The vessel arrived the next day in Venice, Louisiana, where the barges were dropped off. On the evening of 22 August, the crew members began assembling another tow and then moored at a nearby dock for the night.

The next morning, on 23 August, a new captain and deckhand came aboard the *Gracie Claire*. At 0720, before resuming assembly of the tow, the crew members got the vessel under way to load fuel and water. According to one crew member's statement and video evidence, the vessel's freeboard was only a few inches at the stern.

About 20 minutes later, the towboat arrived at the fuel dock of John W. Stone Oil Distributor (Stone Oil) in Tiger Pass, an outlet from the Lower Mississippi River to the Gulf of Mexico. The captain moored the towboat bow-in, perpendicular to the seawall (bulkhead), with two lines extending directly ahead from a centerline bitt at the bow to a single mooring bollard ashore. Soon afterward, at 0750, using the starboard-side fill pipes, the captain began loading the fuel tank forward of the deckhouse, while the deckhand filled the water tank aft of the deckhouse.

The deckhand, who had just begun working in the maritime industry a month prior, told investigators that the towboat appeared to be sitting lower in the water before

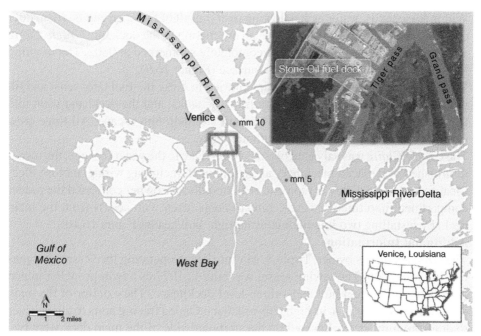

FIGURE 16.8   Location where *Gracie Claire* sank in Tiger Pass while the vessel was refueling at John W. Stone Oil Distributor dock in Venice, Louisiana. *Source*: Photo adapted from original NTSB report.

he began loading the water into the tank; however, he did not consider the condition unusual compared to the vessel's draft during other fueling evolutions he had observed. The *Gracie Claire* began listing to starboard as fuel was loaded through the starboard fill pipe at a rate of 140–150 gal (£972–1041) per minute. The new load was in addition to the estimated 1200 gal (£8640) of fuel already in the tank. According to the captain, listing to one side during fuel loading was not uncommon. He told investigators that the *Gracie Claire* would often list to angles that submerged the deck edge. When the *Gracie Claire* listed to one side during refueling, his practice was to stop fueling and then resume filling the tank on the opposite side to bring the towboat to an even keel because he believed that there were two separate fuel tanks, one on the starboard side and one on the port side. In fact, there was only one fuel tank with a swash bulkhead on the centerline of the vessel that allowed fuel to flow from one side of the tank to the other.

As loading continued, video captured by the camera on the dock showed a disturbance on the surface of the water at the stern of the *Gracie Claire*. The water motion indicated a current acting on the starboard side.

The video footage showed that at 0755 an unidentified crew boat was traveling in the middle of the channel. When the crew boat passed the *Gracie Claire,* its wake washed onto the starboard side of the towboat's stern. The towboat then slowly rolled to starboard as more of the starboard-side main deck became submerged.

Within 90 seconds, water began flowing over the sill of the open engine room door on the starboard side. As the crew began evacuating to the dock, the *Gracie Claire* continued to roll until it was almost completely submerged at 0802. The mooring lines held the bow above the water's surface (Figure 16.9).

While the vessel was sinking, the captain tried to close the fuel tank vents and fill pipes to prevent oil from discharging from the tanks, and the deckhand went into the deckhouse to awaken the relief captain and help him escape. All three crew members abandoned the towboat without injury.

The *Gracie Claire* was salvaged two days later. Due to the water damage, the owner refurbished or replaced machinery and equipment, including wiring, motors, electrical fixtures, and galley equipment. Additionally, the owner converted the rudder compartment into three separate compartments, each with access from the main deck, by installing two longitudinal watertight bulkheads (Figure 16.10).

Additional Information

The *Gracie Claire*'s owner, Triple S. Marine, began operating the vessel after purchasing it in 2006. The towing vessel was built in 1977 as a twin-propeller inland push boat with a steel hull and a three-level deckhouse. The exterior of the main deck was fitted with a 14-in-high bulwark that included freeing ports to allow water to drain from the deck. Located on the starboard side of the main deck were two doors, one to the galley, forward, and one to the engine room, amidships. There was another engine room door on the port side. The bottom coaming of each door was elevated 20 in above the deck.

Beneath the main deck, four watertight bulkheads divided the hull into five compartments. In the fuel and water tanks, centerline swash bulkheads were installed to improve the towboat's stability by preventing the quick shift of the weight of fuel and water from one side to the other when the tanks were slack (not full). The first opening in these bulkheads was 24 in above the bottom of the tank.

The *Gracie Claire* was dry-docked on 1 August, three weeks prior to the accident. Employees from Triple S. Marine inspected the underwater portion of the hull, and shipyard workers replaced the port shaft and propeller. No alterations or new equipment were added that would have increased the weight of the towboat and thereby changed its stability.

Following the accident, a damage assessment of the vessel during dry dock revealed a 1-in diameter hole in the bottom of the rudder compartment. The marine surveyor performing the assessment told investigators that the hole was likely present before fuel was loaded onto the towboat.

A company representative stated that, among measures to ensure stability of the *Gracie Claire*, Triple S. Marine's policy required the captain to check the rudder compartment for leaks each week and did not allow the vessel to be moored

FIGURE 16.9   Screenshots of *Gracie Claire*, starting
a minute after loading began through sinking, from
video captured from the southwest view of the dock's
camera. *Source*: Photo adapted from NTSB report.

FIGURE 16.10 *Gracie Claire* during salvage. *Source*: Photo adapted from NTSB report.

"bow first" (bow-in) if "excessive current was present." Investigators estimated the current's speed at the time of the accident was 1–2 knots (1.1–2.3 mph).

The captain obtained his mariner's license in 1983. He told investigators that he had served as a captain for Triple S. Marine for over five years before the accident, had worked on the *Gracie Claire* often, and had fueled the company's towboats more than 100 times. The deckhand started working in the maritime industry and with Triple S. Marine only four weeks prior to the capsizing.

Analysis

Several factors affecting the stability of the *Gracie Claire* led to its capsizing. Before the vessel arrived at the Stone Oil dock to take on fuel, the rudder compartment likely had been flooding from the hole in the bottom of the hull, a water ingress that the crew was not aware of (Figure 16.11). The water level in the compartment would have risen until it equaled the water level outside the towboat, nearly filling the compartment and creating the equivalent of a slack tank. The lost buoyancy from the water in the compartment would have resulted in the towboat having less freeboard. A lower freeboard would have decreased the towboat's stability, because the main deck edge would be submerged at lesser angles of heel. The lower freeboard would also have made it easier for waves to wash across the main deck.

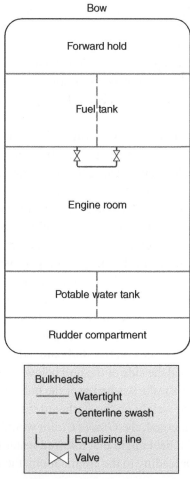

FIGURE 16.11   Plan view of *Gracie Claire*'s hull, not drawn to scale. *Source*: Drawing adapted from NTSB Report.

By mooring perpendicular to the seawall and not parallel to the waterway, the *Gracie Claire* was subjected to a heeling moment caused by the river current. When the captain and deckhand began loading fuel and water through the starboard fill pipes, the vessel already had a starboard list. Although the fuel and water tanks ran the full breadth of the hull, the centerline swash bulkheads would have contained the loaded liquids on the starboard side of the tanks until they reached the height of the 24-in openings above the tank's bottom. Because the liquids would have filled only the starboard side of their respective tanks, the vessel began listing further to starboard.

At the time, the equalizing line valves for the fuel oil tank were open, which was intended to balance the level of the tank's contents on each side of the centerline swash bulkhead. Based on 24 in of fuel oil above the pipe, investigators calculated that the line would transfer approximately 70 gal min^{-1}. However, considering a fill rate of 140–150 gal min^{-1} and the vessel's continued heel to starboard, the captain was filling the tank at a rate that exceeded the capacity of the equalizing line to level the tank.

Aware of the increasing starboard list during filling, the captain decided to shift fuel filling to the port side of the tank to counter the starboard list. He believed that a reduction in the list would occur because the fuel being loaded would accumulate on the port side, or high side, of the tank. However, with the vessel already listing to starboard, once the fuel reached the holes in the swash bulkhead on the port side of the fuel tank, the added fuel would shift by spilling over to the starboard side through the swash bulkhead openings, therefore exacerbating the starboard list. At the same time, the floodwater in the rudder compartment and in the water tank would also shift further to starboard. Had the water in the rudder compartment been discovered, the *Gracie Claire*'s owner or captain presumably would have taken measures to eliminate the water in the space before taking on fuel and water. While the captain was filling the tank from the port side, the crew boat that passed astern added to the heeling forces on the starboard side. First, the displaced water of the passing vessel pushed the side of the hull below the waterline. Second, as the waves from the crew boat's wake washed over the bulwark and onto the main deck of the towboat, they added weight on the starboard side of the main deck. With the starboard deck edge submerged, the *Gracie Claire* would have rolled more easily. Once the vessel reached a heel angle that allowed water to reach above the 20-in coaming of the open door leading to the engine room, the vessel down-flooded and rapidly sank.

Probable Cause

The National Transportation Safety Board determines that the probable cause of the capsizing and sinking of the *Gracie Claire* was the towing vessel's decreased stability and freeboard due to undetected flooding through a hull leak in the rudder compartment, which made the vessel susceptible to the adverse effects of boarding water from the wake of a passing vessel.

Vessel Particulars

Vessel	*Gracie Claire*
Owner/operator	Triple S. Marine LLC
Port of registry	Morgan City, Louisiana
Flag	United States
Type	Towing vessel

Vessel	*Gracie Claire*
Year built	1977
Official number	584 694
IMO number	N/A
Classification society	N/A
Construction	Steel
Length	45.1 ft (13.7 m)
Draft	6.6 ft (2.0 m)
Beam/width	21 ft (6.4 m)
Gross registered tonnage	62
Engine power; manufacturer	900 hp (671 kW) (2) 12 V71; Detroit Diesel
Persons on board	3

NTSB investigators worked closely with our counterparts from Coast Guard Sector New Orleans throughout this investigation. For more details about this accident, visit www.ntsb.gov and search for NTSB accident ID DCA17FM025.

5. Develop an event tree for the following accident.
National Transportation Safety Board
Pipeline Accident Brief
TransCanada **Corporation Pipeline (Keystone Pipeline) Rupture**
Amherst, South Dakota
The Accident
On 16 November 2017, at 4 : 34 a.m. mountain daylight time, a TransCanada Corporation, Keystone Pipeline (Keystone) ruptured near Amherst, South Dakota, between the Ludden, North Dakota, (Ludden) and Ferney, South Dakota, (Ferney) pump stations. Keystone's Operational Control Center (OCC), in Calgary, Alberta, Canada, was monitoring Keystone's Supervisory Control and Data Acquisition system that detected the leak and shut down the pipeline. Keystone's field staff traveled to the indicated leak location, confirmed that the pipeline had ruptured, and initiated their spill response plan. The approximate spill area was comprised of about 5000 barrels of crude oil (Figure 16.12).

When notified about the accident, the National Transportation Safety Board (NTSB) coordinated with the Pipeline and Hazardous Materials Safety Administration (PHMSA) to conduct a metallurgical examination of the pipe to establish a probable cause of the pipeline rupture. As part of its comprehensive investigation of Keystone's operations and practices leading up to the pipeline rupture, PHMSA provided oversight of Keystone's excavation, removal, and shipment of the failed pipe section to the NTSB laboratory in Washington, DC.

FIGURE 16.12 Accident scene in Amherst, South Dakota, taken on 16 November 2017. *Source*: https://twitter.com/TransCanada.

At the time of the pipeline rupture, Keystone was conducting an in-line inspection (ILI) using a cleaning tool (pig) and an acoustic leak detection tool (SmartBall). Prior to the rupture, both the cleaning pig and the SmartBall leak detection tool had traveled past the nearest block valve downstream of the rupture location (Ludden +28). The tools were run in a light-sweet blend batch of crude oil, followed by a batch of sour crude. In preparation for the tools to bypass the next downstream pump station in Ferney, a bypass operation of the Ferney pump station was initiated at 5 : 03 a.m. The bypass of the Ferney pump station was fully executed at 5 : 24 a.m. The pump station bypass resulted in a gradual and anticipated pressure increase at the discharge of the upstream station in Ludden. The pressure records from Ludden indicated that the pressure had increased from 1170 pounds per square inch gage (psig) to 1352 psig when the rupture occurred. Since the pipeline installation, no ILI tools for detecting cracks had been used.

The first indication of the pipeline rupture occurred at 5 : 33 a.m., when an abrupt drop in discharge pressure and a corresponding increase in flow rate were observed at the Ludden pump station. In addition, at 5 : 34 a.m., a corresponding pressure drop was observed at the Ferney pump station. The controller at Keystone's OCC initiated an emergency shutdown of the pipeline at 5 : 36 a.m. and commenced isolation of the pipeline. By 5 : 45 a.m., the failure location had been isolated by using remotely operated valves. Keystone personnel were dispatched to investigate the pipeline right-of-way for signs of a release and confirmed oil on the ground about 9 : 15 a.m. Keystone personnel then notified Marshall County emergency services;

the Britton Fire Department and Marshall County Sheriff responded within minutes of notification. The incident commander established and maintained a 1-mile safety zone during the response.

The Pipeline System

The Keystone Pipeline originates from Hardisty, Alberta, Canada, and delivers crude oil to terminals in Patoka, Illinois, and Cushing, Oklahoma. The rupture occurred downstream of the Ludden pump station, at milepost 234.2, in Marshall County, South Dakota. The pipe at the rupture location was constructed during the fall of 2008 and was commissioned in 2010. The ruptured pipe was 30 in in diameter with a 0.386-in wall thickness and was manufactured by Berg Pipe to American Petroleum Institute Specification 5L grade X-70 product specification Level 2, using a double submerged arc-welded (DSAW) longitudinal weld seam and fusion bonded epoxy (FBE) coating. The pipeline was constructed and operated under a special permit, which allowed operation at pressures up to 80% of specified minimum yield strength (SMYS).

When the pipe was excavated, concrete weights installed during construction were found at the rupture location. The ruptured pipe piece along with a portion of one concrete weight were shipped to the NTSB materials laboratory for examination. Figure 16.13 shows the ruptured pipe after excavation and prior to shipment to the NTSB.

At the NTSB materials laboratory, the ruptured pipe piece was cleaned and visually examined. The metallurgists found midway along the length of the fracture face (a region of 5.52 in long), ratchet marks, crack arrest lines, and other features consistent with a multiple-origin fatigue fracture that originated at an external groove in the pipe wall (Figure 16.14). The fatigue origin area was located near the top of the pipe about 6.5 in away from the longitudinal seam. The pipe exterior exhibited areas of exposed metal and grooves formed by sliding contact that aligned nearly parallel to the pipe axis on the exterior surface coinciding and adjacent to the length of the fracture. Coating material was present and intermixed with the sliding contact marks, and many edges of the remaining coating adjacent to individual contact marks were curled and rounded consistent with sliding contact deformation. As shown on the right side of Figure 16.14, a larger cluster of grooves was present toward the upstream end of the fracture.

Compositional analysis of the grooved surfaces and of cross sections through the grooves near the fatigue crack origin area revealed that smeared metal in the trough of the sliding contact grooves contained higher concentrations of chromium than the pipe material (as determined by energy dispersive spectroscopy and backscattered electron microscopy). This was consistent with the presence of a deposited layer.

The composition of the deposited metal in the grooves and the morphology of the groove colonies are consistent with damage that would result from the pipe being run over by a metal-tracked construction vehicle. The composition of the concrete weights was examined and ruled out as a source of the mechanical damage to the

FIGURE 16.13 Ruptured pipe.

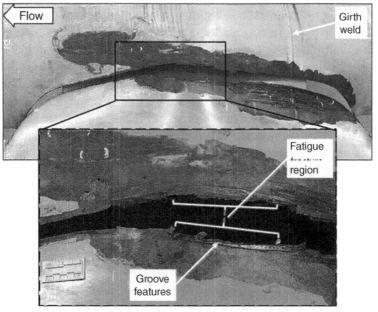

FIGURE 16.14 Overview of the pipe fracture with mating sides of the fracture placed adjacent to each other. The inset image shows a closer view of the fatigue facture initiation region and a colony of sliding contact grooves.

pipe. A full report of the metallurgical examination can be found in accident docket PLD18LR001.

Probable Cause

The NTSB determines that the probable cause of the failure of the Keystone Pipeline was a fatigue crack, likely originating from mechanical damage to the pipe exterior by a metal-tracked vehicle during pipeline installation, that grew and extended in-service to a critical size, resulting in the rupture of the pipeline.

For more details about this accident, visit http://www.ntsb.gov/investigations/dms .html and search for NTSB accident identification PLD18LR001.

REFERENCES

1. US Nuclear Regulatory Agency (2010). Resolution of Generic Safety Issues (Formerly entitled "A Prioritization of Generic Safety Issues"). NUREG-0933, Main Report with Supplements 1–34.
2. NTSB (2011). Hazardous materials accident report release and ignition of hydrogen following collision of a tractor-semitrailer with horizontally mounted cylinders and a pickup truck near Ramona, Oklahoma, 1 May 2001, NTSB/HZM-02/02 PB2002-917003 National Transportation Safety Board Notation 7371A 490 L. Enfant Plaza, S.W., Adopted 17 September 2002, Washington, DC. https://www.ntsb.gov/Publictn/2002/HZM0202.htm (accessed 10 January 2011).
3. NASA (1986). Report to the President: Actions to Implement the Recommendations of the Presidential Commission on the Space Shuttle Challenger Accident. http://history.nasa.gov/rogersrep/actions .pdf (accessed February 2019).
4. National Highway Traffic Administration. Drunk Driving. https://www.nhtsa.gov/risky-driving/ drunk-driving (accessed February 2019).
5. Detwarasiti, A. and Shachter, R. (2005). Influence diagrams for team decision analysis. *Decision Analysis* 2 (4): 207–228.
6. IAEA Report INSAG-7 Chernobyl Accident (1991). Updating of INSAG-1 Safety Series No. 75-INSAG-7. IAEA Vienna, p. 73.
7. Chernobyl Gallery. Radiation levels. http://www.chernobylgallery.com/chernobyl-disaster/ radiation-levels (accessed February 2019).
8. McMah, L. (2016). Tourists flock to radioactive site of the Chernobyl nuclear explosion. *News.com.au.* https://www.news.com.au/travel/world-travel/europe/tourists-flock-to-radioactive-site-of-the-chernobyl-nuclear-explosion/news-story/2f44aa479e90e8a573560455b0768121 (accessed February 2019).
9. Cahill, E. (2001) *Irish Times.* http://www.ratical.org/radiation/Chernobyl/Belarus2001.html (accessed February 2019)

CHAPTER 17

Probabilistic Risk Assessment

17.1 DESCRIPTION

Risk assessment has been a tool/technique used by humanity for much of recorded history. Whether to fight or flee in situations where survival is at stake, the decision to run or engage the threat in many cases would involve some assessment as to whether or not the struggle would be winnable. In Thucydides *History of the Peloponnesian War*, he attributes the Greek general Pericles with stating the advantage that the Athenian army had was their ability to predetermine risk and consequences prior to engaging an enemy. In recent years, with the significant advances in technology, the need to provide decision makers with tools to define risk and minimize the consequences of undesirable events, equipment, and human failure new tools have been required to make these decisions. Probabilistic risk assessment is one of these tools.

Used initially by aviation and nuclear power industry, its use has been incorporated by environmental regulators and industry in making cleanup decisions, in computer security analysis, and as a decision-making tool for increasing safety during all aspects of design, operations, and upkeep of facilities. It also provides an analysis to make cost saving decisions by demonstrating where efforts should be exerted to gain the most benefit or where to avoid the expenditure of large amounts of time and financial resources pursuing efforts of little or no practical gain (1).

17.2 REQUIREMENTS OF THE RISK ASSESSMENT

Any assessment of risk requires an in-depth technical knowledge of the systems/processes being evaluated. The ability to define the system(s) affected as well as the response both with and without operator intervention is imperative to a valid assessment. Also important is the ability to recognize the adverse consequents in the event of a failure or misoperation of the facility and its component systems. If utilized to meet a regulatory requirement, any requirements of that agency need to be identified and verified in the final product. Rigorous analysis and documentation of that analysis will make it more defensible as well as facilitating peer review of the final product. Peer review should be conducted on all risk assessments utilized for decision making.

17.3 SIMPLIFIED PRA PROCEDURE

Dr. Micheal G. Stamatelatos (National Aeronautics and Space Administration [NASA]) presentation, "Risk Assessment and Management, Tools and Applications," summarized a methodology flow for PRA. These steps are summarized below:

1. Identification of end states of interest.
2. System familiarization and data collection.
3. Identification, selection, and screening of initiation events.
4. Definition and modeling of all scenarios linking each initiating event to the end states.
5. Modeling of pivotal events.
6. Risk quantification for each pivotal event and scenario; risk aggregation for all like end states.
7. Full uncertainty analysis; sensitivity analysis as needed.
8. Risk importance ranking (2).

17.4 HAZARD IDENTIFICATION AND EVALUATION

Why do a risk assessment? In many modern endeavors the consequence of misperformance and casualties resulting from internal and external events can lead to dire consequences and liabilities both civil and criminal. In order to protect people and their environment and thus avoid these liabilities, some methods of identifying the potential hazards (consequences) and the causes within the operation that could lead to these hazards/consequences must be utilized. In many cases a hazards analysis is performed to identify and classify the hazards from an initiating event. There are several methods that have been developed to perform this analysis utilizing different schemes to perform and document the analysis. The individual technique utilized is not as important as consistent and defensible application of the technique used in identification and classification. For example, an analyst would probably not want to classify one of two events leading to human casualties as severe and the other event as insignificant. Another possible error would be to classify a short-term release exceeding EPA regulatory standards with no quantifiable adverse environmental effects as critical while characterizing an event that leads to an occupational exposure that hospitalizes several workers and neighbors as negligible (3).

17.5 QUALITATIVE RISK ASSESSMENT

In a simple model of risk assessment, the two factors comprising risk are evaluated. Probability and consequence are analyzed and assigned non-numerical values. These may include high, medium, low, and intermediate. Values assigned to consequence may include none, undesirable, unacceptable, and effect (no, little, significant). A simple evaluation tool can be constructed by forming a grid with probability and consequence forming the x- and y-axes. While relatively simple in concept, it has been demonstrated to be useful for decision makers. However, its simplicity limits its functionality for more complex decisions such as modeling and event over time, incorporating several actions with multiple outcomes. Additionally, the modeling of uncertainty would be impossibly difficult. So, in many cases a more powerful tool is needed.

17.6 QUANTITATIVE RISK ASSESSMENT

The performance of quantitative risk assessment is a more detailed process than the qualitative assessment process. Expert system/process knowledge is a prerequisite to begin the assessment process. This expert knowledge is gained through system knowledge, event/process knowledge, knowledge of normal, abnormal condition and casualty operating procedures, and research on specific failure information. Failure information is not merely the rate at which the equipment fails. Other factors include modes of failure, the sample sizes (how many failures, how many of a particular component are utilized), environments that were utilized, operating time/cycles, repair time, frequency of periodic maintenance, or testing. Sources for information on the failure can come from a wide variety of sources including manufacturer and government testing, maintenance databases, and databases on existing or suspected failures from the Nuclear Regulatory Commission (NRC), NASA, and National Transportation Safety Board (NTSB), among others. Since these analyses are often performed on complex systems, this process is often costly and time consuming. Because of these factors this type of assessment is only performed when required by regulation and liability management or when an appreciable cost benefit is expected to be achieved.

The risk assessment process consists of determining possible events that could occur with detrimental consequences; for example, loss of coolant at a nuclear reactor, loss of insulation on the wing of the space shuttle, and a cyberattack on the infrastructure of the internet are some of the possibilities. Another possibility is to determine an adverse consequence and to reverse engineer (deductive reasoning) the events to determine possible causes of the event. After the initiating events are identified, the event is modeled to determine possible sequences and outcomes of the event. These processes are used to develop the flow path of the event or casualty. NASA uses the term master logic diagram for this process. At this point in the development, the process is still qualitative in nature. With the determination of the flow path, it is possible to determine the set of events, which will lead to the undesired or casualty event. This set of events is known as the cut set. The minimum cut set is defined as the set of events, which cannot be reduced in number, whose occurrence causes the casualty (or undesired event) to occur (2).

The first two parts of this process are utilized to develop failure mode and effect analysis (FMEA) and fault tree analysis (FTA). In these processes the events are diagramed to show the logical sequences with possible outcomes. One key difference between the two methods is that FMEA is an inductive logic process, while FTA is primarily a deductive process. As part of the analysis of the fault tree or failure mode, the possible causes of event are introduced. During this process Boolean algebra is utilized to refine the logic enabling the event to be modeled mathematically. It is at this point the process begins to convert to a quantitative process. The application of Boolean algebra allows multiple probabilities to be combined into a total probability based on the interrelation of the events. For example, a final event can occur; if a multiple set of events occur, then the probability of that event occurring is the sum of the probability of the individual events. If in order for an event to occur, several events must occur. The probability is the product of the probabilities of the requisite events. Of course, this is a simple analysis, in that time-sequenced requirements are not accounted for. An additional benefit is the use of standardized symbols for the mathematical logic.

The next step in quantification of risk is determination of the probabilities of individual events happening. At this time the uncertainty (often identified as error factor) associated with event probability will be required to be identified. Since all probability has a certain amount of uncertainty associated with its derivation due to measurement and calculation techniques, there should be uncertainty associated with all probability values. Factors that affect uncertainty include sample size, difference between laboratory determinations, and operating conditions in installed environments.

Another factor affecting equipment failure is reliability, availability, and maintenance. If operation of a certain piece of equipment is part of the analysis, then factors that affect whether the component is available to respond form a valid part of the analysis. Answers to questions such as what percentage of time the equipment is available form a vital part of the analysis. If a fan belt fails once in a hundred attempts to start and requires several hours to repair, that could impact the analysis of risk. A system that is routinely maintained and restored will have different failure rates than one where the components are run to destruction (no maintenance other than corrective is performed).

The failure probability of many components analyzed will be a function as to whether it is a discrete or continuous function. The most commonly utilized discrete distributions are the binomial distribution and the Poisson distribution. For probabilities that may be determined with a continuous distribution, the gamma, log-normal, exponential, and Weibull distributions are among the most common distributions. Trial-and-error analysis to develop curve fitting techniques has also been utilized. Also, the synthetic mixed distributions have been utilized to demonstrate mixed distributions of different distributions. These mixed distributions are achieved by combining the distribution curves for different failure curves. For example, a component with a high failure rate for a short period of time after initial installation may have an exponential distribution curve for that mode of failure and a gamma distribution for a longer-term failure mode such as a belt or bearing failure. By combining the two distributions, an integrated curve can be developed.

Determination of the statistical method is dependent on the conditions of operation and failure. Generally, a continuous distribution determination is made based on comparing the underlying assumptions for the distribution with the physical nature of the problem.

Another factor introduced during this process is human performance or human reliability analysis (HRA). Since events leading to and subsequent to an event can be caused, mitigated, or exacerbated by the actions and non-actions of the operators, an analysis of human performance is necessary to adequately account for all possible outcomes and probability of occurrences. Human errors are classified as errors of commission, errors of omission, mistakes, lapses, etc. Several sources of data are available based on industry studies. These studies address a wide variety of possible errors and need to be carefully considered for applicability to the given situation.

After identification and quantification of the individual events, the aggregation of the risk can occur. During this process the risks and associated uncertainties are calculated for the cut sets. Because much of the information used to generate statistic input are from sources other than test data, often Bayesian analysis techniques are used. The use of Bayesian statistics is required because the probabilities utilized are considered variables rather than precise values. Because of this approach confidence interval testing becomes essential in providing validity to the calculations.

Uncertainty can be calculated by several methods, the most common being Monte Carlo testing. This technique utilizes random number generation from numbers associated with the distribution of all the variables and follows them through the fault tree to find the value of all independent variables for the combination of dependent variables associated with the components. This method was developed during World War II to assist in modeling nuclear weapons design and is used today for many purposes.

Another factor that should be analyzed during the aggregation of risk is sensitivity. Sensitivity of the model is analyzed to ensure that uncertainties in the data do not lead to wide, and perhaps unacceptable, variations in the calculated result. This may lead the analyst to recommend certain parameters be constrained to preclude these results. These constraints may require implementation of operational and engineering controls in order for the risk assessment to be valid. If implementation of these constraints is beyond the assessor's control, then any boundaries of the assessment should be clearly identified.

While this process may be performed on a simple problem, the complexity of most problems will drive most analysts to use a computer software program specifically written for that purpose. Software is available from governmental sources and private (proprietary) sources.

17.7 USES OF PRA

As stated earlier, the reason to perform a risk assessment is to demonstrate safety to both the workers and the public, as well as save cost by utilizing available funds in the most efficient manner. For the latter purpose the PRA becomes a decision-making tool to assign resources such as component maintenance and replacement or to let equipment with low consequence of failure operate with less upkeep. Implementation of engineered and administrative controls taking decisions from operators under duress, or to minimize/prevent misoperation, may be determined or warranted. The possibilities are unlimited and depend on the specific operation.

If utilized to demonstrate safety to obtain and/or maintain an operating permit, there may be certain regulatory and policy requirements that must be met to rely on the risk assessment. In such an environment the validity of any decision-making tool must be scientifically defensible, and the regulatory agencies have issued requirements for risk assessments used in permit issuance (4).

17.8 EXAMPLE OF PRA USED IN THE NRC

The NRC uses PRA by computing real numbers to determine what can go wrong, how likely is it, and what are its consequences. By completing the process in this way, they are able to gain insights into the strengths and weaknesses of the design and operation of a nuclear power plant. The PRA tool consists of elements that provide both qualitative and quantitative insights and assessments of risk. This is done by addressing the "risk triplet": (i) What can go wrong? (ii) How likely is it? and (iii) What are the consequences? (5)

PRAs for nuclear plants can vary depending on their use; therefore, the scope is defined by (i) radiological hazards, (ii) population exposed to hazards, (iii) plant operating states, (iv) initiating event hazards, and (v) level of risk characterization. PRA is used on different types of nuclear plants in the United States to estimate three levels of risk:

- Level 1
- Level 2
- Level 3

They define these levels as Level 1 estimating the frequency of accidents that cause damage to the nuclear reactor core. Level 2 uses the Level 1 core damage accidents by estimating the frequency of accidents that release radioactivity from the nuclear power plant. Finally, Level 3 uses the Level 2 radioactivity release accidents. The consequences of injury to the public and damage to the environment are then determined from those estimates. Figure 17.1 shows an example of a PRA from the NRC.

A Level 1 PRA models the various plant responses to an event that challenges plant operation. The plant response paths are called *accident sequences.* A challenge to plant operation is called an *initiating event.* There are numerous accident sequences for a given initiating event. The various accident sequences result from whether plant systems operate properly or fail and what actions operators take. Some accident sequences will result in a safe recovery, and some will result in reactor core damage. The accident sequences are graphically represented with event trees. Each event in the event tree (called a top event) generally depicts a system that is needed to respond to the initiating event. An analysis is performed for each top event in the event tree. This analysis is graphically represented with a fault tree.

The frequency for each core damage accident sequence is estimated, and the frequencies for all core damage sequences are summed to calculate the total core damage frequency. In that way, the Level 1 PRA provides the first measure of risk – core damage frequency – which is the input to the Level 2 PRA.

A Level 2 PRA models the plant's response to the Level 1 PRA accident sequences that resulted in reactor core damage. Such core damage sequences are typically referred to as *severe accidents.* Toward that end, a Level 2 PRA analyzes the progression of an accident by considering how the containment structures and systems respond to the accident, which varies based on the initial status of the structure or system and its ability to withstand the harsh accident environment. Thus, a Level 2 PRA must consider the key phenomena that affect accident progression. The following are two of the countless examples that illustrate the phenomena that must be considered:

- Do tubes in the steam generator rupture?
- Is the reactor core debris in a coolable configuration?

Once the containment response is characterized, the analyst can determine the amount and type of radioactivity released from the containment. Thus, the Level 2 PRA estimates the second measure of risk – radioactivity release – which is the input to the Level 3 PRA.

A Level 3 PRA is often called a *consequence analysis.* Consequences result from the radioactive material released in a severe accident. A Level 3 PRA estimates those consequences in the following terms:

- Health effects (such as short-term injuries or long-term cancers) resulting from the radiation doses to the population around the plant.
- Land contamination resulting from radioactive material released in the accident.

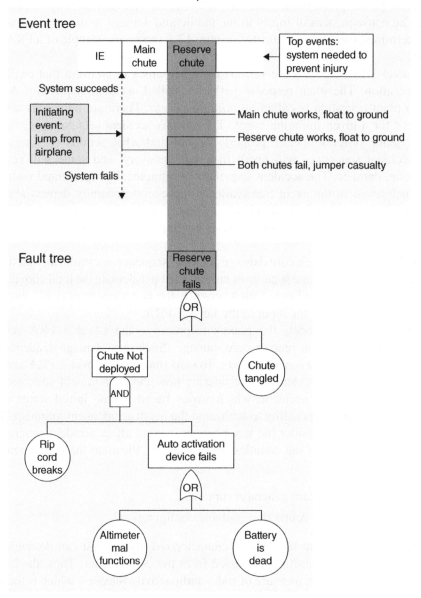

FIGURE 17.1 Sample PRA. *Source*: Courtesy of the NRC.

Consequences are estimated based on the characteristics of the radioactivity release calculated by the Level 2 PRA. Those consequences depend on several factors. For example, health effects depend on the population in the plant vicinity, evacuation conditions, and the path of the radioactive plume. The plume, in turn, is affected by wind speed and direction, as well as rainfall or snowfall. Similarly, land contamination depends on the characteristics of the radioactivity release and how the land surrounding the plant is used.

The Level 3 PRA estimates the final measure of risk by combining the consequences with their respective frequencies. (What can go wrong, how likely is it, and what are the consequences?) For instance, a Level 3 PRA might estimate that an accident would create one chance in a million that a person living near the plant would experience radiation exposure equivalent to a chest X-ray and one chance in a billion that some people would develop cancer over the next 50 years. Figure 17.2 depicts the three levels of PRA from the NRC.

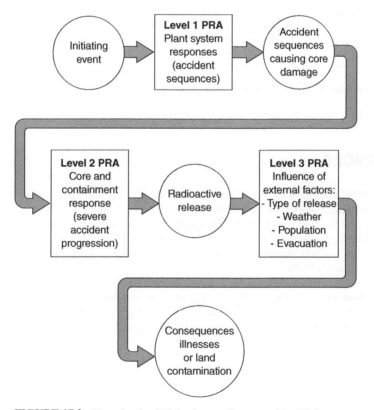

FIGURE 17.2 Three levels of PRA. *Source*: Courtesy of the NRC.

17.9 CONCLUSION

Probabilistic risk assessment can be a valuable tool in the identification, documentation, and management of risk. If developed and implemented properly, the techniques can help save costs, damages, and even lives. However, the utilization of PRA presupposes a certain level of consequence that some people find difficult to understand and accept. This condition can be even more pronounced when the person does not perceive a real benefit from the operation while the operation creates a risk and the consequences are life threatening. An example is a physician once told a patient while informing him of a diagnosis of cancer that it really didn't matter that the probability of getting the cancer was only 1 in 15 000; he was the one.

Self-Check Questions

1. What needs to be done if a PRA is used to meet a regulatory requirement?

2. Why do a risk assessment?

3. In what industries would you use a PRA?

4. What does the NRC use to measure risk?

5. How is human performance used in PRAs?

6. What is the process used in a quantitative risk assessment?

REFERENCES

1. Fullwood, R.R. and Hall, R.E. (1988). *Probabilistic Risk Assessment in the Nuclear Power Industry Fundamentals and Applications*. New York: Pergamon Press.
2. Stamatelatos, M.G. (1988). *Risk Assessment and Management, Tools and Application*, presentation by Dr Micheal G. Stamatelatos. New York: Pergamon Press.
3. McCormick, N.J. (1981). *Reliability and Risk Analysis, Methods and Nuclear Power Applications*. New York: Academic Press.
4. Modarres, M. (ed.) (1999). Volumes I and II. *Proceedings of PSA'99, International Topical Meeting on Probabilistic Safety Assessment*. Washington, DC: American Nuclear Society.
5. U.S. Nuclear Regulatory Commission. Probabilistic Risk Assessment (PRA). https://www.nrc.gov/about-nrc/regulatory/risk-informed/pra.html (accessed 1 October 2018).

CHAPTER 18

Probabilistic Risk Assessment Software

18.1 INTRODUCTION

The objective of this chapter is to provide an overview of the various probabilistic risk assessment (PRA) software packages that are available. Some of these packages were developed through funding from a federal agency and private companies developed some. It is not the intent of this chapter to provide recommendations for one package over another. An older article by Li and Pruessner (1) discusses attributes of some of the PRA software packages. The descriptions of these packages come from report abstracts and/or easily obtainable information from the web. Also, this is not all of the packages that are available. The packages that will be discussed are:

- SAPHIRE
- RAVEN
- IRIS
- SimPRA
- RELAP 7
- iQRAS

18.2 SAPHIRE

The Systems Analysis Programs for Hands-on Integrated Reliability Evaluations (SAPHIRE) is a software application developed for performing a complete PRA using a personal computer (PC) running the Microsoft Windows operating system (2).

Risk Assessment: Tools, Techniques, and Their Applications, Second Edition. Lee T. Ostrom and Cheryl A. Wilhelmsen.
© 2019 John Wiley & Sons, Inc. Published 2019 by John Wiley & Sons, Inc.
Companion website: www.wiley.com/go/Ostrom/RiskAssessment_2e

SAPHIRE version 8 is funded by the US Nuclear Regulatory Commission (NRC) and developed by the Idaho National Laboratory (INL). INL's primary role in this project is that of software developer and tester. However, INL also plays an important role in technology transfer by interfacing and supporting SAPHIRE users, who constitute a wide range of PRA practitioners from the NRC, national laboratories, the private sector, and foreign countries.

SAPHIRE can be used to model a complex system's response to initiating events and quantify associated consequential outcome frequencies (or probabilities). Specifically, for nuclear power plant applications, SAPHIRE 8 can identify important contributors to core damage (Level 1 PRA) and containment failure during a severe accident, which leads to releases (Level 2 PRA). It can be used for a PRA where the reactor is at full power, low power, or shutdown conditions. Furthermore, it can be used to analyze both internal and external initiating events and has special features for managing models such as flooding and fire. It can also be used in a limited manner to quantify risk, using PRA techniques, in terms of release consequences to the public and environment (Level 3 PRA). In SAPHIRE 8, the act of creating a model has been separated from the analysis of that model in order to improve the quality of both the model (e.g. by avoiding inadvertent changes) and the analysis. Consequently, in SAPHIRE 8, the analysis of models is performed by using what are called workspaces. Currently, there are workspaces for three types of analyses: (i) the NRC's Accident Sequence Precursor program, where the workspace is called "Events and Condition Assessment (ECA)," (ii) the NRC's Significance Determination Process (SDP), and (iii) the General Analysis (GA) workspace. Workspaces are independent of each other, and modifications or calculations made within one workspace will not affect another. In addition, each workspace has a user interface and reports tailored for their intended uses.

18.3 RAVEN

Risk Analysis Virtual Environment (RAVEN) is a generic software framework to perform parametric and probabilistic analysis based on the response of complex system codes (3). The initial development's objective was to provide dynamic risk analysis capabilities to the Reactor Excursion and Leak Analysis Program v.7 (RELAP-7) Thermo-Hydraulic code, currently under development at the INL. Although the initial goal has been fully accomplished, RAVEN is now a multi-purpose probabilistic and uncertainty quantification platform, capable to agnostically communicate with any system code. This agnosticism includes providing application programming interfaces (APIs). These APIs are used to allow RAVEN to interact with any code as long as all the parameters that need to be perturbed are accessible through input files or via python interfaces. RAVEN is capable of investigating the system response as well as the input space using Monte Carlo, Grid, or Latin Hyper Cube sampling schemes, but its strength is focused toward system feature discovery, such as limit surfaces, separating regions of the input space leading to system failure, using dynamic supervised learning techniques.

18.4 RELAP-7

Reactor Excursion and Leak Analysis Program-7 (RELAP-7) is the nuclear reactor system safety analysis code currently under development at the INL for the Risk Informed Safety Margin Characterization (RISMC) Pathway as part of the Light Water Reactor Sustainability Program (4). It is an evolution in the RELAP-series reactor systems safety analysis applications. The RELAP-7 code development is taking advantage of the progresses made in the past three decades to achieve simultaneous advancement of physical models, numerical methods, coupling of software, multi-parallel computation, and software design. RELAP-7 uses the INL's open-source Multiphysics Object Oriented Simulation Environment (MOOSE) framework for efficiently and effectively solving computational engineering problems. Unlike the traditional system codes, all the physics in RELAP-7 can be solved simultaneously (i.e. fully coupled), resolving important dependencies and significantly reducing spatial and temporal errors relative to traditional approaches. This allows RELAP-7 development to focus strictly on systems analysis-type physical modeling and gives priority to the retention and extension of RELAP5's system safety analysis capabilities. In addition to the mechanistic calculations that are performed in RELAP-7 to represent plant physics, it has been designed to be integrated into probabilistic evaluation using the RISMC methodology. The RISMC methodology can optimize plant safety and performance by incorporating plant impacts, physical aging, and degradation processes into the safety analysis.

18.5 IRIS

Integrated Risk Information System (IRIS) was developed for the US Federal Aviation Administration for risk-informed safety oversight (5). IRIS software is a platform to perform probabilistic risk analysis (PRA) based on the hybrid causal logic (HCL) methodology. The HCL methodology employs a model-based approach to system analysis. The framework contains a multilayer structure that integrates event sequence diagrams (ESDs), fault trees (FTs), and Bayesian belief networks (BBNs) without converting the entire system into a large BBN. This allows the most appropriate modeling techniques to be applied in the different individual domains of the system. The scenario or safety context is modeled in the first layer using ESDs. In the next layer, FTs are used to model the behavior of the physical system as possible causes or contributing factors to the incidents delineated by the ESDs. The BBNs in the third layer extend the causal chain of events to potential human and organizational roots.

The connections between the BBNs and ESD/FT logic models are formed by binary variables in the BBN that correspond to basic events in the FTs, or initiating events and pivotal events in the ESDs. The probability of the connected events is determined by the BBN. In order to quantify the hybrid causal model, it is necessary to convert the three types of diagrams into a set of models that can communicate mathematically. This is accomplished by converting the ESDs and FTs into reduced ordered binary decision diagrams (ROBDD). BBNs are not converted into ROBDDs; instead, a hybrid ROBDD/BBN is created. In this hybrid structure, the probability of one or more of the ROBDD variables is provided by a linked node in the BBN. IRIS provides a framework to identify risk scenarios and contributing events, calculates probabilities of the various risk scenarios, calculates event risk (probabilities of the undesired events) and identifies the impact of specific changes, and ranks risk scenarios and risk contributors by their probabilities. In addition, IRIS provides tools for hazard identification such as highlighting functions. Trace and drill down functions are provided to facilitate the navigation through the risk model. IRIS also includes a risk indicator feature that allows the user to monitor system risk by considering the frequency of observation and risk significance of particular events (indicators) in the model. All IRIS features can be implemented with respect to one risk scenario or multiple scenarios, e.g. all of the scenarios leading to a particular category or type of end state.

18.6 SIMPRA

Developed for risk-based design of complex hybrid systems under a grant from NASA (5), SimPRA is an adaptive-scheduling simulation-based dynamic PRA (DPRA) environment developed at the University of Maryland under NASA funding. SimPRA provides an extensive and multilayered risk model building capability to capture engineering knowledge, design information, and any available information from operating experience, simplifying (and in part automating) the tasks typically undertaken by the risk analysts. In the SimPRA framework, the estimation of end state probabilities is based on the simulation of system behavior under stochastic and epistemic uncertainties. A new scenario exploration strategy is employed to guide the simulation in an efficient and targeted way. The SimPRA environment provides the analysts with a user-friendly interface and a rich DPRA library for the construction of the system simulation model. In SimPRA, a high-level simulation scheduler is constructed to control the simulation process, generally by controlling the occurrence of the random events inside the system model. To stimulate the desired types of scenarios, the input to the simulation model is also controlled, using scheduling algorithms. Rather than using a generic wide-scale exploration, the scheduler is able to pick up the important scenarios, which are essential to the final system risk, thus increasing the simulation efficiency. To do that, a high-level simulation planner is constructed to guide the scheduler to simulate the scenarios of interest. Therefore the SimPRA environment has three key elements: planner, scheduler, and simulator. The planner serves as a map for exploration of risk scenario space. The scenarios of interest are highlighted in the planner. The scheduler manages the simulation process, including saving system states, deciding the scenario branch selection, and restarting the simulation. The scheduler guides the simulation toward the plan generated by the planner. The scenarios with high importance would be explored with higher priority, while all other scenarios also have a chance to be simulated. Scheduler would favor the events with higher information and importance values. This is done with an entropy-based algorithm.

18.7 IQRAS

iQRAS is a user friendly software tool with a fully integrated environment for constructing and analyzing risk models (6). A powerful aid for conducting PRA, you can construct and quantify risk scenario models, estimate numerical risk levels, and identify major risk contributors. Although initially developed for NASA, the features of iQRAS allow it to be applied to a wide range of applications, including aerospace, military, transportation, and medical procedures. iQRAS was designed to be used by a wide range of engineers and analysts. It is suitable for use by safety engineers who are seeking to enhance their safety analyses. iQRAS is the latest addition to Item Software's suite of risk and reliability software tools. The latest version of the software completes the transformation from the original software as developed by the University of Maryland and NASA. The new environment provides an extensive set of editing and reporting capabilities, in line with those found in ITEM ToolKit, our flagship suite of tools for reliability, availability, maintainability, and safety analysis.

On the computational side, iQRAS offers state-of-the-art algorithms that are both highly efficient and, unlike conventional solution methods used in most other tools, are not subject to often significant approximation errors. With iQRAS you can develop risk models in the form of ESDs and FT models. ESDs describe the possible risk scenarios following potential perturbations of normal system operations. Pivotal events in the risk scenarios are further detailed using FT models or other distributions. This defines the occurrence of those pivotal events as logical combinations of one or more basic events. Each scenario eventually leads to an end state and consequence that designates the severity of the outcome of the particular scenario. The ESDs are organized using a system hierarchy, consisting of structural or functional decomposition. A mission timeline allows the breakdown of the overall mission into multiple mission phases. This organizing capability makes for easy navigation between the potentially large numbers of event sequence diagram models and tailors the risk scenarios to the particular conditions of a mission phase.

REFERENCES

1. Li, M. and Pruessner, P. (2008). Software tools for PRA. https://ieeexplore.ieee.org/document/4925775 (accessed August 2018).
2. US NRC (2011). Systems Analysis Programs for Hands-on Integrated Reliability Evaluations (SAPHIRE) Version 8, NUREG/CR-7039. https://www.nrc.gov/reading-rm/doc-collections/nuregs/contract/cr7039/v2 (accessed February 2019).
3. Alfonsi, A., Rabiti, C., Mandelli, D. et al. (2013). RAVEN as a Tool for Dynamic Probabilistic Risk Assessment: Software Overview, INL/CON-13-28291. https://inldigitallibrary.inl.gov/sites/sti/sti/5806429.pdf (accessed February 2019).
4. Idaho National Laboratory (2018). Relap 7. https://relap7.inl.gov/SitePages/Overview.aspx (accessed August 2018).
5. University of Maryland (2018). IRIS. http://crr.umd.edu/software (accessed August 2018).
6. ITEM Soft (2018). iQRAS. http://www.itemsoft.com/qras.html (accessed August 2018).

CHAPTER 19

Qualitative and Quantitative Research Methods Used in Risk Assessment

19.1 WHAT IS QUALITATIVE RESEARCH?

Qualitative research is not very simple to define. Sure, there are definitions from Wikipedia, psychologists, and anthropologists to name a few, but it depends on what the researcher is trying to capture. The medical world views qualitative research as a type of scientific research, which looks for answers to questions by way of predefined procedures. "Qualitative research is especially effective in obtaining culturally specific information about the values, opinions, behaviors, and social contexts of particular populations" (1).

Qualitative research explores topics in more depth and detail than quantitative and is usually less expensive. You can also use it in trying to understand the reasons behind the results of quantitative research. You do not perform any statistical test with qualitative research. Focus groups and individual in-depth interviews are common tools used in qualitative research. Therefore, I guess this could simply be defined as methods that at least attempt to capture life as it is lived.

Go back to what the researcher is trying to focus on. Different researchers focus on different sources of data. A researcher could use his/her own experiences or others' experiences through behaviors, speaking, writing, technology, artwork, and so on. There are variations in how the researchers collect their data as well. Some look at the "past" and collect artifacts, life histories, or literature; others look at the "present" through observation of what is happening now.

Risk Assessment: Tools, Techniques, and Their Applications, Second Edition. Lee T. Ostrom and Cheryl A. Wilhelmsen.
© 2019 John Wiley & Sons, Inc. Published 2019 by John Wiley & Sons, Inc.
Companion website: www.wiley.com/go/Ostrom/RiskAssessment_2e

In John W. Creswell's book, he uses five approaches to qualitative research in both the inquiry and design phases. These are listed as follows:

1. Narrative research
2. Phenomenological research
3. Grounded theory research
4. Ethnographic research
5. Case study research (2)

19.1.1 Narrative Research

Creswell defines narrative research as "the term assigned to any text or discourse, or, it might be text used within the context of a mode of inquiry in qualitative research (Chase, 2005), with specific focus on the stories told by individuals (Polkinghorne, 1995)" (2). Webster's dictionary defines it as "something that is narrated, e.g. story, account or the representation in art of an event or story; *also*: an example of such a representation" (3). There are many different definitions depending on the particular study of interest. To make it simple think of it as a method on knowing how and why we make meaning in our lives, a way to create and recreate our realities. It is a study of human beings by verbal or written acts of someone telling someone else that something happened.

There are different forms in narrative research such as:

1. *Biographical study*, which is a study where the researcher writes and records the experiences of another person's life.
2. *Autobiography*, which is written by the individual himself or herself.
3. *Oral history*, which is a compilation of events and causes, found in folklore, private situations, and single or multiple episodes (2).

In terms of looking at risk assessment, this would be valuable in writing up observations, performing some comparisons, and developing a risk assessment plan to mitigate against the risk. Narrative research is a way of understanding experience. Mishler describes reference and temporal order as being the relationship between the order in which events actually happened and the order in which they are told in narration. Again, this is valuable in determining risk by looking at an event such as an airplane crash. We look backward in trying to determine the cause of the crash. We look at data from the airlines, from personal observation of the event, and from any video or pictures that might exist of the event. We write everything down, compare, and look at the relationship between the order of the actual crash and the order in which they have been narrated. This helps in understanding how the event occurred and why, so we can again mitigate our risks.

Narrative research can also help in developing pilot studies to gather information that helps design objective research tools. It is used in applied research and basic research and used in the social sciences as well as medicine.

19.1.2 Phenomenological Description

All experiences have both an objective and subjective component; therefore, to understand a phenomenon, you need to understand both sides. In doing so, you look at it from all perspectives, using all your senses, thoughts, and feelings. When using your senses, you need to be prepared to listen. It is easy to talk and reason, but we sometimes really forget to listen. You cannot hear anything if you are too busy telling what it is. This also means we need to put aside all our biases and even our common sense so that we can be open to the phenomenon and accept it. This is not very easy to do. We tend to want to explain everything and use science to help. For example, a psychologist looks for how to measure knowledge, and cause and effect, so they may say that anger is really just nervous system activation or that our thoughts are just neural activities. These explanations are not found in experiences. Creswell states, "A phenomenological study describes the meaning for several individuals of their lived experiences of a concept or a phenomenon" (2). So, in essence, you look for what is common among several individuals in experiencing a phenomenon. One example might be grief. We all experience grief at one point in our lives and many books are written on the subject. Think about a time in your life you have felt grief. Try to recall it fully and then write it down. Then look for a book on others' experiences with grief, or other individuals that have experienced grief, and find the similarities. What makes grief? In order to understand the feelings the individuals are experiencing, you need to listen to how they "feel." Spiegelberg puts it nicely, "Phenomenology begins in silence" (4).

Pain is another phenomenon that individuals experience differently and the same. We tend to migrate to those individuals who may be experiencing similar problems, so we can relate our experiences with them. We can experience or recall the phenomenon and hold it in our awareness. We live it or are involved in it. All of these are aspects of a phenomenon. It is important to write everything down and relate your experience to similar experiences.

A good way to understand phenomenological research is to relate an experience such as drinking hot chocolate. You must drink the beverage, savoring the experience and describing the sensual experience of the taste, smell, feel, and so on. You could do the same with sounds such as music or listening to a bird.

19.1.3 Grounded Theory Research

The phenomenological research emphasizes the meaning of an experience, and the grounded theory generates, or discovers, a theory. The theory is generated or "grounded" from the data collected in the narrative or phenomenological research. Now the researcher can generate a theory or explanation of a particular action, process, or inter-actions that have been shaped by the views of the participants (2).

First, you need to understand what is happening and the roles of the players involved. You can do this through observation, conversation, and interview processes. Take notes and record, or code, the key issues. A comparison of the data begins, and a theory emerges. Then compare the data to the theory. Links may emerge between categories and a category may appear to be central to the study. Keep adding to the sample with more theoretical sampling searching for any different properties. When you reach saturation, it is time to start the sorting process where you group memos with similar memos and then sequence them in an order to clarify your theory. You are ready to write as the theory emerges.

There are two types of grounded theory research:

1. Systematic procedures, which "seeks to systematically develop a theory that explains process, action, or interaction on a topic (e.g. the process of develop-ing a curriculum, the therapeutic benefits of sharing psychological test results with clients)" (2).

   Basically, you conduct 20–30 interviews to collect data to saturate the cat-egories; you continue to collect data and analyze and then collect more data and analyze.

2. Constructivist approach, which is where you look at multiple realities and the complexities of particular worlds, views, and actions.

   This theory depends on the researcher's view through relationships, situa-tion communications, and opportunities. This research relies more on values and beliefs, feelings, and assumptions. The researcher has more control of the study, which could be labeled as suggestive, incomplete, and inconclusive (2).

19.1.4 Ethnographic Research

Ethnographic research goes further than the grounded study research in the fact that it collects data from more than the 20–30 individuals and focuses on an entire cul-tural group. This study describes learned and shared patterns of the group's behaviors, beliefs, and language. Again, the process involves observation but more into the daily lives of the people. The following are three definitions of ethnographic research:

When used as a method, ethnography typically refers to fieldwork (alternatively, participant-observation) conducted by a single investigator who "lives with and lives like" those who are studied, usually for a year or more. (5)

Ethnography literally means "a portrait of a people." An ethnography is a written description of a particular culture – the customs, beliefs, and behavior – based on information collected through fieldwork. (6)

Ethnography is the art and science of describing a group or culture. The description may be of a small tribal group in an exotic land or a classroom in middle-class suburbia. (7)

The three types of data collection methods used in ethnography research are as follows:

1. Interviews
2. Observations
3. Documents

These three types of collection methods then lead to three types of data:

1. Quotations
2. Descriptions
3. Excerpts of documents

Does this sound familiar? It is a narrative description that includes diagrams and charts to tell the story. This is a difficult research method as you need to be a participant and yet an observer to describe the experience to the outsiders. There are debates as to whether a researcher really needs to be a participant to understand the group being studied. However, if the researcher does immerse themselves into the group, then the goals must be made clear to the members of the group, and they must consent to the study beforehand. The researcher cannot harm or exploit the group.

The analysis begins with compiling the raw data to get a total picture. The data are assembled in order and organized into patterns or categories. "The analysis process involves consideration of words, tone, context, non-verbal's, internal consistency, frequency, extensiveness, intensity, specificity of responses and big ideas." Data reduction strategies are essential in the analysis (8).

19.1.5 Case Study Research

"Case study research is a qualitative approach in which the investigator explores a bounded system (a case) or multiple bounded systems (cases) over time, through detailed, in-depth data collection involving multiple sources of information (e.g. observations interviews, audiovisual material, and documents and reports), and reports a case description and case-based themes" (2).

This method has been used for many years and within many different disciplines to examine real-life situations. There are many debates about how valid and useful this research method is. Some believe it is only useful as an exploratory tool, while others believe the intense exposure biases the findings. However, many successful studies have used this case study method. Robert K. Yin has done a great deal of work in the field of case study research and in essence has proposed a six-step process for performing case study research:

1. Determine and define the research questions.
2. Select the cases and determine data gathering and analysis techniques.

3. Prepare to collect the data.
4. Collect data in the field.
5. Evaluate and analyze the data.
6. Prepare the report (9).

The research questions start with either "why" or "how" to help limit conditions and how they interrelate. A literature review is conducted to see what is already in the literature on the topic. This helps define the questions about the problem and thus determines the method of analysis. The researcher can then determine what evidence they want to gather and what analysis techniques they will use (e.g. surveys, observation, interviews, or documentation review). The study must be constructed to ensure validity, both internal and external, as well as reliability.

A large amount of data is generated with case study research; therefore, it is important to organize the data systematically to prevent oversaturation and forgetting the purpose of the study. A good way of organizing the data would be the use of databases, which stores the data, so you can sort, categorize, and retrieve it when you are ready for the analysis phase. In essence, case study research is very complex with the large amounts of data to analyze and the multiple sources of data. Researchers may use this method to build on a theory, dispute a theory, or even develop a new theory. They do relate to everyday experiences and help in understanding the complexity of real-life situations.

19.2 QUANTITATIVE

Quantitative research is the opposite of qualitative research. A hypothesis is generated that needs to be proved or disproved usually through statistical and mathematical means. Randomize the study groups and where you can, include a control group. You should only manipulate one variable at a time to keep the analysis less complex. The research should be able to be conducted again and receive similar results. Quantitative research works well with qualitative research and after the facts, and the data have been collected.

Quantitative research usually involves numbers that can be statistically analyzed to achieve the results. Hard sciences use quantitative research to answer specific answers. There are four types of quantitative research methods:

1. *Descriptive* involves collecting data to answer questions or test a hypothesis.
2. *Correlational* tries to determine whether there is a relationship between two or more quantifiable variables. There is no cause and effect relationship, only a correlation coefficient number between 0.00 and 1.00.
3. *Cause–comparative* establishes a cause and effect relationship and compares the relationship without manipulating the cause.

4. *Experimental* again establishes a cause and effect relationship, but this time during the comparison the cause is manipulated. In experimental, the cause or independent variable makes the difference, and the effect is dependent on the independent variable (10).

Figure 19.1 illustrates the steps followed in conducting a quantitative research study. Most people have experienced some sort of a quantitative inquiry. The data are collected in the form of a survey or questionnaire where the questions are used to measure important factors on a numerical level. Some examples are as follows:

- Telephone surveys
- Mail surveys
- Online surveys
- Panel research

19.3 RISK ASSESSMENT PERSPECTIVE

Now that we have looked at both, qualitative and quantitative research methods, we now talk about the part they play in dealing with risk assessment. Risk is applied across various organizations and varying circumstances, which does not have the same meaning across the various disciplines, organizations, or even individuals. In other words, there is no a "one size fits all" risk assessment method. One thing they do agree on is the term *risk* carries a negative connotation such as destruction, harm, or some undesirable event. Risk, as we have talked about in other chapters, possesses unknown and unpredictable results or consequences, so there is that "uncertainty" element. With qualitative research, we try to piece the event or story back together through the different forms of research such as narrative research. Once we reconstruct the pieces, we can then analyze those pieces, and that is where the quantitative research methods come into play. Without the gathering of information and piecing together, we could not perform any statistical analysis because the data just would not exist.

Some form of risk assessment is used to make decisions and in deciding how much effort should be directed toward avoiding undesirable events. A valid risk analysis should have a procedure in place to determine the appropriate consequence and likelihood levels. When talking about qualitative analyses, this would mean having adequate descriptions for each of the levels of consequence and likelihood. Here is an example that helps illustrate how qualitative and quantitative methods can be used to identify the cause of an undesirable event.

There had been an accident involving two aircraft colliding in air. The Federal Aviation Administration (FAA) and the National Transportation Safety Board (NTSB), along with various agencies, are called to investigate what happened. They cordon off

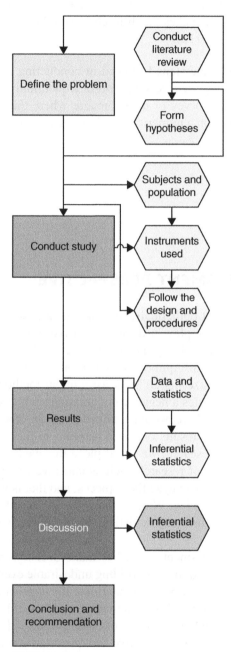

FIGURE 19.1 Quantitative research step process.

the crash site and then identify potential hazards. There are many hazards involved with this type of accident such as toxic chemical releases from the fuel and materials that make up the aircraft, as well as possible buildings or industries that were damaged in the accident. You need to look at those hazards and decide who might be harmed and how. You decide what the potential risk is, for example, illness or injury. This allows you to come up with the best plan for preventing those risks. Then develop precautions to minimize the risk such as protective clothing and equipment. By putting safety policies in place, it can help reduce any harmful incidents in the investigative process.

Once they have the scene cordoned off and the potential hazards identified, they need to piece together what happened and why it happened. Interviews from eyewitnesses are conducted in which they ask the individuals to tell them everything they saw and heard. Does this sound familiar? It is telling a story as to what that person saw and heard. You may have 20 eyewitnesses, or you may only have a few. Either way you record every detail those individuals tell you. You may get conflicting stories and so you have to look for the similarities and identify those. You then ask the eyewitnesses to retell their account of the event again going through this type of process repeatedly. The eyewitnesses are using their senses to try to recall what they saw, heard, and even felt during the incident. This is utilizing the narrative research portion of qualitative research.

Now, the investigators have narrowed down the similarities and start to piece a scenario together. There will most likely be holes in the research. Possible data recorders or Global Positioning Systems (GPS) within the aircraft, if found, could be analyzed. This would help collaborate the accounts of the eyewitnesses and validate the data you have collected. Usually, data from recorders are things such as latitude or longitude, speeds, and directions. These can be quantified, and using a technology, the accident can be recreated. It is important to recreate what exactly happened so as to avoid such accidents in the future. Figure 19.2 illustrates this process.

19.3.1 Aviation Study

Various industries use both qualitative and quantitative methods to collect data needed in determining what risks their companies and employees face. One particular industry that uses both qualitative and quantitative methods in determining risk factors is the aviation industry. This case study involves a qualitative/quantitative or what is commonly referred to as a *mixed method study* to assess the risks and reliability of visual inspections of aircraft.

There have been several studies in determining estimated inspector reliability for finding structural cracks in commercial aircraft, and a set of probability of detection (POD) curves were developed in the 1980s (11). These curves were used as part of the basis of the damage tolerance-rating concept used by a major airline to help determine inspection and maintenance intervals. These curves indicated that the inspectors have a relatively small probability of finding cracks of less than 1 in length. Data that are more current needed to be collected in determining how well the inspectors actually

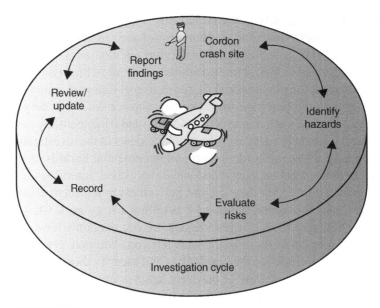

FIGURE 19.2 Investigation cycle.

estimated these lengths visually. Therefore, a visual line length measurement study was developed to determine inspector reliability (12). This study was needed because it would impact the estimated POD for the cracks. For instance, if inspectors always underestimated the crack lengths, then the POD would be less than what would be predicted from the actual data. If the inspectors overestimated the crack lengths, the POD would be more than what would be predicted from the actual data. In addition, airline structural parts are usually painted with gray or green paint. The question existed as to whether the color of the part affected the visual line length estimates. There was very little research available to compare the literature findings; however, a paper on a basic visual training course was helpful in defining some of the processes and skills an inspector follows while performing his/her duties. These included perception, recognition, and attention.

Perception processes are used to organize, interpret, and extract sensory information. Knowledge about the situation can influence the perception. In other words, a person's knowledge of the situation in terms of prior experiences and formal education can influence what is perceived and how it is interpreted. "At an early age, we fail to understand that changes in distance and lighting do not alter the perceived object. As we age, we develop the capacity to perceive the object as constant even if we are a short or long distance away or if lighting changes occur." "This is a learned skill that allows us to perceive an object as constant, even though the environment may suggest the object has changed in size (distance), and in color (lighting)" (13).

Recognition represents the ability to process complex stimuli to identify it as a class of objects. Therefore, when you see a friend's face, listen to a Celine Dion song, or taste tuna fish, you can recognize each one as something you previously experienced. We combine stimulus information and previous experience to produce recognition.

The inspection process involves observing and recognizing cracks on a part of the airplane. Since perceptions are based on prior experiences, knowledge, and expectations, the inspector's experience plays an important role. The difference in the expectations and knowledge contributes to differences in perception (13). One of the comments made by the inspectors at the airline was that if they knew that a particular part was prone to having more cracks, they inspected that area more carefully.

"Attention to sensory cues is at the control of the observer. Research has shown, however, that with highly practiced and routine tasks, attention can be automatized (performed with little conscious awareness)" (13). Three types of attention are used in inspection activities depending on the task involved. Some require focused attention, some selective attention, and some divided attention. The selective attention allows multiple inputs to be processed. Divided attention gives the ability to monitor two sources of input simultaneously and the ability to produce two different responses. A resource intensive task, such as visual inspection, requires focused attention on a single source of input and processing.

Referring back to phenomenological concepts, you can see the inspectors use their senses as they see, hear, feel, think, and judge how long the crack may be and where it may be located.

The qualitative portion of this study involved a written survey before the actual experiment took place. The inspectors met with the researchers and a questionnaire was completed. The following is an example of the survey used in this particular case study to help determine the inspectors experience level and percentage cracks they felt they detected in performing their task:

1. How long have you been with this airline? _____
2. How many years have you been an inspector? _____
3. Before becoming an inspector what was your occupation? _____
4. What percentage of cracks do you measure? _____
5. What percentage of cracks do you eyeball? _____
6. Do you wear eyeglasses? _____
7. Are the glasses bifocals or trifocals? _____
8. Do you wear contacts? _____
9. How many inspections do you complete in a day? _____

When conducting an inspection check the lighting conditions: Good Adequate Poor (14).

The results from the questionnaire indicated that the inspectors were well experienced, and most cracks were measured visually rather than measured with tools.

The quantitative method involved collecting measurable data. It is usually good practice to include protocol instructions, so everyone gets the same information before answering the questions. The following is an example of the protocol used in this particular case study.

19.3.2 Protocol Instructions

The purpose of the study is to determine how well do inspectors determine crack length visually.

We have been looking at crack propagation data from xyz airlines and actual inspection data from the airlines. In our analysis, we need to statistically bound the data by knowing how accurately inspectors conduct a visual inspection. From this, we will better predict inspector reliability. This is to the benefit of the airlines.

You will be looking at a series of crack mockups made up of a skin panel, a lug, and a floor beam. The crack lengths range from 1/8 to 2 in. Please give us your first impression of the crack length. There is no time limit, but we are looking for your first impression.

19.3.3 Design

The study was designed with 3 different mocked-up parts, 2 colors, and 16-line lengths: 1/8 to 2 in (the range of cracks we primarily saw in the data). The three different mocked-up parts used in this study consisted of drawings of a lug, a beam, and a flat panel. The two colors were gray and green. They are the main colors used in aircraft parts. To help determine consistency, seven measurements of each of the gray parts were repeated in random order. The measurements repeated were 5/8 to 1 and 3/8 in. Therefore, the gray lug, beam, and panel had additional data to analyze.

The mocked-up parts were drawn on an 8.5 × 11 in colored cardstock with the cracks drawn on the parts. These drawings were developed in AutoCAD to the specific measurements, and each was hand measured to ensure accuracy. Each mocked-up part was laminated, and the back of the card contained a code number, which corresponded with the correct measurement. Therefore, the cracks were true measurements. The design for the study is illustrated in Figure 19.3.

Twenty inspectors completed the experiment; the spreadsheet contained $96 \times 20 = 1920$ data points. The data analysis showed that inspectors primarily underestimate the length of the small lines and begin to overestimate the length of the larger lines (14).

This is only a portion of the actual study, but it illustrates how using both quantitative and qualitative methods to collect the data can help in determining risk assessments in any industry. This particular mixed method approach determined the following:

1. Inspectors overestimate the length of 1 in or less and begin to underestimate the length of lines longer than 1 in. The relationship between the estimates and line length can be represented by a linear regression equation.

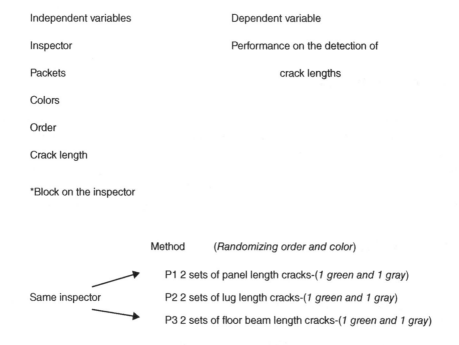

FIGURE 19.3 Example of the design for the visual crack study given to inspectors.

2. Mocked-up part type did have an impact on the inspectors' ability to estimate the line length. Lines drawn on the parts without any points of reference such as the flat panel appear to influence the inspectors' estimate more than when a reference point is available. For instance, the mocked-up floor beam with the holes of standard size allowed the inspectors a reference point; therefore, they appeared to estimate slightly more accurately the length of the line. (This aspect could be further studied for risk assessment involving the importance to safety and quality in both industrial and military operations.)

3. The shift the inspectors were on also affected the estimates. The swing shift was the least accurate and the graveyard shift was the least consistent.

4. The color of the mocked-up parts used in this study did not affect the estimates.

5. Gender differences could not be determined.

The above case study illustrated that the best way to determine risk assessment is to use a mixed method study. A combination of both the qualitative and the quantitative information assesses risk assessment from all aspect (13, 15, 16).

19.4 CONCLUSION

Qualitative and quantitative research methods are best used in tandem. Various organizations and disciplines use risk assessment to determine the extent of potential threats and risks associated with their organization. Both qualitative and quantitative research methods can be used in helping to assess those risks through observation, questionnaires, interviews, surveys, and so on. Then the data can be derived from the data collected. The results can be input and can be used to mitigate and reduce risks.

Self-Check Questions

1. What is qualitative research? How is it used in risk?

2. How many forms of grounded theory research does the chapter discuss, and what are they?

3. Describe case study research. How is the data collected?

4. What is quantitative research? How is it used in risk?

REFERENCES

1. Mack, N. and Family Health International, United States Agency for International Development (2005). *Qualitative Research Methods: A Data Collector's Field Guide*. Research Triangle Park, NC: Family Health International https://www.fhi360.org/sites/default/files/media/documents/Qualitative%20Research%20Methods%20-%20A%20Data%20Collector's%20Field%20Guide.pdf (accessed 3 October 2018).
2. Creswell, J. (2007). *Qualitative Inquiry and Research Design – Choosing Among Five Approaches*, 2e. Thousand Oaks, CA: Sage.
3. Chase, S. (2005). Narrative inquiry: multiple lenses, approaches, voices. In: *The Sage Handbook of Qualitative Research*, 3e (ed. N.K. Denzin and Y.S.Lincoln), 651–680. Thousand Oaks, CA: Sage.
4. Polkinghorne, D.E. (1995). Narrative configuration in qualitative analysis. *Qualitative Studies in Education* 8 (1): 5–23.
5. Webster M. (2011). Merriam-webster.com. Retrieved 20 September 2011. http://www.merriam-webster.com/dictionary/narrative
6. Spiegelberg, H. (1982). *The Phenomenological Movement*. The Netherlands: Kluwer Academic Publishers.
7. Van Maanen, J. (1996). *Ethnograhy*, 2e. London: Routledge The Social Science Encyclopedia.
8. Harris, M. (2000). *Cultural Anthropology*, 5e. Needham Heights, MA: Allyn and Bacon.
9. Fetterman, D.M. (1998). *Ethnography*, 2e. Thousand Oaks, CA: Sage.

10. Krueger, A.R. (1994). *Focus Groups: A Practical Guide for Applied Research*. Thousand Oaks, CA: Sage.

11. Yin, R.K. (1984). *Case Study Research: Design and Methods*. Newbury Park, CA: Sage.

12. Ostrom, L.T., Wilhelmsen, C., Valenti, L., and Hanson, E. (2002). Probability of crack detection: a unique approach to determining inspector reliability. Presented at the System Safety Society Meeting, Denver, CO (August 2002).

13. Toquam, J.M. (1995). *Basic Visual Observation Skills Training Course Appendix A*. The International Atomic Energy Agency, Department of Safeguards.

14. Wilhelmsen CA. (2002). Aviation visual crack measurement. Presented at the System Safety Society Meeting, Denver, CO (August 2002).

15. Boardman, J. (1999). Risk: a four letter word. Speech presented at a Luncheon at the 39th Australian Petroleum Production and Exploration Association (APPEA) Conference in Perth, Australia (20 June 1999).

16. Ostrom, L.T. and Wilhelmsen, C.A. (2002). *Guide for Developing and Evaluating Procedures Using Risk Assessment Techniques*. Idaho Falls, ID: University of Idaho.

CHAPTER 20

Risk of an Epidemic

20.1 INTRODUCTION AND BACKGROUND

A public health risk is anything that makes the wider population vulnerable to illness or injury. This can happen in a variety of ways. As a global society, people are constantly traveling and exposed to pathogens that otherwise they never would have previously encountered. Many people are foregoing safety measures like vaccinations out of ignorance or willful indifference (1). There are pathogens that some people are immune to that are being transmitted unknowingly to vulnerable populations. Tools to perform appropriate risk assessments are valuable to help stratify public safety threats and to guide agencies in the type of response to mount based on the situation.

The World Health Organization (WHO) released a manual in 2012 to help guide the assessment of risk for public health events (2). Figure 20.1 shows the general flow of the WHO risk assessment process. The general purpose of this manual is to guide users in taking appropriate steps in a timely fashion to rapidly identify and control public health risk. When thinking about the risk of a public health event, it is important to keep an open mind, to think outside the box, and to avoid prematurely narrowing down the list of possible culprits. Those performing the risk assessment need to consider several different factors:

- Is the source endemic to a defined area and always present, like malaria throughout parts of Africa?
- Is the source some novel, not yet elucidated agent similar to the current eruption of acute flaccid myelitis in the United States?
- Is the agent susceptible to becoming used by bioterrorists similar to the anthrax mailed to the members of the press during the 2001 attack (3)? Is the agent an environmental or industrial toxin mimicking symptoms of a biological disease?
- Are there public health protections in place like vaccinations to protect the larger populations?

Risk Assessment: Tools, Techniques, and Their Applications, Second Edition. Lee T. Ostrom and Cheryl A. Wilhelmsen.
© 2019 John Wiley & Sons, Inc. Published 2019 by John Wiley & Sons, Inc.
Companion website: www.wiley.com/go/Ostrom/RiskAssessment_2e

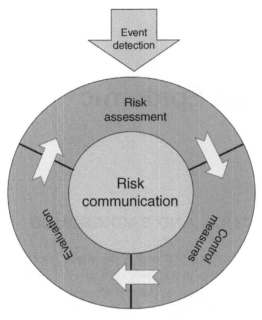

FIGURE 20.1   WHO epidemic risk assessment flow.

There are many ways agencies target the public to reduce risk of transmission of public health threats. Early mandatory reporting of communicable diseases to the health department has led to reduction in spread of diseases like chlamydia and syphilis through identification and treatment of infected individuals and others they may have encountered. The use of antibiotic monitoring programs led by community health departments has led to reduction in tuberculosis, particularly in populations that previously would have had difficulty completing treatment, such as the mentally ill or homeless.

Now, where do we begin when we are evaluating the risk of a public health threat?

20.2 WORLD HEALTH ORGANIZATION GUIDELINES

Let's consider the WHO recommendations. Initially, all available information about an outbreak needs to be considered. Small details can often be overlooked, so it is essential to be as detailed as possible when starting our risk assessment. What some-one had for lunch can be the single linking factor among affected individuals. Consider that in tracking multidrug-resistant Klebsiella infections in hospitals, there was initial difficulty in identifying how patients that were never in contact with each other were becoming infected. A systematic review of all processes ultimately led to identify-ing the sink traps in individuals' rooms as the culprit. They were not being properly sanitized, and this small, easily overlooked detail was the key to preventing further transmission of this illness.

Obviously, details need to be ranked in order of priority, and the reasoning behind why some details are more important than others should be documented.

Good recordkeeping is essential in tracking possible flaws in protective mechanisms. Identification of who was present, what was their capacity, when were they engaged in the process, and where did contact take place provides key components in assessing risk after possible exposure to a pathogen.

During an epidemic, these details should be reviewed repeatedly. This will allow for real-time updates to identify potential further exposures and adjust methods to contain and control the ongoing public health threat. If a method is clearly not working and, despite use of developed protocols, the public health event continues, the method can be identified as a failure and adjusted or eliminated.

The WHO recommends development of terminology that is common among all frontline personnel when developing these control strategies. This should allow for seamless communication between departments and allow for more rapid responses as well as rolling reassessment of the ongoing management.

These events are tracked based on predetermined criteria. Certain events when recorded by public health agencies trigger closer monitoring, and from this, a response is generated. For example, a relatively healthy person is admitted to a hospital with acute bloody diarrhea, nausea, and abdominal pain. The stool is tested and indicates *Escherichia coli* O157:H7. Two days later, this information is sent to the health department and triggers a warning to other hospitals to be on the lookout for similar presentations. The person in the hospital is interviewed in detail to identify a source. Now we need to consider if this is because of an isolated individual behavior or part of a larger issue. Do they have a penchant for eating raw beef, or have they been eating at the local burrito restaurant regularly? This is a type of event-based surveillance. Indicator-based surveillance is monitoring of specific disease processes like eruptions of clusters of seasonal flu. As more flu victims are identified, the trend is tracked: if the trend is moving up with identification of new cases, a response is amplified, or if the trend is moving down, the response is lessened and resources are allocated elsewhere.

In the example of *E. coli* O157:H7 cited above we have a known organism that will trigger the health department to look for other similar cases. On further questioning it is discovered that the infected individual was eating raw beef. This makes this much less likely to represent the beginning of an epidemic. Alternatively, a person infected from the local burrito restaurant would represent a public health risk. The detailed interview should examine the ingredients on the individuals' burrito, the exact day they visited, and the time of day to narrow down the source of infection and to generate a response that is as wide or narrow as necessary.

Now that we have identified a threat to the public in the burrito restaurant, let's follow along and identify the rest of the process. We need to establish the team to help manage this case. As this is most likely a food-borne illness, we should involve the local agency that regulates food safety, obtain proper testing of the ingredients in the restaurant, and assess the cleanliness of the restaurant. We should also consider that the burrito restaurant could be a red herring and investigate other potential causes of illness though, for the sake of this example, this is the only potential cause.

Now, we need to examine how broad of an investigation we need to pursue:

- Is this a local burrito restaurant or a national chain with ingredients from many different sources?
- How many potential victims are there? Is this a very popular restaurant with rapid turnover of inventory, or is this a slow-paced restaurant that could have allowed a batch of food to sit for prolonged periods of time under less than ideal conditions?
- Has this restaurant been the subject of similar events in the past, and if so what was the cause at that time?
- Are people aware of this risk, and does the restaurant need to be shut down during the investigation?

Often, in performing the risk assessment, as questions are answered, many more are generated. Appropriate and cost-effective utilization of resources is important, so the data gathering phase is the most important piece of the risk reduction puzzle.

WHO describes three components to every episode that should be considered: the assessment of the hazard, the route of exposure, and the context that the episode is occurring in. This is not completed sequentially but rather unfolds as new information is learned about the event.

20.3 HAZARD ASSESSMENT

In our example described above, *E. coli* O157:H7 can be readily identified in stool samples of those affected. Other outbreaks are not always as easily identified. In the case of the emerging epidemic of acute flaccid myelitis, to date medical researchers have not yet identified a causative agent (4). It is not currently known if this is viral, bacterial, environmental, or something else despite extensive testing of spinal fluid, blood, stool, and urine specimens from impacted patients. Some information can be predicted in this particular scenario because the illness is similar to polio, so similar viral strains are implicated. Additionally, as this is causing neurological symptoms, it is reasonable to consider that the culprit is likely to be discovered in the spinal fluid. It is primarily impacting children, so we can infer that there is something about this population that makes them more susceptible. It has emerged in a variety of environments, so locations are not particularly useful information, but it has flared up seasonally.

A big barrier to hazard assessment with an unknown agent is identifying populations that may be affected. Spreading information to relevant agencies to be on the lookout for a similar spectrum of symptoms is essential in identifying clusters of disease and then, through detailed interviews, narrowing down potential sources. For example, if all of those affected swam in the same lake next to a chemical plant, that fact implicates the lake or the chemical plant. If all of those affected happened to be on the same flight back from Mali, that fact implicates something unique to the flight or something all the afflicted individuals encountered in Mali.

It is possible to identify a public health threat without knowing the causative agent but makes the response more ill-defined. That's why identifying all similar features becomes so important.

Once a potential public health hazard has been identified, steps can be taken to determine who is at greatest risk and how can this risk be minimized.

Next we need to review the routes of exposure. Identification of those that may have encountered the potential health hazard is essential in containing the event. This should start as broad as possible and narrow down accordingly. For example, Joe has been working in Mali with those impacted by Ebola; he feels unwell but does not reveal this to anyone and boards his flight back to Texas. During the flight he develops a fever and upon landing takes a bus to the local hospital where he is promptly quarantined and diagnosed with Ebola. Everyone from the airport in Mali to the main entrance of the hospital was potentially infected. It is unreasonable though to quarantine everyone, so we must use what we know about Ebola to guide our management. We know that Ebola is transmitted through blood or body fluids. We know that Ebola cannot spread to others until signs or symptoms are showing; in our example when Joe developed a fever, we can presume he was contagious. We know that Ebola can remain on surfaces for several hours. So those at highest risk are those on the airplane in the immediate vicinity of Joe as well as those that would have contact with his seat on the plane after he had left. The people that sat next to him on the bus and the hospital workers that met Joe when he first arrived are likely at the highest risk.

To summarize, when performing the exposure assessment, start broad and identify the facts of the agent if known: how the agent is spread (aerosol, vector borne, droplets, direct transmission), how long symptoms take to develop, when those infected are the most contagious, and the community protections such as vaccines that are available.

20.4 CONTEXT ASSESSMENT: WHO IS GETTING THE DISEASE AND IN WHAT SETTING?

Is this public health threat occurring in a specific population, or is it indiscriminate? An example of this would be the emergence of multidrug-resistant Klebsiella infections in the ICU at a hospital in Baltimore. This was a specific population – a specific set of patients in the ICU rooms. We can infer that the spread of this disease occurred because of some overlooked area when cleaning and sterilizing the rooms. The context assessment in this case would be specifically the ICU at the specific hospital in Baltimore. People admitted to the ICU in a hospital in Dallas are not at risk from this same strain of bacteria, though we can infer that they could be at risk for a similar event if their sterilization process is similar.

So, let's put it all together. Peter traveled to Brazil for a much-needed vacation. While there, Peter neglected to use mosquito netting or bug spray and was bitten by numerous mosquitos. He developed joint and body pain and eye redness and generally did not feel well for about a week after returning from his vacation and now presents to his local primary care physician. Now, let's perform our risk assessment.

His savvy physician knows that numerous mosquito-borne diseases are endemic in Brazil. He considers that Peter's symptoms could be due to dengue fever, malaria, Zika, or chikungunya. He also considers other etiologies that are more common such as the seasonal flu. Due to the physicians detailed history, Peter's numerous mosquito bites and blasé attitude for self-preservation numerous samples are sent off for testing, and Zika is diagnosed.

Now we need to look at the vector or how is our pathogen spread. We must also consider the virulence of the virus and who is the most vulnerable. Can Peter our patient zero spread this to others? Peter happened to have gone with a large tour group from all over the country. The individuals from this tour group should be identified. Medical records should be surveilled, and positive testing for Zika should flag public health authorities that there may be a new public health event. The appropriate health authorities in Brazil will investigate where the tour group traveled and ideally capture mosquitos carrying the virus. They can then alert other tour agencies and restrict travel into those areas.

Consider instead that upon returning from Brazil and prior to his diagnosis, Peter, who for some reason does not mind mosquito bites and tends to not use bug spray, goes fishing in Minnesota. The public health authorities would evaluate the possibility that Peter's illness came from there rather than Brazil. They will also consider that Peter could have unwittingly spread his Zika to the mosquitos in Minnesota. Luckily, in this example the mosquito that transmits Zika has never been identified in Minnesota, but state entomologists would perform testing on the local mosquitos to ensure that this was still the case. The general population of Minnesota is therefore inferred to not be at risk.

Press releases warning travelers going to the affected areas of Brazil will be generated, and those at highest risk will be warned prior to travel. In Brazil, areas with the largest numbers of the vectors (mosquitos in this case) will be monitored, and pesticides will be used to control the vector population. Vulnerable areas with large amounts of standing water that can attract mosquitos will be identified, and the standing water conditions rectified. Travelers will be cautioned of the risk and will hopefully utilize prevention methods like mosquito netting and bug sprays. Researchers will start looking for a vaccine to aid in prevention.

Once we have performed the initial risk assessment, let's determine the level of risk to the population at large. The WHO recommends use of a risk matrix evaluating the consequences of a public health event in relation to the likelihood of such an event occurring.

We need to determine now how useful is the information in our risk assessment. This is called the confidence in the assessment and is based on the quality and breadth of data we can obtain from all of our sources. Low confidence would be in the case of a poorly understood or emerging disease process with little corollary information to guide assessment, whereas high confidence would be determined in a case of a well-known disease process that is well elucidated.

The consequences of a public health threat can be determined by considering the risk assessment as a whole. In poor Peter's Zika example, the level of risk is relatively low. He is minimally contagious to others, and the vector that carries the disease is not present in his home state in Minnesota. In this example the consequences are low to moderate, the risk is low, and the confidence in the risk assessment is high.

In the Ebola example earlier in the chapter, however the risk is high. Joe was likely actively shedding virus during his flight with numerous others that were likely continuing their travels. The virus has an incubation period between 2 and 21 days. Ebola in the United States is uncommon and would not be on many healthcare providers' radar initially. So, Joe's contacts would not necessarily be diagnosed correctly early on. In this example, the consequences are severe, the risk is high, and the confidence of the risk assessment is moderate to high.

Now, in the example of the cases of acute flaccid myelitis, the confidence of the risk assessment would be considered low. This is an emerging illness impacting primarily children. The consequences are severe and as the etiology is unknown, the risk is high.

20.5 SUMMARY

Upon conclusion of the public health event, those involved with the response should be debriefed. This allows for identification of areas that could be handled more efficiently. Safeguards can then be installed. For example, after the Ebola scare in 2015, many healthcare facilities will now routinely ask if there has been any foreign travel. This then triggers the healthcare provider to be aware of potential unusual illnesses immediately and can then rule them out early rather than rule them in later.

Developing protocols for risk assessment with public health threats presents challenges in that often, due to variety of different events potentially encountered, a single algorithm cannot be generated. There may be a huge amount of extraneous and irrelevant information, especially at first. Minor details may be overlooked but may present a single unifying piece between affected individuals.

Multiple factors can throw a wrench in the best efforts of public health agents when trying to control an emerging public health threat.

Financial limitations and working across different agencies, and potentially with foreign governments, may present challenges with keeping up standards. Cultural differences may result in increased ease of transmission of diseases. Mutations or development of antibiotic resistance can lead a previously well-controlled pathogen to become a threat. Failure to complete vaccination series or avoidance of vaccines can lead to not only infection of those unvaccinated individuals but also reduction in herd immunity. A civil war could limit the ability of healthcare staff to reach afflicted individuals.

All of these factors need to be considered when performing a risk assessment. It is much less cut and dry than in other fields, but with a critical eye and some basic awareness, risk managers can help control public health threats:

- What steps would you take if an employee presented to you, his risk manager, with reports of acute wheezing and shortness of breath after exposure to loose debris at a work site?
- What questions would you ask?
- How would you narrow down the cause of his symptoms?

Case Study

Read the following case study and discuss the questions at the end.

Food Risk Assessment Case Study

Chipotle Outbreak:

In 2015 there were a string of food-borne outbreaks throughout the Chipotle restaurant franchises.

Beginning in July a small *E. coli* O157:H7 infected less than 10 people at a single location in Seattle. They were unable to determine the source of the outbreak. Even though it was localized and did not gain media attention, it was foreboding for what was to come.

The second outbreak of the norovirus occurred concurrently with the Salmonella outbreak happened at a single location in Simi Valley, California – initially believed to have infected more than 200 customers including 18 employees. The source of the contamination is unknown.

http://www.foodsafetynews.com/2015/12/a-timeline-of-chipotles-five-outbreaks/#.W0t72qdKiM8

The third of the 2015 Chipotle outbreak occurred throughout the state of Minnesota. Nearly a hundred cases were found to be linked to a Salmonella infection. The source of the outbreak is believed to have been a result of contaminated tomatoes imported from Mexico.

The fourth outbreak occurred in October. This outbreak is believed to have been caused by a STEC O26 strain of E. coli. First being reported in the states of Washington and Oregon, this period more cases were discovered in seven more states. In the end more than 50 customers were reported as infected with at least 20 being hospitalized. The source of the infection was never discovered, but it is believed to have been from contaminated fresh produce. Due to the cases of infecting having occurred in multiple locations, the contaminated items would have occurred before reaching the restaurants.

https://www.cdc.gov/ecoli/2015/o26-11-15/index.html

The fifth outbreak was also caused by a different unrelated strain of *E. coli*. This small outbreak infected less than 10 people from locations in Kansas and Oklahoma. The source of the contamination is unknown.

The sixth and final outbreak happened in December. At a single Chipotle restaurant, 80 students at Boston College in Massachusetts were infected with norovirus. In the month nearly 150 cases of norovirus were reported from patients that had never eaten at Chipotle. This is believed to have been a result of a secondary outbreak due to the contaminated Chipotle victims in contact with others who would sequentially get infected. The cause of this outbreak is believed to be the result of an infected employee who was allowed to work while noticeably ill.

There were several economic results from such a massive scale. This included class action lawsuits from investors, lawsuits from the individuals that were infected, and the temporary national shutdown of all stores.

Reevaluations of Chipotle safety protocols were implemented. Senior corporate staff were replaced. Several food supply chain distributors were replaced. The company centralized the processing and began pretreating several food products that were once performed freshly at the restaurant sites. Employees have been retrained in food safety protocols. Employees are now required to dip the raw produce into hot water to aid in the removal of possibly harmful bacteria on the exterior of the produce and are now required to wash their hands every 30 minutes.

What additional steps would you take to decrease the risk of future outbreaks?

What could have Chipotle done to avoid the outbreaks from occurring?

What avoidable mistakes did Chipotle do that could have prevented or mitigated these outbreaks?

What actions do you believe Chipotle and regulative agencies did correct that may have lessened the results of the outbreaks?

Should Chipotle as a multinational organization be principally responsible for each of these outbreak? Explain your reasoning.

Would the general population soon forget about the food issues at a major chain restaurant?

REFERENCES

1. WHO Six common misconceptions about immunization. http://www.who.int/vaccine_safety/initiative/detection/immunization_misconceptions/en/index1.html (accessed August 2018).
2. WHO (2012). *Rapid Risk Assessment of Acute Public Health Events*. Geneva, Switzerland: WHO Press.
3. NPR.Org (2011). Timeline, how the Anthrax terror unfolded. https://www.npr.org/2011/02/15/93170200/timeline-how-the-anthrax-terror-unfolded (accessed August 2018).
4. Centers for Disease Control (2018). Acute flaccid myelitis. https://www.cdc.gov/acute-flaccid-myelitis/index.html (accessed August 2018).

OTHER SOURCES

English, J.F., Cundiff, M.Y., Malone, J.D. et al. (1999). APIC Bioterrorism Task Force). *Bioterrorism Readiness Plan: A Template for Healthcare Facilities*. Washington, DC: Associationfor Professionals in Infection Control and Epidemiology.

Centers for Disease Control and Prevention, Fort Collins; Colorado Department of Public Health and Environment, Denver, Colorado; New Mexico Departments of 3Health and Environment, Santa Fe, New Mexico; and California Department of Health Services, California. Retrieved 3 September 2011 http://cid.oxfordjournals.org at Serials Section Norris Medical Library.

CHAPTER 21

Vulnerability Analysis Technique

21.1 INTRODUCTION

Vulnerability assessment and risk assessment are in essence the same. They both seek to determine risks to a system, a building, a plant, a ship, an airplane, a country, or people. However, vulnerability assessment is usually more interested in determining vulnerabilities in a system, building, plant, ship, airplane, country, or persons from persons, organizations, or countries with intent on doing harm. It is also common to call a risk assessment a vulnerability assessment if it concerns natural disasters, such as earthquakes, hurricanes, tornados, floods, or strong storms.

In this regard, the initiating event is someone, an organization, or a country that wants to harm the system. In the case of a natural disaster, the initiating event is an earthquake or tornado. The probability of the initiating event, therefore, is 1.0 or 100%. The subsequent analysis determines where in the system the vulnerabilities reside. The same tools can be used to conduct vulnerability assessments that are used to conduct a risk assessment. The outcome of a vulnerability assessment is used by analysts to modify the system to reduce the probability of a vulnerable component or to eliminate it.

Common parts of a vulnerability assessment are:

1. Background
2. Purpose
3. Scope
4. Assumptions

Risk Assessment: Tools, Techniques, and Their Applications, Second Edition. Lee T. Ostrom and Cheryl A. Wilhelmsen.
© 2019 John Wiley & Sons, Inc. Published 2019 by John Wiley & Sons, Inc.
Companion website: www.wiley.com/go/Ostrom/RiskAssessment_2e

5. Description of system

    (a) System attributes

    (b) System sensitivity

6. Systems security

    (a) Administrative security

    (b) Physical security

    (c) technical security

    (d) Software security

    (e) Telecommunications security

    (f) Personnel security

7. System vulnerabilities

    (a) Technical vulnerability

    (b) Personnel vulnerability

    (c) Telecommunications vulnerability

    (d) Environmental vulnerability

    (e) Physical vulnerability

Another common nomenclature difference between a risk assessment and a vulnerability assessment is that usually the first step in a vulnerability assessment is called a threat assessment. This is analogous to a hazard assessment in a risk assessment.

> The federal government has implemented *The Risk Management Process for Federal Facilities: An Interagency Security Committee Standard*, which states,
>
> Risk is a function of the values of threat, consequence, and vulnerability. The objective of risk management is to create a level of protection that mitigates vulnerabilities to threats and the potential consequences, thereby reducing risk to an acceptable level. A variety of mathematical models are available to calculate risk and to illustrate the impact of increasing protective measures on the risk equation.
>
> Facility owners, particularly owners of public facilities, should develop and implement a security risk management methodology that adheres to the Interagency Security Committee (ISC) standard while also supporting the security needs of the organization. Landlords who desire to lease space to federal government agencies should implement the ISC standard in the design of new facilities and/or the renovation of existing facilities (1).

Two case studies are used to demonstrate how vulnerability assessments are performed. The first uses a preliminary hazard analysis (PHA) approach to analyze a facility for vulnerabilities associated with a potential intruder attack. The second is a much broader assessment examining the vulnerabilities of a multipurpose academic building.

21.2 CASE STUDY 1: INTRUDER

One of the scariest types of events that have become far too common is an armed intruder who intends to do harm to the occupants of a building. A vulnerability assessment can be used to help identify those vulnerabilities of the building that can contribute to the event. These types of assessments can be performed using PHA, failure mode and effect analysis (FMEA), tree analysis techniques, and even probabilistic risk assessment techniques. No matter what the technique is, the vulnerability assessment assumes an intruder will attempt to gain access to a building. The scenarios that are exercised during the assessment must be realistic and reasonable. For instance, a realistic and reasonable approach is to assume that one or two armed individuals will seek to kill and injure as many employees in a corporate office as possible. It is unreasonable and unlikely that a squad of ninjas with intent of world domination will invade a local car dealership.

The types of intruder scenarios exercised for a community hospital vulnerability assessment might include:

- Disgruntled employee seeking revenge.
- Disgruntled patient or family member seeking revenge.
- Estranged father trying to abduct or harm a new baby.
- Drug-related invasion of a pharmacy.
- Mental patient wreaks havoc.
- Patient has a bad reaction to a drug and wreaks havoc.
- Random violence.

Hospitals are especially vulnerable because of the mix of patients, staff, family members, and drugs. Most hospitals, until the present time, were designed to be open and allow free movement between floors and areas of the hospitals. With increased potential threats, hospitals, as well as schools, will become more secure. The Federal Emergency Management Agency (FEMA) has a series of publications concerning improving the security and safety of facilities such as hospitals and schools (www .fema.gov) (2).

Table 21.1 shows an example of PHA type of analysis for a hospital in response to the threat of an armed intruder. Note that this is not a complete analysis.

21.3 CASE STUDY 2: MULTIPURPOSE ACADEMIC BUILDING

The purpose of this vulnerability assessment was to determine the risks associated with a multipurpose academic building. The building contains classrooms, offices, and

TABLE 21.1
Partial Vulnerability Assessment of a Community Hospital

Threat	Location	Current situation	Potential events	Proposed mitigating factor
Armed intruder	Main lobby	Open lobby, with three volunteers at a receiving desk. Three corridors, 60-in wide, radiate from the lobby. There are two cameras that monitor the lobby. One security guard monitors these cameras, along with 22 more cameras	Intruder can attack volunteers or can walk past volunteers and enter other parts of the hospital	Lobby is manned by two armed security guards. Doors are placed on the corridors, with electronic locks that are opened by the guard or by code. Add a security guard to the camera monitoring task
Armed intruder	Neonatal ward	The ward contains 16 beds for new infants. It is a classic neonatal ward, with a large viewing window. The room is separated from the hallway by a door with an electronic lock that requires a code	Intruder could walk in with a staff person or watch a staff person enter the code and then use the code to gain entry to the ward when no staffs are in the room. A camera does not monitor the door	Change the code pad with a card reader. Update the access codes on the cards at least on a monthly basis. Consider having a staff person in the neonatal ward at all times. Add a camera to monitor the door
Armed intruder	Pharmacy	The pharmacy has three entrances. There is an outside door that is used to receive supplies from delivery trucks. There is an entrance from the main corridor that is used by pharmacy staff during the course of the day. There is a side door to the corridor that leads to the operating rooms. The outside entrance has a keypad lock. The other two entrances have card readers. The doors are alarmed. The main entrance door is a hollow-core door. The dispensing window is made from plate glass. There is a camera at the dispensing window, but no other cameras. One to two pharmacists are on duty from 9 : 00 a.m. till 6 : 00 p.m. There are at least two pharmacy technicians on duty at all times between 9 : 00 a.m. and 6 : 00 p.m. Response time to the pharmacy from the security office is five minutes	An armed intruder could gain access to the pharmacy during nonworking hours. An intruder could break through the main entrance door and security guards could not respond before the intruder has time to grab pharmaceuticals and vacate the facility. An armed intruder could force the pharmacists to produce drugs during working hours. The main entrance door and dispensing window are not made of strong enough materials. There is no camera monitoring the main door	Replace the main entrance door with a stronger door. Replace the glass in the dispensing window with toughened polycarbonate or bullet proof glass. Increase the number of cameras to monitor all entrances to the pharmacy

chemical, biological, and engineering laboratories. The types of hazards analyzed for this risk assessment fall into three basic categories: building, natural, and man-made. An analysis of the building hazards due to the building itself and its associated systems will determine the impact on all personnel who work within the building. An analysis must also be performed on the impact to those who are not working in the building such as personnel in other university buildings and the public as well as to the environment in the event of a release of hazardous materials. Natural and man-made hazards will be analyzed for the same reasons. However, in this case, the analysis must also address the hazards outside of the university and their impact on the building. The basic requirement for conducting a risk assessment (or hazard assessment) comes from 29 CFR 1910.1200, "Hazard Communication," and the "Superfund Amendments Reauthorization Act (SARA) of 1986, Title III, The Emergency Planning and Community Right-To-Know Act."

The purpose of this risk assessment was to identify materials at risk such as chemicals and potential energy sources such as mechanical and electrical energy. Security of the building is included in this risk assessment. Once all hazards have been identified, a qualitative assessment was performed on the risk of each hazard that may cause harm. Their modes of failure and their associated controls were then identified. A PHA was not performed since the building was already built. This risk assessment was conducted as a follow-up to construction. It is assumed that the event would occur regardless of cause, probability, and frequency. Many mitigations and controls are mandated by law and codes regardless of their chances of occurring.

There are many requirements governing the hazards and their controls. The Code of Federal Regulations (CFRs) set the minimum standards for them. However, there are consensus codes established by societies that implement and improve on the CFRs. Subject matter experts (SMEs) are important to consult in order to determine these codes and requirements. The following sources should be consulted, but not limited to, in order to determine which consensus codes to apply:

1. American Conference of Governmental Industrial Hygienists (ACGIH).
2. American Glovebox Society (AGS).
3. American National Standards Institute (ANSI).
4. American Nuclear Society (ANS).
5. American Society of Heating, Refrigerating and Air-Conditioning Engineers (ASHRAE).
6. American Society of Mechanical Engineers (ASME).
7. Compressed Gas Association (CGA).
8. National Fire Protection Agency (NFPA).
9. National Institute for Occupational Safety and Health (NIOSH).
10. Scientific Apparatus Manufacturer Association (SAMA).

11. Sheet Metal and Air Conditioning Contractors' National Association (SMACNA).

12. State and local planning, zoning, and building codes.

The fire department has resources on hand to assist in proper emergency response. These can assist in determining hazard controls during emergency response. These sources would include the following, which are available online:

1. Chemical Transportation Emergency Center (CHEMTREC).

2. Chemical Hazards Response Information System (CHRIS).

3. Computer-aided management of emergency operations (CAMEO).

4. Emergency Response Guidebook (ERG).

21.3.1 Methods of Collecting Information

The following is a list of the various methods used to collect information for the risk assessment.

1. *Task Analysis*

    (a) The task analysis will aid in identifying the tasks performed and the associated hazards with performance of those tasks. This will identify the hazards the personnel will be exposed to, which require mitigation and controls plus any training required. A task analysis is typically a tabletop analysis performed by the analyst and a group of SMEs.

2. *Observation of Personnel*

    (a) Direct observation of personnel performing duties at the job site supports the task analysis. It identifies the hazards they will be exposed to while working or by their being in the proximity to hazards. Observation will assist in determining those tasks performed that a tabletop task analysis may miss.

3. *Walk Down of Building and Processes*

    (a) This will allow the analyst to actually see and identify the actual hazards that are currently in place in the facility. Applicable drawings and procedures should be used to assist in identifying hazards. A walk down of the surrounding area will identify the areas and people that the facility hazards may affect and identify outside hazards that may affect the facility. This will identify hazards that will affect the personnel in the facility, the public, and the environment.

4. *Interviews with SMEs*

    (a) There are many types of hazards and controls associated with any facility. No one person can be an expert in multiple fields or know all requirements.

Consulting with SMEs for the various areas (chemical, biological, radiological, fire, electrical, industrial hygiene, etc.) is necessary to ensure all hazards are identified and the appropriate code or standard is applied. The following are those SMEs consulted for this analysis:

1. Chemistry and biology lab managers.
2. Research lab manager.
3. Maintenance supervisor.
4. Public safety manager and officers.
5. Local fire chief.
6. Local police department.
7. Local records department.
8. Local building inspector.
9. The architectural firm that designed the building.
10. County emergency planner.
11. Various SMEs for various functions such as industrial hygiene, fire safety, industrial safety, radiological controls, chemical coordinators, and hazardous waste managers. The fire marshal and building contractor would make valuable resources.

5. In addition to the regulatory requirements and consensus code resources mentioned above, a review of existing documentation, policies, and statutes such as the following but are not limited to:

(a) Building drawings
(b) Area maps
(c) Local ordinances
(d) Idaho statutes
(e) University policies

21.3.2 Task Analysis

The following is a list of the various tasks performed in the building by faculty and students:

1. Handle chemicals in laboratories (acids, bases, salts, flammables, metals).
2. Grow biological cultures.
3. Handle radioactive materials.
4. Operate Bunsen burners.
5. Operate laboratory natural gas shutoffs.
6. Operate compressed gas bottles.
7. Operate cryogenic systems.

8. Operate emergency eyewashes and showers.

9. Dispose of hazardous waste.

10. Operate fire extinguishers.

The following is a list of tasks performed by maintenance personnel in the building:

1. Operate the building ventilation system.

2. Operate electrical breakers and disconnects.

3. Operate natural gas shutoffs (laboratory and building).

4. Operate compressed gas bottles.

5. Operate cryogenic systems.

6. Operate and perform maintenance on emergency eye washes and showers.

7. Dispose of hazardous waste.

8. Operate fire extinguishers.

9. Operate and perform maintenance on boilers.

10. Mix and add water treatment chemicals.

11. Access cooling water sump and perform maintenance on cooling water pumps.

12. Access cooling tower and perform maintenance on cooling tower components.

13. Perform maintenance on electrical systems.

14. Perform maintenance on plumbing systems.

15. Perform maintenance on building structures.

16. Perform maintenance on elevator.

17. Perform maintenance on natural gas lines.

The following is a list of tasks performed by public safety in the building:

1. Maintain lab and building security.

2. Enforce student code of conduct.

3. Enforce state and federal laws.

4. Render emergency medical aid.

21.3.3 Risk Assessment

The risk assessment will include a description of the building and systems and the surrounding area. It will also include the hazards, both internal and external to the building, the risks and consequences of those hazards, their likelihood, and their controls. Hazard classification for each hazard reflects the credible worst-case potential consequence. Hazard classifications range from class I to IV. Class I hazards are negligible. Class II hazards have marginal effects with minor consequences. Class III hazards are critical in nature, which can cause severe injury or damage to equipment or the facility. Class IV hazards have catastrophic consequences leading to death, disabling injuries, and loss of a facility or system. For the purposes of this analysis, the hazards are basic and straightforward. Failure is assumed regardless of the cause or mode. The controls will be the same regardless. Therefore, a quantitative analysis, a fault tree analysis, and a human reliability analysis event tree need not be performed. It is assumed that there will be equipment failure or human error.

21.3.4 General Description

The multipurpose academic building was designed in 1991; construction began in 1992 and was completed in 1994. It is located on a hill approximately 100 ft from a major western river. The geographical area is a high plains desert with an annual rainfall of 14.21. The desert contains grasses and sage brush with lava flows. Farming is predominant in the surrounding area. The building is located next to a city park. There are office buildings, residential areas, and research laboratories in the vicinity. The local regional airport is located across the river nearby to the west next to an interstate. Major highways are to the east a short distance away. Two railways are located to the north as well as to the east in close proximity. The campus where the building is located includes other buildings. The building conducts research for various non-university groups in the science, mechanics, and materials (SIMM) laboratory for the university. The building also has a chemistry and biology lab for students. The following sections describe in more detail the building.

21.3.5 Building and Systems Design

1. *Multipurpose Academic Building*
   (a) 65 336 ft3.
   (b) Built to 2B seismic code.
   (c) Built to withstand 70 mph wind.
   (d) 33 lb/ft2 snow loading.
   (e) 30-in frost depth.
   (f) Located outside 100-year flood plain on hill.
   (g) Construction type 2 for occupancy: one-hour fire rating. A-3 and B-2 occupancy for the number of people to occupy building, number of exits, size of corridors, and number of sprinklers.
   (h) The steel framework and other structures are built to a three-hour fire rating. Interior materials are rated for one hour. Stairs and walls around laboratories rated for two hour. Stairs are designed for fire safe area for rescue. Wall materials are fire rated and tested per Underwriters Laboratories

(UL), Gypsum Association (GA), International Conference of Building Officials (ICBO), and ANSI standards.

(i) Elevator installed for third floor access for the disabled.

(j) Building rated for importance factor of one (low).

(k) Roofing built to class A standards. It includes gypsum thermal barrier, insulation, and single-ply material for roofing membrane.

(l) Parking lot size built to accommodate faculty and students and is lit at night.

(m) Stormwater pollution prevention is not required, but there is a drainage pit at the end of the parking lot for rainwater runoff.

(n) The building is not constructed to prevent flooding. However, the foundation has been sealed for seepage. Planter boxes in the front of the building have water weep holes installed.

(o) Access to the facility is by a park access road. Exiting the facility can be by the normal access route or by the local laboratory office building parking lot (access normally restricted) or by the side entrance to park that leads to another entrance/exit.

2. *Ventilation*

   (a) 16 ventilation fans located on top of the building.

   (b) 100% air exchange rate every 10 minutes.

   (c) Laboratory ventilation is separate from the rest of the building ventilation. Systems are redundant.

   (d) Ventilation is monitored by computer.

3. *Natural Gas*

   (a) Enters building on north side by the north exit. Supplies the laboratories. Main shutoff is located where the gas line meters the building. Laboratory shutoffs located at the laboratory exits.

   (b) Natural gas supplies two maintenance buildings that are next to the building just to the west. Shutoffs are located next to each building.

4. *Cooling Tower*

   (a) A cooling tower for the ventilation heat recovery system is located outside of the building next to the building on the west side.

   (b) The cooling tower's cooling water system has a sump with pumps in the mechanical room.

5. *Fire Water and Fire Systems*

   (a) Fire water enters the building through the mechanical room. All rooms and hallways have sprinkler systems. The piping is filled with water (wet pipe). The system is monitored by an electronic fire alarm panel.

   (b) Fire alarm pull boxes are located at the exits.

(c) Smoke detectors will activate the fire alarm.

(d) Fire extinguishers are located in the hallways near the exits and halfway down the hallways. Fire extinguishers are located in each laboratory.

(e) Fire hydrants are located near the main entrance next to the parking lot and at the southwest corner of the building.

6. *Plumbing*

   (a) Supplies, bathrooms, water fountains, and laboratories. (Note: This is considered a non-hazard; however, for the analysis, it will be considered as a source for contamination of laboratory chemicals entering the environment via sinks. If piping were to break, there are no hazards severe enough to cause any problems).

7. *Electrical*

   (a) Electrical power to the building is supplied by city power via a transformer next to the south side of the parking lot. Supply breakers for equipment are located in the mechanical room.

8. *Boilers*

   (a) There are four 1 M BTU boilers located in the mechanical room for building heating.

9. *Fume Hoods*

   (a) In each laboratory.

   (b) Ventilation separate from main ventilation. Each hood has an HEPA filter.

10. *Glove Boxes*

    (a) In the SIMM laboratory.

    (b) Ventilation separate from main ventilation.

11. *Public Safety*

    (a) The university employs its own security force. Public safety is not law enforcement but will ensure all federal and state laws are being enforced as well as local policies.

    (b) Public safety officers have law enforcement background, typically hired from the law enforcement academy. Some are currently serving duty as law enforcement or have in the past.

21.3.6 Hazards

1. *Chemicals (Class IV Hazards)*

   (a) Assorted acids, bases, salts, flammables, and metals. Inventory and quantities are considered restricted on a need-to-know basis. No more than 55 gal of materials in total is present.

    (b) Satellite accumulation areas (SAA) for the laboratories and a temporary accumulation area (TAA) in the SIMM laboratory for disposal of hazardous waste.

    (c) Water purification chemicals for cooling tower.

2. *Compressed Gases* (*Class IV Hazards*)

    (a) Compressed gases in bottles such as nitrogen, argon, helium, and hydrogen in laboratories.

    (b) Argon dewar in SIMM laboratory.

3. *Radiological* (*Class II Hazards*)

    (a) Small quantities of radiological materials are present. Types and amounts considered restricted on a need-to-know basis.

4. *Biological* (*Class II Hazards*)

    (a) Small quantities of Escherichia coli and *Salmonella*.

    (b) Located in the biology laboratory's refrigerator.

5. *Systems* (*Class IV Hazards*)

    (a) Building

        1. Third floor not wheelchair accessible other than by elevator.

    (b) Natural gas

        1. Supplied to laboratories for Bunsen burners. Shutoffs to laboratory gas supply located at exits. Building supply and shutoff located north end of building.

    (c) Cooling tower sump

        1. In mechanical room. Deep sump. Has lid with wench. Access door to room locked.

    (d) Electrical

        1. Wiring.

        2. Flooding.

        3. Laboratories (chemicals).

        4. Power outages.

    (e) Boilers

        1. Steam.

        2. Electrical.

    (f) Fire

        1. Electrical.

        2. Chemical (laboratories).

        3. Lightning.

        4. Hydrogen.

     5. Natural gas.

     6. Combustibles (microwave oven cooking, cigarettes).

6. *External Hazards*

  (a) River/flooding (class I hazards)

     1. The river is located approximately 100 ft from the building. The building is on a hill and is approximately 25 ft above the elevation of the river. The building is located just outside of the 100-year flood plain. The building area did not flood in 1976 when a local dam broke nor did it flood during the 1997 flood.

     2. Sources of flooding include water from rapid snow melt and storms. Another source is from the Palisades and Ririe Dams if they were required to release excess water from snow melt or if they fail.

  (b) Wildfires (class IV hazards)

     1. The desert surrounding the area, mainly to the west and southwest, contains grasses and sage brush. There is an open field adjacent to the building. These are fire hazards with the potential to cause wildfires. The area has frequent droughts, which increase the risk of wildfires.

  (c) Weather (class II hazards)

     1. Thunderstorms can cause flooding. Wildfires, structural fires, and power outages are caused by lightning strikes. Thunderstorms can create high winds that can damage structures. Power outages may occur.

     2. Tornadoes/high winds may damage structures and cause power outages.

     3. Winter storms and blizzards can create high winds and also heavy snow packs on roofs. With power outages, there may be freezing temperatures inside the building, which can rupture water pipes.

  (d) Earthquake (class IV hazards)

     1. The town is located in a 2B seismic zone. Yellowstone National Park, located 110 miles from the town, is seismically active. The high desert has volcanic buttes. Two of the largest earthquakes in the continental United States in the past 40 years have occurred within 150 miles of the town.

  (e) Electrical (class II hazards)

     1. Power outages can result in the loss of building ventilation and loss of heating. Loss of heating in the winter can result in water pipes freezing and bursting.

     2. Electrical shock (class IV hazard) can severely injure or cause death.

(f) Local national laboratory (class I hazards due to location)

1. There is a national laboratory west of the building. It conducts research with nuclear reactors and materials. The facility also has hazardous chemicals. The closest nuclear facility is located 30 miles west of the building.

2. The facility has a research laboratory in the town approximately 1 mile away to the east. It may contain small quantities of biohazards at any given time.

(g) City (class III hazards)

1. The city has water pump houses located throughout the city. Two are within 1 mile of the building. Each pump house has 2–150 lb chlorine bottles for water treatment. Southeast of the city is the water treatment plant that has a 1500 lb container of chlorine. It is located approximately 2.5 miles from the building.

2. The regional airport is located 1100 ft to the west of the building across the river and the interstate.

(h) Interstate, highways, and railways (class IV hazards)

1. Two railways, an interstate, and two highways are in close proximity to the building. A railway to the northeast is 800 ft away. Another is located to the east and is 1.5 miles away. The interstate is to the west across the river 1000 ft away. The highways are to the east 3500 ft and 2.25 miles away.

2. This is considered by the local fire department and the county director of emergency management to be the most severe source of hazards in the county. Trains and trucks transport many types of hazardous materials. Depending on the material and the amount and leak rate, the emergency planning zone can be 5 miles.

(i) Palisades and Ririe Dams (class II hazards due to location)

1. Palisades Dam is located on the Snake River 56 miles upstream from the town.

2. Ririe Dam is located on Willow Creek, a tributary of the Snake River, 18 miles upstream of the town.

(j) Security (class III hazards)

1. Typical threats may include disorderly conduct, stalking, threats, and robbery. Violations of the university student code of conduct are possible.

(k) Administration building (class II hazards)

1. The administration building is constructed mainly of wood. It is next to the building and connected by a small walkway.

21.3.7 Risks

1. Tables 21.2–21.4 are an assessment of the risks associated with each hazard. This includes the consequences and frequency of each hazard.
2. The hazards listed are building hazards, natural hazards, and man-made hazards. Natural and man-made hazards are the external hazards listed in the Section 21.3.6.
3. An airplane or jet crash is not analyzed because it is beyond all likelihood, and if it did occur, actions would be similar to a fire or seismic event. There is no credible way to protect against such an event.
4. The administration building is not included but is considered a source of a fire, and the actions for a fire are the same regardless of the source.
5. Table 21.2 describes the building hazards. Table 21.3 describes the natural hazards. Table 21.4 describes the man-made hazards.

TABLE 21.2
Building Hazards

	Consequence	Frequency	Risk
Chemical	Splash chemicals on body (burns) resulting in injury, spills into city sanitary drain system, fire/explosion with incompatible materials	Moderate	High
Biological	*Escherichia coli* and *Salmonella* contamination/illness	Extremely low	Low
Radiological	Radiation exposure/contamination	Extremely low	High
Fire	Death or injury to personnel, release of hazardous chemicals and biohazards, explosions from gas bottles and natural gas lines	Low	High
Electrical	Electric shock/electrocution, fires, explosions from natural gas or chemicals	Low	High for maintenance, low for faculty and students
Natural gas	Gas leak, fire/explosion, asphyxiation	Very low	Moderate
Compressed gases	Asphyxiation, fire due to hydrogen leak, explosion during fire, injury due to falling bottle or flying parts	Very low	Moderate
Cryogenics	Frost bite, asphyxiation	Very low	Moderate
Confined space	Injury and/or drowning due to falling in	Very low	Moderate
Heat	Burns due to contact with boiler and piping surfaces or hot water	Low	High

TABLE 21.3

Natural Hazards

	Consequence	Frequency	Risk
Flooding	Flooding of building, electrical components; release of chemicals, and biohazards	Extremely low	Very low
High winds/ tornadoes	Damage to building, windows; loss of electrical power and ventilation, release of hazardous chemicals, radioactivity, and biohazards	High for winds, very low for tornadoes	Low for damage, moderate for loss of power and ventilation
Winter storms	Snow loading collapsing roof, damage due to high winds	High	Moderate for snow loading, low for building damage
Wildfires	Damage to building, fire/explosion, release of hazardous chemicals, radioactive material, biohazards, damage to compressed gas bottles	Moderate	Very low
Volcanic/ seismic activity	Damage to building, fire/explosion, release of hazardous chemicals, radioactive material, biohazards, damage to compressed gas bottles, dam failure	Low	Very low

21.3.8 Hazard Controls

1. Hazard controls are derived from a number of sources. The general requirements (references) for each hazard control will be included. SMEs and management will be responsible for the correct implementation of the requirements and controls. Since the building is under a university, it is considered a state entity and is not required to follow federal laws. However, as appropriate, the federal laws will be referenced as a resource. State laws will typically reference consensus codes (International Building Code [IBC], NFPA, ANSI, etc.) and hand off to them. The most restrictive codes or statutes will be used.

2. To ensure the building is built to applicable codes, the city building inspector will use software, "plan analyst," to enter building design criteria. A report is then generated, which outlines all applicable specifications for building construction in accordance with the IBC.

TABLE 21.4

Man-made Hazards

	Consequence	Frequency	Risk
Power outage	Loss of ventilation and release of hazardous chemical vapors, buildup of inert gases from compressed gas bottles, freezing of water pipes	Low	Moderate
Dam failure	Flooding, loss of power, release of hazardous chemicals, radioactive material, biohazards, fire/explosion	Very low	Low for Palisades Dam, moderate for Ririe Dam due to time of flood water arrival. Palisades requires 22 hours. Ririe requires 3.5 hours
Idaho National Laboratory	Release of radioactive materials, biohazards	Very low	Very low
Hazardous materials from trains and highways	Release of hazardous chemicals	Very low	Low
Chlorine gas from city of Idaho Falls water system	Release of chlorine gas, death/injury to public	Very low	Moderate
Diesel storage tanks	Explosion, toxic fumes	Very low	Very low
Public safety	Robbery, violence, stalking, intoxication, terrorism, medical emergencies	Low	Low

3. The IBC addresses structure design, fire safety, hazardous material safety (in building design), and electrical, mechanical, plumbing, and elevator systems. The IBC also hands off to the International Fire Code.

4. According to Intermountain Gas Company, natural gas lines are installed up to the gas meter by Intermountain Gas. Lines inside the building are installed by contractor personnel. Intermountain Gas uses numerous codes from the Office of Pipeline Safety, Public Utilities Commission, and the Department of Transportation. The city inspector says contractors will use the International Fuel Gas Code.

5. The following are the hazards and their associated controls. Note: Training will typically be a control for these hazards and will not need to be mentioned below.

6. Table 21.5 discusses building hazards. Table 21.6 discusses natural hazards. Table 21.7 discusses man-made hazards.

TABLE 21.5

Building Hazards and Associated Standards/Regulations

Hazard	Controls	Requirement
Chemical	Ventilation, PPE, flammable storage containers and segregation, glove boxes/ventilation hoods, chemical neutralization in drain system, MSDS, spill kits, emergency eyewash/shower, satellite accumulation areas and temporary accumulation area, electrical systems constructed for laboratories, sprinkler system, fire alarm system and fire extinguishers, locked rooms, preventive maintenance, evacuation. Note: Neutralization in drain system removes the requirement for a stormwater pollution prevention plan. Drains will still be monitored	• International Building Code (IBC) • 29 CFR 1926 • 29 CFR 1910.94/1000/1200/1450 • 40 CFR 262 • IDAPA 07, Title 07, Chapter 1 • NFPA 1, 10, 30, 45, 91, 101 • NEC • Chemical Hygiene Plan • Hazardous Waste Management Policy and Procedures Manual • AGS-G001 • ANSI Z9.5 • ERG • SARA Title I, III • City Code Title 3 Chapter 2
Biological hazards	PPE, approved storage, cleanliness, locked rooms, evacuation	• National Institute of Health "Biosafety in Microbiological and Biomedical Laboratories"
Radiological	In accordance with NRC license, posting, locked rooms, evacuation	• NRC license, which implements 10 CFR 19 and 20 • Radiation Safety Policy Manual • SARA Title I, III • City Code Title 3 Chapter 2
Fire	Fire-resistant materials rated for one hour, construction type 2, A-3 and B-2 occupancy, sprinkler system, fire alarm system and fire extinguishers, approved smoking area, fire department access to building and list of hazards, electrical systems built to code, housekeeping, preventive maintenance, evacuation. Note: These controls apply to the TAB building catching on fire	• IBC, chapters • 29 CFR 1926 • NFPA 1, 10, 101 • NFPA 70 (NEC) • SARA Title I, III • City Code Title 3 Chapter 2 • The fire department has a list of hazards and access to the Building
Electrical	Electrical systems built to code for building and laboratories, lockout/tagout for maintenance, PPE for live systems, maintenance room locked	• 29 CFR 1926 • 29CFR 1910.147 • NFPA 70 (NEC) • NFPA 45, 70E

TABLE 21.5
(*Continued*)

Hazard	Controls	Requirement
Natural gas	Gas systems built to code, shutoffs at exits, ventilation, evacuation	• 29 CFR 1910
Compressed gas	PPE, approved bottles by vendor, approved storage (racks and chains), ventilation, room locked, evacuation	• 29 CFR 1910.101 • CGA P-1 • NFPA 45, 55 • ISU Chemical Hygiene Plan • SARA Title I, III • City Code Title 3 Chapter 2
Cryogenics	PPE, approved storage, ventilation, room locked, evacuation	• 29 CFR 1910.1450 • Chemical hygiene plan
Confined space	Access cover, room locked	• 29 CRF 1910.146
Heat	Pressure vessels inspected and certified, PPE, preventive maintenance, lockout/tagout for maintenance, room locked	• 29 CFR 1926 • 29 CFR 1910.147 • NFPA 70 (NEC) • ASME B.31.1 • IDAPA 17.06.01

MSDS, material safety data sheet; NRC, Nuclear Regulatory Commission; PPE, personnel protective equipment.

TABLE 21.6
Natural Hazards and Associated Standards/Regulations

Hazard	Controls	Requirement
Flooding	Built outside flood zone, importance level 1, evacuation	• IBC • SARA Title I, III • City Code Title 3 Chapter 2
High winds/tornado	Built to withstand 70 mph winds, take shelter, and evacuate as appropriate	• IBC • SARA Title I, III • City Code Title 3 Chapter 2
Winter storms	Built class A roof, snow loading to 33 lb/ft3, evacuation	• IBC • SARA Title I, III • City Code Title 3 Chapter 2
Wildfires	Evacuation	• SARA Title I, III • City Code Title 3 Chapter 2
Volcanic/seismic activity	Built to 2B seismic code, evacuation	• IBC • SARA Title I, III • City Code Title 3 Chapter 2

TABLE 21.7

Man-made Hazards and Associated Standards/Regulations

Hazard	Controls	Requirement
Power outage	Evacuation, supplemental heat in the winter for pipes if ventilation maintained	• Chemical Hygiene Plan • City Code Title 3 Chapter 2
Dam failure	Evacuation	• SARA Title I, III • City Code Title 3 Chapter 2
National laboratory	Evacuation	• SARA Title I, III • City Code Title 3 Chapter 2
Hazardous materials from trains and highways	Shelter in place or evacuation	• SARA Title, III • City Code Title 3 Chapter 2 • ERG
Chlorine gas from city of Idaho Falls water system	Shelter in place or evacuation	• SARA Title, III • City Code Title 3 Chapter 2
Diesel storage tanks	Shelter in place or evacuation	• SARA Title III • City Code Title 3 Chapter 2 • ERG
Public safety	Public safety surveillance in day, Idaho Falls police department surveillance at night, shelter in place or evacuation, call 911, citizen's arrest followed by police arrest, enforcement of student code of conduct, emergency phones stationed thought campus, increase security surveillance consistent with Homeland Security Advisory System threat level, render first aid until EMS arrives	• Idaho Statute Title 19 Chapter 6 • City Code Title 3 Chapter 1 • Public Safety Mission Statement

Public safety is a best management practice dictated by the Idaho State Board of Education. There are no requirements to have public safety. However, the above requirements outline derivation of authorities.

21.4 OTHER RELEVANT CODES AND STANDARDS

- Executive Order 12977, "Interagency Security Committee"
- FEMA
  - FEMA 386–7 *Integrating Manmade Hazards into Mitigation Planning*
  - FEMA 452 Risk Assessment – *A How-To Guide to Mitigate Potential Terrorist Attacks Against Buildings*
- ISC *Security Design Criteria* – Defines Threat/Risk classifications and resultant federal protective design requirements (Official Use Only)
- Unified Facilities Criteria (UFC) – UFC 4-010-01 *DoD Minimum Anti-Terrorism Standards for Buildings* – Establishes prescriptive procedures for Threat, Vulnerability and Risk assessments and security design criteria for DoD facilities (Official Use Only) (1).

21.5 CONCLUSION

The building has numerous hazards, which can cause serious bodily harm or death. The fact that it is a university setting does not make the building and processes inherently safe. The identified hazards must be mitigated or controlled in a manner consistent with established laws, statutes, consensus codes, and best practices established by the federal government, state and local governments and authorities, and recognized associations. With the numerous regulations and standards that must be followed, an analyst must consult with appropriate SMEs in order to properly identify the hazards and the requirements for mitigating and controlling the hazards. Without the proper assistance, it would be impossible to adequately mitigate and control hazards sufficiently to protect personnel, the public, and the environment. A task analysis of the different duties of personnel is required to ensure that the tasks they perform are identified so that the appropriate mitigations and controls are in place. A physical walk down of the facility using drawings, procedures, and SMEs is necessary to properly identify and document the hazards. Hazards inside the building and on the outside are identified. Observation of work being performed is another requirement. Not all personnel are subjected to the same hazards. It depends on their job assignments. Once this part of the analysis is performed, the hazards are organized, and their risks are analyzed. Mitigations and controls are then determined.

Sample sources of information to perform a study of this type are contained in Table 21.8.

TABLE 21.8
Sample Sources of Information for a Vulnerability Study

Organization	Source(s)
Emergency management, fire, and security	FEMA. Available at http://www.fema.gov/plan/prevent/rms National Fire Protection Agency (2009). *NFPA 1, Fire Code*. Quincy, MA: NFPA. National Fire Protection Agency (2009). *NFPA 101, Life Safety Code*. Quincy, MA: NFPA. National Fire Protection Agency (2009). *NFPA 484, Standard for Combustible Metals*. Quincy, MA: NFPA. National Fire Protection Agency (2008). *NFPA 30, Flammable and Combustible Liquids Code*. Quincy, MA: NFPA. National Fire Protection Agency (2010). *NFPA 72, National Fire Alarm and Signaling Code*. Quincy, MA: NFPA. United States Department of Transportation (2008). *Emergency Response Guide*. Washington, DC. http://www.ehso.com/ehsodot.php?URL=http%3A%2F%2Fphmsa.dot.gov/staticfiles/PHMSA DownloadableFiles/Files/erg2008_eng.pdf 10CFR1910 OSHA Standard County Hazards Vulnerability Analysis City and state codes
Laboratory safety	American Glovebox Society (2007). *AGS-G001, Guideline for Gloveboxes*, 3e. Centers For Disease Control and Prevention, National Institutes of Health (2009). *Biosafety in Microbiological and Biomedical Laboratories*, 5e. http://www.cdc.gov/biosafety/publications/bmbl5/index.htm National Fire Protection Agency (2011). *NFPA 45, Standard on Fire Protection for Laboratories Using Chemicals*. Quincy, MA: NFPA. National Fire Protection Agency (2010). *NFPA 91, Standard for Exhaust Systems for Air Conveying of Vapors, Gases, Mists, and Noncombustible Particulate Solids*. Quincy, MA: NFPA.
Building requirements	Uniform Building Code
Compressed gases	American Society of Mechanical Engineers (2007). *ASME B31.1, Power Piping: ASME Code for Pressure Piping*. Compressed Gas Association (2000). *CGA P-1, Safe Handling of Compressed Gases in Containers*. 29CFR1910 OSHA Standard National Fire Protection Agency (2010). *NFPA 55, Compressed Gases and Cryogenic Fluids Code*. Quincy, MA: NFPA. State and Local Codes and Standards
Hazardous materials and waste	Environmental Protection Agency. 40 CFR 262, Standards Applicable to Generators of Hazardous Waste. http://ecfr.gpoaccess.gov/cgi/t/text/textidx?c=ecfr&sid=aab68f7aa9c461f1ae41bd67332dd4c3&rgn=div5&view=text&node=40:25.0.1.1.3&idno=40 Superfund Amendments and Reauthorization Act of 1986 (SARA). SARA Title III—Emergency Planning and Community Right-to-Know. http://epw.senate.gov/sara.pdf 29CFR1910 OSHA Standard
Radiation protection	Nuclear Regulatory Commission Regulations: 10CFR33 "Specific Domestic Licenses of Broad Scope for By-Product Material" 10CFR Part 35 "Medical Use of Byproduct Material" 10CFR73 "Physical Protection of Plants and Materials" Department of Energy Orders and Guidelines

Self-Check Questions

1. What does a vulnerability analysis technique seek to determine?

2. What are the parts of a vulnerability assessment?

3. What methods are used to collect information for the risk assessment?

4. What are some of the standards/regulations used when assessing risk?

REFERENCES

1. WBDG (2018). Threat/vulnerability assessments and risk analysis. https://wbdg.org/resources/threat-vulnerability-assessments-and-risk-analysis (accessed October 2018).
2. Federal Emergency Management Agency (FMEA) (2018). Risk Management Series (26). Created 26 July 2013. https://www.fema.gov/media-library/resources-documents/collections/3# (accessed October 2018).

CHAPTER 22

Developing Risk Model for Aviation Inspection and Maintenance Tasks

22.1 INTRODUCTION

Risk assessment has been used to analyze a wide range of industries to determine vulnerabilities with the ultimate purpose of eliminating the sources of risk or reducing them to a reasonable level. The purpose of this chapter is to show how risk assessment tools can be used to develop risk models of aviation maintenance tasks. Two tools will be discussed in this chapter, though many other methods exist. The tools discussed in this chapter are:

- Failure mode and effect analysis (FMEA)
- Event and fault tree analysis

22.2 FAILURE MODE AND EFFECT ANALYSIS

FEMA is discussed detail in Chapter 11. A procedure analysis will be used to demonstrate how an FMEA can be conducted. An FMEA is conducted on a step-by-step basis. Table 22.1 shows an example of an FMEA table.

The first step is to create a flow diagram of the procedure. This is a relatively simple process in which a table or block diagram is constructed that shows the steps in the procedure. Table 22.2 shows the simple steps checking an engine chip detector. Note that this is a simple example and not an exhaustive analysis. Table 22.3 lists the major, credible failures associated with each step in the process. Table 22.4 shows the effect of the potential failures. Table 22.5 shows the complete FMEA for the task.

Risk Assessment: Tools, Techniques, and Their Applications, Second Edition. Lee T. Ostrom and Cheryl A. Wilhelmsen.
© 2019 John Wiley & Sons, Inc. Published 2019 by John Wiley & Sons, Inc.
Companion website: www.wiley.com/go/Ostrom/RiskAssessment_2e

TABLE 22.1
Example FMEA Table

Item	Potential failure mode	Cause of failure	Possible effects	Probability	Criticality (optional)	Prevention

Step in procedure, part, or component	How it can fail:
	• Pump not working
	• Stuck valve
	• No money in a checking account
	• Broken wire
	• Software error
	• System down
	• Reactor melting down
		What caused the failure:
		• Broken part
		• Electrical failure
		• Human error
		• Explosion
		• Bug in software
			Outcome of the failures:
			• Nothing
			• System crash
			• Explosion
			• Fire
			• Accident
			• Environmental release
				How possible is it:
				• Can use numeric values:
				  ○ 0.1, 0.01, or 1.00E−05
				
				• Can use a qualitative measure:
				  ○ negligible, low probability, or high probability
					How bad are the results:
					• Can use dollar value:
					  ○ $10., $1 000., or $1 000 000
					
					Can use a qualitative measure:
					  ○ nil, minimal problems, or major problems
						What can be done to prevent either failures or results of the failures?

TABLE 22.2
Process Steps for Checking a Chip Detector

Inspecting chip detector	
Step number	Process steps
1	Cut and remove lock wire from oil drain plug
2	Remove oil drain plug
3	Drain oil
4	Cut and remove lock wire from chip detector
5	Remove chip detector
6	Examine chip detector
7	Clean chip detector
8	Replace chip detector
9	Lock wire chip detector
10	Replace oil drain plug
11	Lock wire oil drain plug
12	Replace oil

TABLE 22.3
Failures Associated with Each Step

Inspecting chip detector	
Process steps	Major failures
Cut and remove lock wire from oil drain plug	No major failures that affect process outcome
Remove oil drain plug	No major failures that affect process outcome
Drain oil	No major failures that affect process outcome
Cut and remove lock wire from chip detector	No major failures that affect process outcome
Remove chip detector	Improper removal can remove debris from chip detector and cause false reading. Chip detector can be damaged if improperly removed
Examine chip detector	Aircraft Maintenance Technician (AMT) fails to notice debris on chip detector
Clean chip detector	AMT fails to properly clean chip detector
Replace chip detector	AMT fails to properly install chip detector
Lock wire chip detector	AMT fails to properly lock wire chip detector
Replace oil drain plug	AMT fails to properly install oil drain plug
Lock wire oil drain plug	AMT fails to properly lock oil drain plug
Replace oil	AMT fails to properly replace oil

FMEA is a relatively simple but powerful tool and has a wide range of applicability for analyzing aircraft maintenance tasks.

22.3 EVENT TREE AND FAULT TREE ANALYSIS

As discussed in Chapters 14 and 16, an event tree is a graphical representation of a series of possible events in an accident sequence (1). Using this approach assumes

TABLE 22.4
Effect of Potential Failures

Inspecting chip detector

Process steps	Potential failure modes	Potential failure effects
Remove chip detector	Improper removal can remove debris from chip detector and cause false reading. Chip detector can be damaged if improperly removed	Engine could fail if chips are not properly detected
		Added cost to replace damaged chip detector
Examine chip detector	Aircraft maintenance technician (AMT) fails to notice debris on chip detector	Engine could fail if chips are not properly detected
Clean chip detector	AMT fails to properly clean chip detector	Debris could be placed back into engine
Replace chip detector	AMT fails to properly install chip detector	Oil could leak past chip detector
		Threads of chip detector could be damaged
Lock wire chip detector	AMT fails to properly lock wire chip detector	Chip detector could become lose and fall out, leading to loss of engine oil
Replace oil drain plug	AMT fails to properly install oil drain plug	Engine oil could leak out
		Oil drain plug could become damaged
Lock wire oil drain plug	AMT fails to properly lock oil drain plug	Oil drain plug could become loose and fall out
		Oil drain plug could become damaged
Replace oil	AMT fails to properly replace oil	Engine could fail

that as each event occurs there are only two outcomes, failure or success. A success ends the accident sequence, and the postulated outcome is either that the accident sequence terminated successfully or was mitigated successfully. For instance, a fire starts in an engine. This is the initiating event. Then the automated system closes fuel feed. If the lack of fuel does not extinguish the fire, the next step is that the fire suppression system is challenged. If the system actuates the fire suppression system, the fire is suppressed and the event sequence ends. If the fire suppression system fails, the fire is not suppressed, and then the accident sequence progresses. Table 22.6 shows this postulated accident sequence. Figure 22.1 shows this accident sequence in an event tree.

As in most of the risk assessment techniques, probabilities can be assigned to the events and combined using the appropriate Boolean logic to develop an overall probability for the various paths in the event. Using our example from above, we will now add probabilities to the events and show how the probabilities combine for each path. Figure 22.2 shows the addition of path probability to the event tree (Table 22.7).

TABLE 22.5
Complete FMEA for Chip Detector Task

Procedure step	Potential failure mode	Cause of failure	Possible effects	Probability	Criticality	Prevention
Cut and remove lock wire from oil drain plug	No major failures that affect process outcome	AMT fails to perform task	Delay in performing task	Very low	Not critical	Ensure AMTs follow work schedule
Remove oil drain plug	No major failures that affect process outcome	AMT fails to perform task	Delay in performing task	Very low	Not critical	Ensure AMTs follow work schedule
Drain oil	No major failures that affect process outcome	AMT fails to perform task	Delay in performing task	Very low	Not critical	Ensure AMTs follow work schedule
Cut and remove lock wire from chip detector	No major failures that affect process outcome	AMT fails to perform task	Delay in performing task	Very low	Not critical	Ensure AMTs follow work schedule
Examine chip detector	AMT fails to notice debris on chip detector	AMT fails to properly perform task	Engine could fail if chips are not properly detected Added cost to replace damaged chip detector	Moderate	Critical	Training, procedures, and inspection oversight
Clean chip detector	AMT fails to properly clean chip detector	AMT fails to properly perform task	Engine could fail if chips are not properly detected	Moderate	Critical	Training, procedures, and inspection oversight
Replace chip detector	AMT fails to properly install chip detector	AMT fails to properly perform task	Debris could be placed back into engine	Moderate	Critical	Training, procedures, and inspection oversight

(continued)

321

TABLE 22.5
(*Continued*)

Procedure step	Potential failure mode	Cause of failure	Possible effects	Probability	Criticality	Prevention
Lock wire chip detector	AMT fails to properly lock wire chip detector	AMT fails to properly perform task	Oil could leak past chip detector; Threads of chip detector could be damaged	Moderate	Critical	Training, procedures, and inspection oversight
Replace oil drain plug	AMT fails to properly install oil drain plug	AMT fails to properly perform task	Chip detector could become lose and fall out, leading to loss of engine oil	Moderate	Critical	Training, procedures, and inspection oversight
Lock wire oil drain plug	AMT fails to properly lock oil drain plug	AMT fails to properly perform task	Engine oil could leak out; Oil drain plug could become damaged	Moderate	Critical	Training, procedures, and inspection oversight
Replace oil	AMT fails to properly replace oil	AMT fails to properly perform task	Oil drain plug could become loose and fall out; Oil drain plug could become damaged; Engine could fail	Low	Critical	Training, procedures, and inspection oversight

TABLE 22.6

Accident Sequence

Event	Description	Possible outcomes
Fire	This is the initiating event	
Fuel feed is stopped	The lack of fuel causes the fire to stop	Success – the fire stops
		Failure – the fire continues
Fire suppression system actuates	The fire suppression system detects the fire and, it actuates	Success – system actuates and controls the fire
		Failure – fire destroys the engine

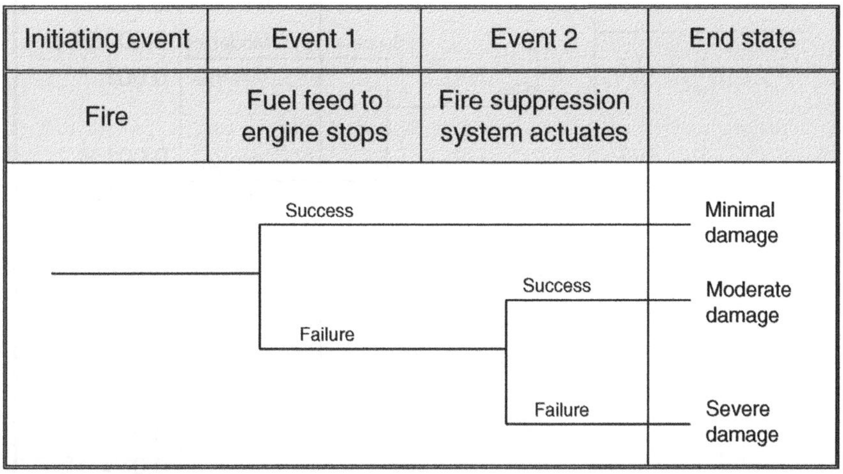

Initiating event	Event 1	Event 2	End state
Fire	Fuel feed to engine stops	Fire suppression system actuates	

Success — Minimal damage
Failure — Success — Moderate damage
Failure — Severe damage

FIGURE 22.1   Event tree.

The result of this analysis tells us that the probability derived for a fire in which the fuel feed system stops fuel supply to engine actuates and the consequence in minimal damage is approximately 1/1000 or 1×10^{-3}. The probability derived for a fire in which the fuel feed system fails to actuate, but the fire suppression system successfully extinguishes the fire and there is only moderate damage is 1E-6 or 1×10^{-6}. Finally, the probability that a fire occurs and both the fuel feed system and fire suppression system fail and severe damage occurs is 1E-8 or 5×10^{-8}.

This approach is considered inductive in nature. Meaning the system uses forward logic. A fault tree, discussed in Chapter 14, is considered deductive because usually the analyst starts at the top event and works down to the initiating event. In complex risk analyses event trees are used to describe the major events in the accident sequence, and each event can then be further analyzed using a technique most likely being a fault tree (2).

Refer to Chapter 14 for a discussion concerning the rules of constructing a fault tree.

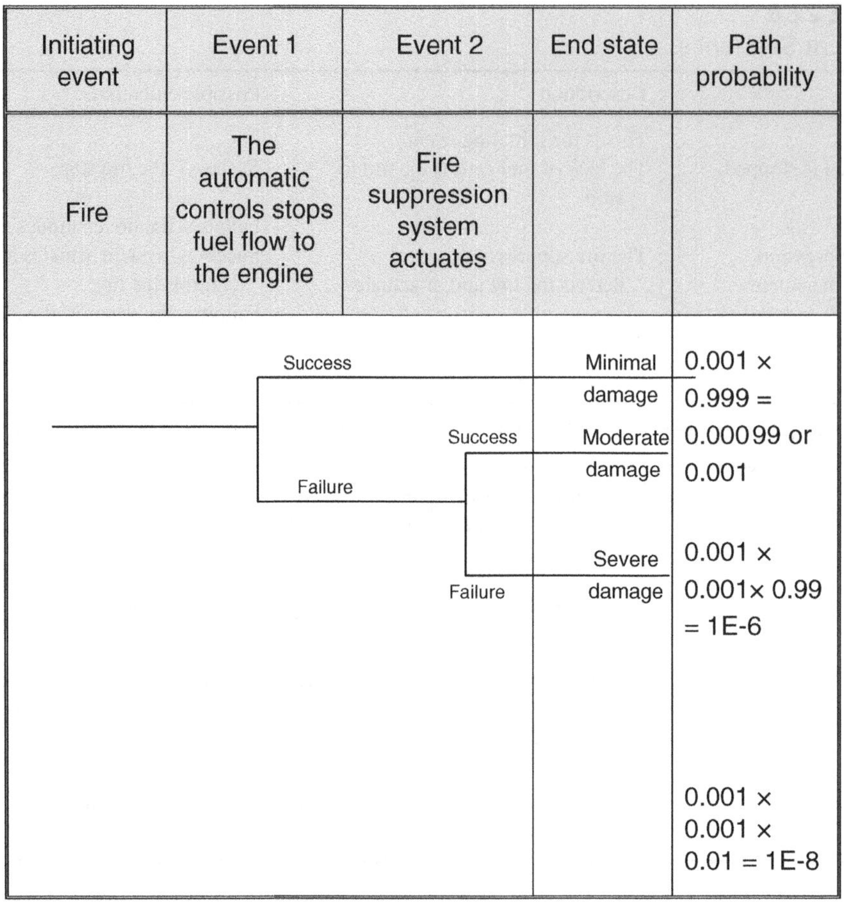

Initiating event	Event 1	Event 2	End state	Path probability
Fire	The automatic controls stops fuel flow to the engine	Fire suppression system actuates		

Success — Minimal damage — $0.001 \times$

$0.999 =$

Success — Moderate damage — 0.00099 or

Failure — 0.001

Severe — $0.001 \times$

Failure — damage — 0.001×0.99

$= 1E-6$

$0.001 \times$

$0.001 \times$

$0.01 = 1E-8$

FIGURE 22.2 Event tree with path probabilities.

The analyst should provide a description of the system analyzed, as well as a discussion of the problem definition, a list of the assumptions, the fault tree model(s), lists of minimal cut sets, and an evaluation of the significance of the minimal cut sets. Any recommendations should also be presented. An example fault tree for the engine fire example is shown in Figure 22.3.

22.4 SUMMARY

This chapter discussed how common risk assessment techniques could be used to perform risk assessments of aviation-related activities. Most all of the risk assessment tools and techniques presented and discussed in this book can be applied to a wide range of aviation maintenance and inspection tasks. Also, they can of course be applied to marine, rail, and other transportation tasks.

TABLE 22.7
Event Sequence with Probabilities

Event	Description	Possible outcomes	Probability
Fire	This is the initiating event		0.001
Fuel feed stops	The automatic controls stops fuel flow to the engine	Success – stopping fuel flow stops fire	0.999
		Failure – fire continues	0.001
Fire suppression system actuates	The fire suppression system detects the fire, and it actuates	Success – system actuates and controls the fire	0.99
		Failure – system fails to control the fire	0.01

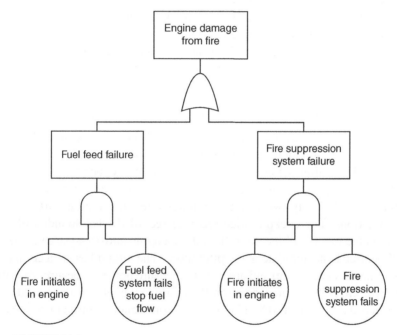

FIGURE 22.3   Example fault tree.

Self-Check Questions

1. Select a task or a procedure and analyze it using the methods discussed above.

2. Select an accident sequence from the literature and develop a risk model.

3. Develop a risk model using the accident scenario below.
   The following description of an accident involving an Air France Airbus A330-203 was adapted from several sources (3–5). This accident occurred on 31 May 2009.

The aircraft involved in the accident was an Air France Airbus A330-203, with manufacturer serial number 660, registered as F-GZCP.

The aircraft departed normally from Rio de Janeiro-Galeão International Airport on 31 May 2009 at 19 : 29 local time (22 : 29 UTC). The scheduled arrival at Paris Charles de Gaulle Airport was at 11 : 03 (09 : 03 UTC) the following day, after an estimated flight time of 10 : 34. The last voice contact with the aircraft was at 01 : 35 UTC, three hours and six minutes after the 22 : 29 UTC departure, when it reported that it had passed waypoint INTOL (1°21′39″S 32°49′53″W), located 565 km (351 miles) off Natal, on Brazil's northeastern coast. The aircraft left Brazilian Atlantic radar surveillance at 01 : 49 UTC and entered a communication dead zone.

The Airbus A330 is of modern design, with a glass cockpit. Though the A330-203 is normally to be flown by a crew of two pilots, this flight was crewed by three pilots, a captain and two first officers. This is because the 13-hour "duty time" (flight duration, plus preflight preparation) for the Rio–Paris route exceeds the maximum 10 hours permitted by Air France's procedures for pilots to operate an aircraft without a break. With three pilots on board, each of them can take a rest during the flight, and for this purpose the A330 has a rest cabin, situated just behind the cockpit.

Captain Dubois had sent one of the copilots for the first rest period with the intention of taking the second break himself. At 01 : 55 UTC, he woke first officer Robert and said: " … he's going to take my place." The captain left the cockpit to rest at 02 : 01 : 46 UTC after having attended the briefing between the two copilots. At 02 : 06 UTC, the pilot warned the cabin crew that they were about to enter an area of turbulence. Probably two to three minutes after this, the aircraft encountered icing conditions. The cockpit voice recorder recorded what sounded like hail or grapple on the outside of the aircraft, and the engine anti-ice system came on. Ice crystals started to accumulate in the pitot tubes. The pitot tubes are a component of the flight instrument system and are used to measure how fast the aircraft is moving through the air. Bonin turned the aircraft slightly to the left and decreased its speed from Mach 0.82 to Mach 0.8. This is the recommended "turbulence penetration speed."

At 02 : 10 : 05 UTC the autopilot disengaged because the blocked pitot tubes were no longer providing valid airspeed information, and the aircraft transitioned from normal rules to alternate rules. The engines' autothrust systems disengaged three seconds later. Without the autopilot, the aircraft started to roll to the right due to turbulence, and Bonin reacted by deflecting his side-stick to the left. One consequence of the change to alternate rules was an increase in the aircraft's sensitivity to roll, and the pilot's input overcorrected for the initial upset. During the next 30 seconds, the aircraft rolled alternately left and right as Bonin adjusted to the altered handling characteristics of his aircraft. At the same time he abruptly pulled up on his side-stick, raising the nose. This action was unnecessary and excessive under the circumstances. The aircraft's stall warning sounded briefly twice due to the angle of attack tolerance being exceeded, and the aircraft's recorded airspeed dropped sharply from 274 knots (507 km h^{-1}; 315 mph) to 52 knots (96 km h^{-1}; 60 mph). The aircraft's angle of attack increased, and the aircraft started to climb above its cruising level of FL350. By the time the pilot had control of the aircraft's roll, it was climbing at nearly 7000 ft min^{-1} (36 m s^{-1}) (for comparison, typical normal rate of climb for modern airliners is only 2000–3000 ft min^{-1} (10–15 m s^{-1}) at sea level and much smaller at high altitude).

At 02 : 10 : 34 UTC, after displaying incorrectly for half a minute, the left-side instruments recorded a sharp rise in airspeed to 223 knots (413 km h^{-1}; 257 mph), as did the integrated standby instrument system (ISIS) 33 seconds later. The right-side instruments are not recorded by the recorder. The icing event had lasted for just over a minute. The pilot continued making nose-up inputs. The trimmable horizontal stabilizer (THS) moved from three to 13° nose up in about one minute and remained in that latter position until the end of the flight.

At 02 : 11 : 10 UTC, the aircraft had climbed to its maximum altitude of around 38 000 ft (12 000 m). There, its angle of attack was 16°, and the engine thrust levers were in the fully forward takeoff/go-around (TOGA) detent. As the aircraft began to descend, the angle of attack rapidly increased toward 30°. A second consequence of the reconfiguration into alternate rules was that stall protection no longer operated. Whereas in normal rules, the aircraft's flight management computers would have acted to prevent such a high angle of attack, in alternate rules this did not happen. The wings lost lift and the aircraft stalled.

In response to the stall, first officer Robert took over control and pushed his control stick forward to lower the nose and recover from the stall; however, Bonin was still pulling his control stick back, lifting the nose further up. The inputs canceled each other out.

At 02 : 11 : 40 UTC, Captain Dubois reentered the cockpit after being summoned by first officer Robert. Noticing the various alarms going off, he urgently asked the two crew members: "What the hell are you doing?" The angle of attack had then reached 40°, and the aircraft had descended to 35 000 ft (11 000 m) with the engines running at almost 100% N1. This is the rotational speed of the front intake fan, which delivers most of a turbofan engine's thrust. The stall warnings stopped, as all airspeed indications were now considered invalid by the aircraft's computer due to the high angle of attack. In other words, the aircraft had its nose above the horizon but was descending steeply. Roughly 20 seconds later, at 02 : 12 UTC, the pilot decreased the aircraft's pitch slightly, airspeed indications became valid, and the stall warning sounded again; it then sounded intermittently for the remaining duration of the flight but stopped when the pilot increased the aircraft's nose-up pitch. From there until the end of the flight, the angle of attack never dropped below 35°. From the time the aircraft stalled until its impact with the ocean, the engines were primarily developing either 100% N1 or TOGA thrust, although they were briefly spooled down to about 50% N1 on two occasions. The engines always responded to commands and were developing in excess of 100% N1 when the flight ended. First officer Robert responded with: "We've lost all control of the aeroplane, we don't understand anything, we've tried everything" and then: "Climb climb climb climb." When Bonin replied: "But I've been at maximum nose-up for a while!" captain Dubois realized that Bonin was causing the stall, causing him to shout: "No no no, don't climb!" However, the aircraft was now too low to recover from the stall. Shortly thereafter, the ground proximity warning system sounded an alarm, warning the crew about the aircraft's now imminent crash with the ocean. Bonin, realizing the situation was now hopeless, said: "F#@#! We're going to crash! This can't be true. But what's happening?" The last CVR recording was Captain Dubois saying: "[ten] degrees pitch attitude."

The flight data recordings stopped at $02:14:28$ UTC, or 3 hours 45 minutes after takeoff. At that point, the aircraft's ground speed was 107 knots (198 km h^{-1}; 123 mph), and it was descending at 10 912 ft min^{-1} (55.43 m s^{-1}) (108 knots (200 km h^{-1}; 124 mph) of vertical speed). Its pitch was 16.2° (nose up), with a roll angle of 5.3° left. During its descent, the aircraft had turned more than 180° to the right to a compass heading of 270°. The aircraft remained stalled during its entire 3 minutes 30 seconds descent from 38 000 ft (12 000 m). The aircraft crashed belly first into the ocean at a speed of 152 knots (282 km h^{-1}; 175 mph), comprising vertical and horizontal components of 108 knots (200 km h^{-1}; 124 mph) and 107 knots (198 km h^{-1}; 123 mph), respectively. The Airbus was destroyed on impact; all 228 passengers and crew on board were killed instantly by extreme trauma.

REFERENCES

1. Vesely, W., Dugan, J., Fragola, J. et al. (2002). *Fault Tree Handbook with Aerospace Applications*. National Aeronautics and Space Administration http://www.hq.nasa.gov/office/codeq/doctree/fthb .pdf (accessed 17 January 2010).
2. Modarres, M. (2006). *Risk Analysis in Engineering: Techniques, Tools, and Trends*, 1e. CRC Press. ISBN: 1574447947.
3. Bureau d'Enquêtes et d'Analyses (2012). Final Report On the accident on 1st June 2009 to the Airbus A330–203 registered F-GZCP operated by Air France flight AF 447 Rio de Janeiro – Paris.
4. Wikipedia. Air France Flight 447. Retrieved August 2018. https://en.wikipedia.org/wiki/Air_France_ Flight_447 (accessed February 2019).
5. BEA. On the accident on 1st June 2009 to the Airbus A330-203 registered F-GZCP operated by Air France flight AF 447 Rio de Janeiro - Paris. Final Report. https://www.bea.aero/docspa/2009/f-cp090601.en/pdf/f-cp090601.en.pdf (accessed February 2019).

CHAPTER 23

Risk Assessment and Community Planning

23.1 INTRODUCTION

On 24 December 2008, a natural gas leak caused an explosion and fire. It killed one person and injured five others, including one firefighter and a utility worker. The explosion also destroyed one house completely and severely damaged two others adjacent to the destroyed house. Several other houses in the neighborhood were damaged. The neighborhood is located in Rancho Cordova, California. Pacific Gas and Electric Company, the utility owner and operator, operate 42% of California's natural gas pipe lines. According to Pacific Gas and Electric Company, the property damage was $267 000 (1).

The incident can be traced back to a phone call that Pacific Gas and Electric Company received on their Customer Contact Center Hotline at 9 : 16 a.m. The phone call was made by a resident of 10716 Paiute Way who reported a gas odor outside of her house. The Customer Contact Center prepared a case ticket and contacted the Pacific Gas and Electric Dispatch Office as part of their normal procedures. A field technician received the message and headed toward Paiute Way to investigate the call. On arriving at the affected aforementioned residence, the technician's portable gas detector detected gas on her initial approach across the yard. The technician, equipped only with gas detectors suitable for indoor gas detection, contacted the dispatch office requesting equipment suitable for outdoor gas detection along with the assistance from the maintenance and construction department.

On meeting with the resident who had reported the leak, the technician learned that she also smelled gas at her neighbor's residence. The technician called her dispatch office to report the smell of gas at this new residence; they dispatched a field man, a leak investigator, and a foreman to the scene at 10 : 28 a.m.

Risk Assessment: Tools, Techniques, and Their Applications, Second Edition. Lee T. Ostrom and Cheryl A. Wilhelmsen.
© 2019 John Wiley & Sons, Inc. Published 2019 by John Wiley & Sons, Inc.
Companion website: www.wiley.com/go/Ostrom/RiskAssessment_2e

The technician then proceeded to 10712 Paiute Way. She did not detect any leaks inside the residence; however, the male resident of the house indicated that he knew where there was a natural gas leak in his neighbor's yard. He escorted the technician to 10708 Paiute Way. Once in the yard, the technician detected a strong natural gas smell originating from a small patch of dead grass.

The technician then proceeded to check the gas meters of all the houses along this stretch of Paiute Way. She did not detect any unusual gas meter readings that showed excessive gas flows. The technician then knocked on the door of 10708 Paiute Way. She did not receive an answer, so she parked her truck out front and waited for the arrival of the field man, leak investigator, and foreman.

The leak investigator needed to stop by the Pacific Gas and Electric service center to pick up the ionization detector (a type of detector used to locate outdoor natural gas leaks) but discovered a problem with his truck brakes. The leak investigator had to transfer all his equipment into another service truck delaying his departure by roughly an hour.

The foreman and leak technician finally arrived at Paiute Way 2 hours and 47 minutes after the technician had requested the assistance from dispatch. The technician relayed the information she had gathered and then left the scene. The field man arrived shortly thereafter.

They proceeded to track down the location of the leak with the ionization detector. They reported a reading of 60 000 ppm and then 80 000 ppm, and finally, the device flamed out and sounded an alarm. This revealed that the gas-to-air mixture was so rich that it could not support a flame, indicating very strong amounts of natural gas. A man from the neighborhood informed them that he remembered Pacific Gas and Electric Company fixing a gas leak in this general area before. The Pacific Gas and Electric employees then noticed two sunken spots in the grass, indicating that the area had indeed been excavated sometime in the past.

At 1 : 35 p.m., the leak investigator knocked on the door of 10708 Paiute Way (the house at which the technician could not earlier get an answer). After speaking with the residence of the house, the leak investigator closed the door and turned to head back to the leak when the house exploded. The explosion killed the resident and injured five others. The explosion was caused by gas leaking from the previous repair done by Pacific Gas and Electric Company in September 2006. The gas then migrated into the home at 10708 Paiute Way and ultimately exploded two years after the faulty repair.

A more recent example occurred on 13 September 2018, around 4 : 00 p.m. Eastern Standard Time. High-pressure natural gas was released into a low-pressure gas distribution, causing a series of explosions and fires in the northeast region of the Merrimack Valley in Massachusetts. This system was owned by Columbia Gas of Massachusetts. This overpressure killed one person and injured at least 21 other individuals, as well as destroying 5 homes in the city of Lawrence, Andover, and North

Andover. Several of these structures were destroyed by the natural gas explosion, and others were destroyed through fires ignited by gas-fueled appliances.

Four evacuation centers were set up for the residents to evacuate the area. The electrical power was shut down and nearby roads were closed. The low-pressure natural gas distribution system was shut down. This system was installed in the early 1900s and upgrades were made in the 1950s.

A work crew was working on a project on this particular day, and the pressure dropped about 0.25 in of water, so the workers increased the pressure in the system. They fully opened the system, which allowed the full flow of high-pressure gas to be released, which exceeded the maximum allowable pressure. Just minutes before the explosion and the fires, the Columbia Gas company received high-pressure alarms but had no way of closing or opening the valves. Residents started calling at 4 : 11 p.m.

Columbia Gas will replace all cast iron and bare steel piping in the affected neighborhood. "The new system will consist of high-pressure plastic mains with regulators at each service meter to reduce the line pressure from the main to the required pressure" (2).

These types of incidents were not all that uncommon between 2010 and 2016; according to FracTracker Alliance, they counted 230 explosions that cost $3.4 billion in damages. "Altogether, there were 470 injuries and 100 deaths. The total number of incidents were 4215. Of those, equipment failure accounted for 1438; corrosion failure made up 752 and, excavation damage was 406. Texas accounted for 1102 and California made up 297. The pipeline operators responsible for the most incidents resulting in death are Pacific Gas & Electric at 15, Washington Gas Light at 9, and Consolidated Edison Co. of New York at 8. The key variable, it notes, is population density" (3).

The US Department of Transportation's Pipeline and Hazardous Materials Safety Administration (PHMSA) in 2018 issued $89.9 million in grants for the safety programs to all 50 states, 5 US territories, and 8 federally recognized tribes and tribal organizations. According to the PHMSA, the risk of pipeline accidents has been declining, and the injuries or deaths due to accidents have fallen to about 10% every three years. The ecological damage due to spills has also decreased by 5% a year (4).

In the United States, they are far less common than in other countries. Some of the other recent natural gas explosions in other countries include the following:

2018: On October 9, a 36-in Enbridge natural gas pipeline exploded 13 km north of Prince George, British Columbia. About 1 million BC customers and 750 000 US customers were affected (5).

2013: A Sinopec Corp oil pipeline exploded in Huangdao, Qingdao, Shandong Province, on 22 November 2013. 55 people were killed (6).

2012: On 18 September 2012, 22 workers died when a gas leak from a Kinder Morgan pipeline at Reynosa, Tamaulipas, Mexico, sparked an explosion, which became a fireball that overtook workers running for their lives. Lead plaintiff Javier Alvarez del Castillo said, "They were engulfed in fire that burnt and singed every inch of skin from their head to their ankles, taking every bit of hair from their head, laying the plaintiffs 'skinless,' like skeletons bare to the bones, with in most cases only their footwear attached to the only portion of their body not reduced to skeleton." He blamed Kinder Morgan for not adding enough of the odorant methyl mercaptan to the gas. (Natural gas is odorless, so energy companies add the sulfur compound to make leaks smelly and therefore noticeable.) "A gas company may be liable if facts show that it fails to act reasonably after having notice of defects in the pipes through which gas flows," the ruling states, citing the Texas appellate court case Entex, a division of NorAm Energy Corp. v. Gonzalez (7). On 13 July 2016, a US Federal Court ruled that only Kinder Morgan and not any of the other companies originally sued by plaintiffs' groups should face the charges of gross negligence and negligence (8). The cause of the leak was a valve that apparently failed as workers performed routine testing, but gaps remain in what is known about the events that led up to the Reynosa explosion (9).

Risk assessment should be a major factor as communities plan for growth and development, as well as for planning for emergencies. Risks at the community level can include but are not limited to the following:

- Chemical releases and catastrophic fires:
    - Chemical plant and other process plant explosions.
    - Train derailments.
    - Truck accidents.
    - Multiple structure fires.
    - Wild land fires.
- Natural disasters:
    - Floods.
    - Tornados.
    - Hurricanes.
    - Dust storms.
    - Earthquakes.
    - High winds.
    - Heavy snow and blizzards.
    - Heat waves or extreme cold.
- Transportation disruption:
    - Highway accidents.
    - Commuter train derailment.

  ◦ Transportation workers' strike.
  ◦ Power outages.
- Widespread power outages.
- Airplane crash.
- Loss of employers.
- Loss of workforce.
- Loss of tax base.

A good example of a repeating natural disaster is the flooding of the Red River. The Red River that flows from North Dakota into Manitoba and regularly floods the towns of Fargo (ND), Oakton (MN), and Winnipeg (Manitoba) is a good example of river system where floods occur frequently. Major floods have affected the river system multiple times over the past 110 years. Floods have affected the city of Fargo, ND, in 1997, 2008, and 2011. However, residents continue to live in Fargo and businesses still remain open in the flood-prone areas (10).

Risk assessment techniques can and should be applied to the planning and development of communities to reduce the risks associated with natural and man-made disasters. The following section presents two examples of how these techniques can be applied to community planning to reduce the potential for disasters.

23.2 EXAMPLE ANALYSIS

In this example, a corporation is planning on locating a new factory in one of several locations within a city. A general map of Medium City is shown in Figure 23.1. Approximately 50 000 people live in this city. The city has the following:

- Three fire stations
- One hospital
- A community college
- Two grade schools
- One high school
- One junior high school
- Three minor factories
- A train yard
- A shopping mall

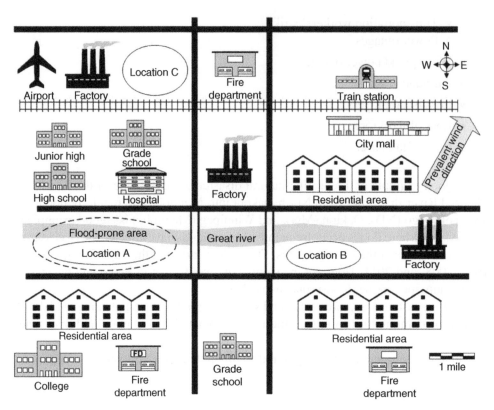

FIGURE 23.1   General layout of the Medium City.

These are located on the map. The bulk of the city sits more than 12 ft above the river, but the western side of the city is only 6 ft above the river. The river flows west to east. In addition to the flood hazard, the city can experience tornados, high winds, thunderstorms, and hail storms and has a moderate earthquake risk. The prevalent wind direction is from southwest to northeast.

The factory will manufacture food flavorings. It will employ 2000 workers. It will use four chemicals to manufacture the flavorings. These chemicals are listed in Table 23.1.

There are several potential risks associated with the placement of the factory. The risks are both to the city from the factory and to the factory from the location in the city.

The risks to the factory from the placement of the factory in the city include but are not limited to the following:

- Flooding hazard.
- Weather hazards.
- Response time from the fire department if the travel distance is too great.

TABLE 23.1

List of Chemicals

Chemical	Hazardous nature	Quantity to be stored on site
Chemical A	Highly flammable, high vapor pressure (gas at room temperature), highly toxic	10 000 gal
Chemical B	Combustible, solid at room temperature, highly toxic	20 000 lb
Chemical C	Combustible, liquid at room temperature, low toxicity	5000 gal
Chemical D	Nonflammable, solid at room temperature, nontoxic	10 000 lb

The risks to the city from the factory include but are not limited to the following:

- Fire and explosion hazards
- Toxic releases
- Traffic increases

Also listed on the map are the three possible sites for the new plant. For each location, risk assessment tools can be used to help elicit the risks for each potential site. The site with the least risks both to the city from the factory and to the factory from the site can be selected once the risks are determined. Preliminary hazard analysis (PHA) is a tool that can be used in this process.

23.2.1 Preliminary Hazard Analysis for Site A

Table 23.2 lists the hazards for site A.

23.2.2 Preliminary Hazard Analysis for Site B

Table 23.3 lists the hazards for site B.

23.2.3 Preliminary Hazard Analysis for Site C

Table 23.4 lists the hazards for site C.

From these analyses, site C is selected. It poses the least amount of risk to the factory from the city and from the factory to the city.

Other risk assessment tools can also be used. For instance, failure mode and effects analysis (FMEA), event trees, and fault trees along with other methods can be used to perform these analyses. Also, a probabilistic analysis can be added, if probabilities can be assigned to the various events.

TABLE 23.2
Preliminary Hazard Analysis for Site A

Hazard	Potential event	Probable cause	Relative probability of occurrence	Preventative actions
Flooding	Factory is flooded	Heavy rain or snow melt causes Great River to flood above 6 ft	Great River exceeds 10 ft very 20 years, but has not been >12 ft	Build dike around proposed plant that is higher than 10 ft
Toxic chemicals	Chemical release from proposed factory and the wind directs the cloud toward schools, hospital, and other factory location	Process upset releases a cloud of chemicals	Chemical releases occur approximately once every five years due to process upsets. Most releases are small	Increase safety measures to reduce potential of process upsets
Toxic chemicals	A liquid or solid chemical spill could inadvertently contaminate the Great River	A spill due to a tank or storage container rupture, process upset, or delivery truck spill could occur, and gravity or rainwater could carry the material to the river	Chemical releases occur approximately once every five years due to process upsets. Most releases are small	Build containment dike around planned facility to contain potential spills
Fire	Fire occurs and spreads quickly through the facility. Fire department response time at this site is approximately five minutes	Fire occurs due to process upset or maintenance actions	Fires occur within similar facilities approximately once every 10 years	Increase number of fixed fire suppression systems and train a company fire brigade
Traffic	Traffic congestion around site hinders movement of workers into and out of the site. Traffic can increase travel time by 15 minutes	Traffic in this area is heavy because of schools and residential area	Traffic studies have shown traffic to be high 200 of 250 work days	Pay city to add capacity to the roads
Transportation of materials in and out of the facility	Material traffic into and out of the plant could be affected by traffic patterns	Traffic in this area is heavy because of schools and residential area	Traffic studies have shown traffic to be high 200 of 250 work days	Pay city to add capacity to the roads

TABLE 23.3

Preliminary Hazard Analysis for Site B

Hazard	Potential event	Probable cause	Relative probability of occurrence	Preventative actions
Flooding	Factory is flooded	Heavy rain or snow melt causes Great River to flood above 12 ft	Great River exceeds 10 ft every 20 years, but has not been >12 ft	Build low dike around proposed plant that is approximately 2 ft in height to further reduce potential for flooding
Toxic chemicals	Chemical release from proposed factory and the wind directs the cloud toward city mall and residential area	Process upset releases a cloud of chemicals	Chemical releases occur approximately once every five years due to process upsets. Most releases are small	Increase safety measures to reduce potential of process upsets
Toxic chemicals	A liquid or solid chemical spill could inadvertently contaminate the Great River	A spill due to a tank or storage container rupture, process upset, or delivery truck spill could occur, and gravity or rainwater could carry the material to the river	Chemical releases occur approximately once every five years due to process upsets. Most releases are small	Build containment dike around planned facility to contain potential spills
Fire	Fire occurs and spreads quickly through the facility. Fire department response time at this site is approximately seven minutes	Fire occurs due to process upset or maintenance actions	Fires occur within similar facilities approximately once every 10 years	Increase number of fixed fire suppression systems and train a company fire brigade
Traffic	Traffic congestion around site hinders movement of workers into and out of the site. Traffic can increase travel time by five minutes	Traffic in this area is relatively heavy because of the residential areas	Traffic studies have shown traffic to be high 100 of 250 work days	Pay city to add capacity to the roads
Transportation of materials in and out of the facility	Material traffic into and out of the plant could be affected by traffic patterns	Traffic in this area is heavy because of the residential areas	Traffic studies have shown traffic to be high 100 of 250 work days	Pay city to add capacity to the roads

337

TABLE 23.4
Preliminary Hazard Analysis for Site C

Hazard	Potential event	Probable cause	Relative probability of occurrence	Preventative actions
Flooding	Factory is flooded	Heavy rain in the area causes localized flooding around the plant	Downpours from thunderstorms result in localized flooding every 10 years	Provide adequate stormwater drainage and catchment basin to collect stormwater
Toxic chemicals	Chemical release from proposed factory and the wind directs the farmland north and east of the proposed site location	Process upset releases a cloud of chemicals	Chemical releases occur approximately once every five years due to process upsets. Most releases are small	Increase safety measures to reduce potential of process upsets
Fire	Fire occurs and spreads quickly through the facility. Fire department response time at this site is approximately three minutes	Fire occurs due to process upset or maintenance actions	Fires occur within similar facilities approximately once every 10 years	Increase number of fixed fire suppression systems and train a company fire brigade
Traffic	Traffic congestion around site hinders movement of workers into and out of the site. Traffic can increase travel time by three to five minutes	Traffic in this area is heavy because of the airport	Traffic studies have shown traffic to be high, 250 of 250 work days, but only during periods when flights are arriving or departing from the local regional airport	Stagger shifts to work around heavy airport traffic
Transportation of materials in and out of the facility	Material traffic into and out of the plant could be affected by traffic patterns	Traffic in this area is heavy because of the airport	Traffic studies have shown traffic to be high, 250 of 250 work days	Stagger delivery times around flight schedules. Also, utilize the readily available railroad for deliveries

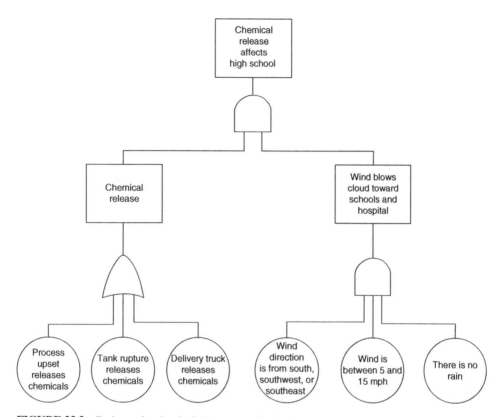

FIGURE 23.2   Fault tree for chemical release event for location A.

Figure 23.2 shows how a fault tree for a potential chemical release event from this proposed factory for location A might look. Figure 23.3 shows the fault tree for a similar event for location C. Using the probabilities in Table 23.5 for the various basic events, one can calculate the risk for a chemical release event that would affect the schools and the hospital shown on the map in Figure 23.1.

Next, the probability of the conditions existing that might cause a chemical release affecting a school or the hospital is calculated for both locations A and B. These calculations are shown in Table 23.6.

These calculations show that there is more than five times less likelihood that the wind would blow a chemical release from location C as from location A.

In this example, both the qualitative and quantitative risk assessments show that location C is the better location from a risk perspective.

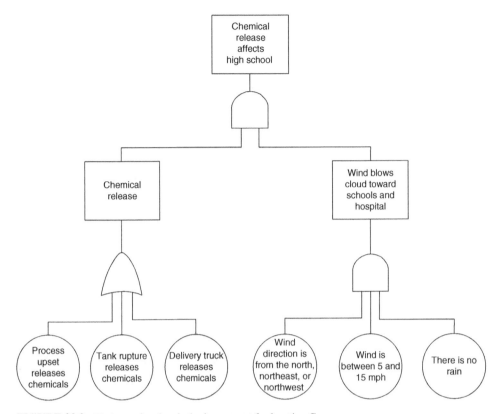

FIGURE 23.3 Fault tree for chemical release event for location C.

TABLE 23.5
Probabilities for Basic Events

Basic events	Probability
Process upset	One in five years or 1 every 1825 days
Storage tank rupture	One in ten years or 1 in every 3650 days
Delivery truck leaking	One in two years or 1 in 730 days
Wind blowing from south, southwest, or southeast	Ten month per year or 1 in 0.83 year (not blowing from this direction = 0.017)
Wind blowing between 5 and 15 mph	One month per year or 1 in 0.08
No rain	Nine month per year or 0.75

TABLE 23.6
Chemical Release Spill Calculations

Location	Probability
Location A	$(0.20 + 0.10 + 0.50) \times (0.83 \times 0.083 \times 0.75) = 0.041$ or 1 in 24 years
Location C	$(0.20 + 0.10 + 0.50) \times (0.16 \times 0.083 \times 0.75) = 0.0079$ or 1 in 126 years

Self-Check Questions

1. Should Risk assessment be a major factor in communities for their growth and development? If so what types of risks should be included?

2. What are the risks associated to the factory in the city? What risks are associated to the city from the factory?

3. What are some of the factors that, if they had been in place at Fukushima, could have prevented so many deaths?

23.3 SUMMARY

Every year hundreds of events occur of which a village, town, or city is affected by man-made or natural disasters. In many of these events, hundreds to thousands of people are affected. For instance, during 26 December 2004 tsunami in Indonesia, over 200 000 people died (11). If the coastal residents had lived away from the coast and not directly on it, tens of thousands of lives could have been saved. If the Fukushima reactors would have been built away from the coast and or better protected from tsunamis, then the potential for the reactors to be damaged would have been significantly less. A risk perspective must be applied to the placement of factories, schools, residential areas, waste facilities, nuclear and conventional power plants, and even highways to minimize the potential for loss of life and destruction of the environment.

REFERENCES

1. CBS San Francisco (2011). NTSB Blames PG&E Litany of Failures in San Bruno Explosion. http://sanfrancisco.cbslocal.com/2011/08/30/ntsb-reveals-final-san-bruno-pipe-explosion-report (accessed October 2018).
2. National Transportation Safety Board (2018). Pipeline Accident Report. https://www.ntsb.gov/investigations/AccidentReports/Reports/PLD18MR003-preliminary-report.pdf (accessed October 2018).
3. FRACTRACKER ALLIANCE (2018). Updated Pipeline Incident Analysis. https://www.fractracker.org/2016/11/updated-pipeline-incidents (accessed October 2018).
4. PHMSA (2018). U.S. Transportation Secretary Elaine L. Chao Announces $90 Million to Advance Pipeline and Hazardous Materials Safety Efforts Across the Nation. https://www.phmsa.dot.gov/news/us-transportation-secretary-elaine-l-chao-announces-90-million-advance-pipeline-and-hazardous (accessed October 2018).
5. Azpiri, J. (2018). Prince George pipeline explosion not 'criminal in nature': RCMP. *Global News* (10 October 2018). https://globalnews.ca/news/4538358/prince-george-pipeline-explosion-not-criminal-in-nature-rcmp (accessed October 2018).
6. REUTERS (2013). Chinese oil pipeline explosion. *China Daily* (22 November 2013). https://www.reuters.com/news/picture/chinese-oil-pipeline-explosion-idUSRTX15OM8#a=1 (accessed October 2018).

7. Langford, C. (2016). Kinder Morgan on hook for blast that killed 22. *Courthouse News Service* (15 July 2016). https://www.workerscompensation.com/news_read.php?id=24257&forgot=yes (accessed October 2018).

8. Hazardex (2016). Kinder Morgan must defend fatal Mexican gas pipeline explosion court case. *Hazardex* (19 July 2016). http://www.hazardexonthenet.net/article/121852/Kinder-Morgan-must-defend-fatal-Mexican-gas-pipeline-explosion-court-case.aspx (accessed October 2018).

9. Stevenson, M. (2012). Mexico blast a blow to Pemex's improving safety. *Deseret News* (20 September 2012). https://www.deseretnews.com/article/765605627/Mexico-blast-a-blow-to-Pemexs-improving-safety.html (accessed 7 January 2018).

10. NDSU (2011). The Fargo Flood Homepage. http://www.ndsu.edu/fargoflood (accessed October 2018).

11. Wikipedia (2011). 2004 Indian Ocean Earthquake and Tsunami. http://en.wikipedia.org/wiki/2004_Indian_Ocean_earthquake_and_tsunami (accessed October 2018).

CHAPTER 24

Threat Assessment

24.1 INTRODUCTION

The majority of this book has dealt with threats of one sort or another. This particular chapter deals with determining external threats to the safety and security of people and facilities. Threats must be eliminated or minimized once the threats are identified. However, this is difficult to do when the threats come from individuals who might still work for the organization and seem rational one day and who are on a killing spree the next day. During the first two months of 2018, 17 school shootings occurred in the United States (1). These are listed below:

4 January 2018 – Seattle, Washington: Shots were fired by an unidentified shooter into an administrative office at New Start High School. The school went into lockdown, though no injuries were reported.

10 January 2018 – Denison, Texas: A student at Grayson College picked up a gun, belonging to an advisor. She discharged the weapon, believing it was not loaded and shot through the wall. There were no injuries and no charges filed.

10 January 2018 – Sierra Vista, Arizona: A 14-year-old student at Coronado Elementary School shot himself in the bathroom at school. The shooting was initially reported as an active shooter, forcing the school into lockdown. The seventh grader was pronounced dead at the scene.

10 January 2018 – San Bernardino, California: Gunshots came through a window at California State University. The university immediately went into lockdown. No suspects were identified, and no injuries were reported.

15 January 2018 – Marshall, Texas: Gunshots were fired from a vehicle in the parking lot of a dorm at Wiley College. A bullet was found to have gone through a window into a dorm room, but none of the residents were injured.

Risk Assessment: Tools, Techniques, and Their Applications, Second Edition. Lee T. Ostrom and Cheryl A. Wilhelmsen.
© 2019 John Wiley & Sons, Inc. Published 2019 by John Wiley & Sons, Inc.
Companion website: www.wiley.com/go/Ostrom/RiskAssessment_2e

20 January 2018 – Winston-Salem, NC: A 21-year-old football player, Najee Ali Baker, was shot and critically injured following an altercation at a Wake Forest University party. Baker was taken to a hospital, where he succumbed to his injury.

22 January 2018 – Italy, Texas: A 16-year-old student at Italy High School opened fire with a semiautomatic handgun in the school cafeteria. The gunman wounded a fellow student, who later recovered, the AP reported. The school was placed under lockdown, following the incident.

22 January 2018 – Gentilly, Louisiana: An unknown person in a pickup truck drove past The NET Charter High School and fired shots at students standing in the campus parking lot. One boy was injured, but not by gunfire. Two students were arrested on suspicion and the school was temporarily placed into lockdown.

23 January 2018 – Benton, Kentucky: A 15-year-old student brought a handgun to Marshall County High School and opened fire in the school's atrium, leaving 2 dead and 17 injured, the Associated Press reported.

25 January 2018 – Mobile, Alabama: A disagreement between two 16-year-old students at Murphy High School escalated when one of them pulled out a handgun. The student with the handgun then fled school administrators, who tried to calm him, firing four or five times into the air. Nobody was injured. The suspect was taken into custody and charged for multiple offenses, including possession of a weapon on school property.

26 January 2018 – Dearborn, Michigan: A fight broke out during a basketball game at Dearborn High School. School officials removed the two persons involved, who were not students. Shots were later fired in the parking lot, but no injuries were reported.

31 January 2018 – Philadelphia, PA: A 32-year-old man was shot and killed outside of Lincoln High School during a basketball game. Police responded to reports of a fight and said at least three different weapons were discharged. The school was put into lockdown.

1 February 2018 – Los Angeles, California: A 12-year-old student at Salvador B. Castro Middle School was charged with negligent discharge of a firearm after a semiautomatic rifle she brought to school went off. Four students were injured, and the school was placed on lockdown for several hours.

5 February 2018 – Oxon Hill, Maryland: An Oxon Hill High School junior was shot twice in a school parking lot, during what police say was an attempted robbery. The 17-year-old later recovered. The shooting took place on school grounds but after school hours.

8 February 2018 – New York, New York: A 17-year-old student fired a gun, hitting the floor of a classroom at Metropolitan High School. Police took the student in custody, but no injuries were reported.

5 February 2018 – Maplewood, MN: A third grader at Harmony Learning Center pressed the trigger on a school liaison officer's gun. Although the weapon was

outfitted with a trigger guard, it discharged and hit the floor. No injuries were reported.

14 February 2018 – Parkland, Florida: A 19-year-old former student entered Marjory Stoneman Douglas High School and opened fire, killing 17 and injuring multiple others. The school was put into lockdown. The incident has been ranked as one of the top 10 deadliest mass shootings in modern US history.

Whether this is a larger or lower than normal is not the point. The point is properly assessing the threats for these types of events and taking action to prevent them from happening. School shootings are not just the problem. Workplace shooting also occurs frequently. At least two occurred during the same period of time.

On January 28, Timothy Smith who was wearing body armor killed four people at a car wash in Melcroft, PA (2).

On February 6, Vernest Griffen killed two people at two separate businesses (3).

During the time of writing this chapter on 28 June 2018, Jarrod Ramos killed five journalists and staff members at the Capital Gazette newspaper offices in Annapolis, Maryland (4).

In fact, Occupational Safety and Health Administration (OSHA) states that 403 of the 4679 fatal accidents at work reported in 2014 were homicides (5). Homicide is currently the fourth leading cause of fatal occupational injuries in the United States.

However, OSHA defines workplace violence much more broadly than just physical harm. OSHA states: "Workplace violence is any act or threat of physical violence, harassment, intimidation, or other threatening disruptive behavior that occurs at the work site. It ranges from threats and verbal abuse to physical assaults and even homicide. It can affect and involve employees, clients, customers and visitors."

24.2 SCHOOL VIOLENCE

School violence is a complex issue, to say the least. The Center for Disease Control (CDC) states that "Most educators and education researchers and practitioners would agree that school violence arises from a layering of causes and risk factors that include (but are not limited to) access to weapons, media violence, cyber abuse, the impact of school, community, and family environments, personal alienation, and more" (6). To say that school violence was not a problem in the past is not true. Growing up there were fights at my schools on a daily basis in the 1960s and 1970s. In fact, my grandfather was kicked out of school for beating up his school principal in the early 1900s. However, the level of violence has increased. Guns are now used more and more and the targets appear to be much more random. In the past two people would arrange a fist fight. Now, it could be one or both have guns and blast away, potentially killing each other and accidently, or purposely, killing others.

24.2.1 Risk Factors for the Perpetration of Youth Violence

The National Association of School Psychologists (NASP) state (6): "There is NO profile of a student who will cause harm. There is no easy formula or profile of risk factors that accurately determines whether a student is going to commit a violent act. The use of profiling increases the likelihood of misidentifying students who are thought to pose a threat. Most students who pose a substantive threat indicate their intentions in some way. Examples include statements to friends, ideas in written work, drawings, and postings on social media that threaten harm."

On the other hand, the CDC states the following concerning factors for youth violence (7):

> Research on youth violence has increased our understanding of factors that make some populations more vulnerable to victimization and perpetration. Risk factors increase the likelihood that a young person will become violent. However, risk factors are not direct causes of youth violence; instead, risk factors contribute to the likelihood of youth violence occurring.

> Research associates the following risk factors with perpetration of youth violence.

Individual Risk Factors

- History of violent victimization.
- Attention deficits, hyperactivity, or learning disorders.
- History of early aggressive behavior.
- Involvement with drugs, alcohol, or tobacco.
- Low IQ.
- Poor behavioral control.
- Deficits in social cognitive or information-processing abilities.
- High emotional distress.
- History of treatment for emotional problems.
- Antisocial beliefs and attitudes.
- Exposure to violence and conflict in the family.

Family Risk Factors

- Authoritarian child-rearing attitudes.
- Harsh, lax, or inconsistent disciplinary practices.
- Low parental involvement.
- Low emotional attachment to parents or caregivers.
- Low parental education and income.
- Parental substance abuse or criminality.

- Poor family functioning.
- Poor monitoring and supervision of children.

Peer and Social Risk Factors

- Association with delinquent peers.
- Involvement in gangs.
- Social rejection by peers.
- Lack of involvement in conventional activities.
- Poor academic performance.
- Low commitment to school and school failure.

Community Risk Factors

- Diminished economic opportunities.
- High concentrations of poor residents.
- High level of transiency.
- High level of family disruption.
- Low levels of community participation.
- Socially disorganized neighborhoods.

24.2.2 Protective Factors for the Perpetration of Youth Violence

Protective factors buffer young people from the risks of becoming violent (7). These factors exist at various levels. To date, protective factors have not been studied as extensively or rigorously as risk factors. However, identifying and understanding protective factors are equally as important as researching risk factors. Studies suggest the following protective factors.

Individual Protective Factors

- Intolerant attitude toward deviance.
- High IQ.
- High grade point average (as an indicator of high academic achievement).
- High educational aspirations.
- Positive social orientation.
- Popularity acknowledged by peers.
- Highly developed social skills/competencies.
- Highly developed skills for realistic planning.
- Religiosity.

Family Protective Factors

- Connectedness to family or adults outside the family.
- Ability to discuss problems with parents.
- Perceived parental expectations about school performance are high.
- Frequent shared activities with parents.
- Consistent presence of parent during at least one of the following: when awakening, when arriving home from school, at evening mealtime, or going to bed.
- Involvement in social activities.
- Parental/family use of constructive strategies for coping with problems (provision of models of constructive coping).

Peer and Social Protective Factors

- Possession of affective relationships with those at school that are strong, close, and pro-socially oriented.
- Commitment to school (an investment in school and in doing well at school).
- Close relationships with nondeviant peers.
- Membership in peer groups that do not condone antisocial behavior.
- Involvement in prosocial activities.
- Exposure to school climates that characterized by:
  ○ Intensive supervision.
  ○ Clear behavior rules.
  ○ Consistent negative reinforcement of aggression.
  ○ Engagement of parents and teachers.

24.3 WORKPLACE VIOLENCE

According to OSHA workplace violence is any act or threat of physical violence, harassment, intimidation, or other threatening disruptive behavior that occurs at the worksite (5, 8). It ranges from threats and verbal abuse to physical assaults and even homicide. It can affect and involve employees, clients, customers, and visitors. Nearly two million American workers report having been victims of workplace violence each year. Unfortunately, many more cases go unreported. Research has identified factors that may increase the risk of violence for some workers at certain worksites. Such factors include exchanging money with the public and working with volatile, unstable people. Working alone or in isolated areas may also contribute to the potential for violence. Providing services and care and working where alcohol is served may also impact the likelihood of violence. Additionally, time of day and location of work, such as working late at night or in areas with high crime rates, are also risk factors that should be considered when addressing issues of workplace violence. Among those with higher risk are workers who exchange money with the public, delivery drivers, healthcare professionals, public service workers, customer service agents, law enforcement personnel, and those who work alone or in small groups.

In most workplaces where risk factors can be identified, the risk of assault can be prevented or minimized if employers take appropriate precautions. One of the best protections employers can offer their workers is to establish a zero-tolerance policy toward workplace violence. This policy should cover all workers, patients, clients, visitors, contractors, and anyone else who may come in contact with company personnel.

By assessing their worksites, employers can identify methods for reducing the likelihood of incidents occurring. OSHA believes that a well-written and implemented workplace violence prevention program, combined with engineering controls, administrative controls, and training, can reduce the incidence of workplace violence in both the private sector and federal workplaces.

This can be a separate workplace violence prevention program or can be incorporated into a safety and health program, employee handbook, or manual of standard operating procedures. It is critical to ensure that all workers know the policy and understand that all claims of workplace violence will be investigated and remedied promptly. In addition, OSHA encourages employers to develop additional methods as necessary to protect employees in high-risk industries.

Healthcare facilities are one of the more workplace violence prone areas (5). Some of the risk factors in healthcare facilities are as follows:

Patient, Client, and Setting-Related Risk Factors

- Working directly with people who have a history of violence, abuse drugs or alcohol, gang members, and relatives of patients or clients.
- Transporting patients and clients.
- Working alone in a facility or in patients' homes.
- Poor environmental design of the workplace that may block employees' vision or interfere with their escape from a violent incident.
- Poorly lit corridors, rooms, parking lots, and other areas.
- Lack of means of emergency communication.
- Prevalence of firearms, knives, and other weapons among patients and their families and friends.
- Working in neighborhoods with high crime rates.

Organizational Risk Factors

- Lack of facility policies and staff training for recognizing and managing escalating hostile and assaultive behaviors from patients, clients, visitors, or staff.

- Working when understaffed – especially during mealtimes and visiting hours.
- High worker turnover.
- Inadequate security and mental health personnel on-site.
- Long waits for patients or clients and overcrowded, uncomfortable waiting rooms.
- Unrestricted movement of the public in clinics and hospitals.
- Perception that violence is tolerated and victims will not be able to report the incident to the police and/or press charges.

24.4 THREAT ASSESSMENT PROCEDURES

Though the basic process for conducting a threat assessment is similar, the type of facility being assessed will dictate the type of threat assessment procedure being used.

24.4.1 School Threat Assessment

The NASP recommends the following procedures for assessing threats (6):

1. Establish district-wide policies and procedures.
2. Creating interdisciplinary assessment teams.
3. Educating the school community.

These are more detailed below:

Establish district-wide policies and procedures. All threats of violence must be taken seriously and investigated, so it is important to have specific policy and established procedures for dealing with student threats. The policy should clarify the role of educators in relation to that of law enforcement, identify the threat assessment team (TAT), and specify the team's training requirements.

Create an interdisciplinary assessment team. Effective threat assessment is based on the combined efforts of a school-based team including representatives from administration, school-employed mental health professionals, and law enforcement. In unusually complex cases, the team might draw upon professionals in the local community. The interdisciplinary team approach improves the efficiency and scope of the assessment process and reduces the risk of observer bias. The TAT model is best for assessing the potential for both school and workplace violence. A description of the TAT model will be provided later in this chapter.

Educate the school community about threat assessment. Implementation of a threat assessment approach hinges on educating the school community about the importance of a positive school climate that focuses on providing help for students before problems escalate into violence. Schools should regularly assess

their climate, with particular emphasis on students' trust in adults and willing-
ness to seek help for problems and concerns. All members of the community,
especially students, must understand the distinction between seeking help to
prevent violence and "snitching," or informing on someone for personal gain.
Written materials should be publicly available, and specific efforts should be
made to explain relevant aspects of the threat assessment policy to staff members, students, and families.

24.4.2 Workplace Threat Assessment

Recognizing the early signs of workplace violence is an important step in minimizing
or avoiding the impact. This article looks at early warning signs and some basic steps
to take in response.

Sometimes, like in cases of profit-motivated violence, there may be no clue that
violence is imminent. However, few violent acts in the workplace are totally unpredictable, the result of a person who just "snapped."

Most individuals who commit violence give some sort of warning. To maintain a
safe and productive work environment, you should learn to recognize these warning
signs.

Some early warning signs of workplace violence are:

- Serious and escalating conflict with clients or colleagues; conflicts in romantic
relationships.
- Romantic obsessions and extreme stress.
- Bizarre or suicidal thoughts or other signs of emotional or mental problems.
- Insulting, discriminatory comments, or behavior directed at specific people.
- Telling others about violent thoughts or fantasies; bringing weapons to work.
- Predicting problem that may lead to warning signs of workplace violence.
  Employee education in recognizing the key warning signs of workplace
  violence helps prepare a company to diffuse a potentially violent situation
  before it occurs.
  Knowing what behaviors to look for and how to report them are two powerful
  means companies have to predict people problems before they happen.

The following information is extracted from the OSHA Guidelines for Preventing
Workplace Violence for Healthcare and Social Service Workers (5). Please consult
the original report for a complete list of references. The guidelines provided here are
similar to how any good safety program is organized and can be expanded for use at
other business types as well.

OSHA recommends a written program for workplace violence prevention. When
incorporated into an organization's overall safety and health program, OSHA offers
an effective approach to reduce or eliminate the risk of violence in the workplace. The

building blocks for developing an effective workplace violence prevention program include:

1. Management commitment and employee participation.
2. Worksite analysis.
3. Hazard prevention and control.
4. Safety and health training.
5. Recordkeeping and program evaluation.

A violence prevention program focuses on developing processes and procedures appropriate for the workplace in question.

Specifically, a workplace's violence prevention program should have clear goals and objectives for preventing workplace violence, be suitable for the size and complexity of operations, and be adaptable to specific situations and specific facilities or units. The components are interdependent and require regular reassessment and adjustment to respond to changes occurring within an organization, such as expanding a facility or changes in managers, clients, or procedures. As with any occupational safety and health program, it should be evaluated and updated, if needed, on a regular basis. Several states have also passed legislation and developed requirements that address workplace violence.

24.4.3 Management Commitment and Worker Participation

Management commitment is essential to any safety and health-related policy and program. Effective management leadership begins by recognizing that workplace violence is a safety and health hazard.

Management commitment, including the endorsement and visible involvement of top management, provides the motivation and resources for workers and employers to deal effectively with workplace violence. This commitment should include:

- Acknowledging the value of a safe and healthful, violence-free workplace and ensuring and exhibiting equal commitment to the safety and health of workers and patients/clients.
- Allocating appropriate authority and resources to all responsible parties. Resource needs often go beyond financial needs to include access to information, personnel, time, training, tools, or equipment.
- Assigning responsibility and authority for the various aspects of the workplace violence prevention program to ensure that all managers and supervisors understand their obligations.
- Maintaining a system of accountability for involved managers, supervisors, and workers.
- Supporting and implementing appropriate recommendations from safety and health committees.

- Establishing a comprehensive program of medical and psychological counseling and debriefing for workers who have experienced or witnessed assaults and other violent incidents and ensuring that trauma-informed care is available.
- Establishing policies that ensure the reporting, recording, and monitoring of incidents and near misses and that no reprisals are made against anyone who does so in good faith.

Management must effectively communicate the policies and goals, provide sufficient resources, and uphold program expectations.

The TAT, to be discussed later in the chapter, should be implemented so that workers from different functions within the organization have a voice in the process (6, 9, 10). The range of viewpoints and needs should be reflected in committee composition. This involvement should include:

- Participation in the development, implementation, evaluation, and modification of the workplace violence prevention program.
- Participation in safety and health committees that receive reports of violent incidents or security problems, making facility inspections and responding to recommendations for corrective strategies.
- Providing input on additions to or redesigns of facilities.
- Identifying the daily activities that employees believe put them most at risk for workplace violence.
- Discussions and assessments to improve policies and procedures – including complaint and suggestion programs designed to improve safety and security.
- Ensuring that there is a way to report and record incidents and near misses and that issues are addressed appropriately.
- Ensuring that there are procedures to ensure that employees are not retaliated against for voicing concerns or reporting injuries.
- Employee training and continuing education programs.

24.4.4 Worksite Analysis and Hazard Identification

The tools described throughout this book can be employed to aid in workplace hazard assessments. A failure mode and effect analysis (FMEA), decision trees, and event trees are all tools that can be used for this purpose. These techniques are discussed in other chapters in this book. A worksite analysis involves a mutual step-by-step assessment of the workplace to find existing or potential hazards that may lead to incidents of workplace violence.

The TAT team should be utilized for conducting worksite analyses. Although management is responsible for controlling hazards, workers have a critical role to play in helping to identify and assess workplace hazards, because of their knowledge and familiarity with facility operations, process activities, and potential threats. Depending on the size and structure of the organization, the team may also include representatives from operations: employee assistance, security, occupational safety and health, legal, and human resources staff. The assessment should include a records review: a review of the procedures and operations for different jobs, employee surveys, and workplace security analysis.

Once the worksite analysis is complete, it should be used to identify the types of hazard prevention and control measures needed to reduce or eliminate the possibility of a workplace violence incident occurring. In addition, it should assist in the identification or development of appropriate training. The assessment team should also determine how often and under what circumstances worksite analyses should be conducted. For example, the team may determine that a comprehensive annual worksite analysis should be conducted but require that an investigative analysis occurs after every incident or near miss. The analyses should also be conducted after building modifications are performed.

Additionally, those conducting the worksite analysis should periodically inspect the workplace and evaluate worker tasks to identify hazards, conditions, operations, and situations that could lead to potential violence. The advice of independent reviewers, such as safety and health professionals, law enforcement or security specialists, and insurance safety auditors may be solicited to strengthen programs. These experts often provide a different perspective that serves to improve a program.

Information is generally collected through (i) records analysis, (ii) job hazard analysis (JHA), and (iii) employee surveys.

Records Analysis and Tracking

In general, personnel records are considered personally identifiable information (PII) and cannot be reviewed by just anyone in an organization. Records of certain types of events might fall under the PII umbrella, and human resource professionals need to be consulted to determine what can and cannot be reviewed by the TAT. Therefore, great sensitivity should be applied to records reviews relating to workplace violence.

Records review is important to identify patterns of assaults or near misses that could be prevented or reduced through the implementation of appropriate controls. Records review should include medical, safety, specific threat assessments, workers' compensation, and insurance records. The review should also include the OSHA Log of Work-Related Injuries and Illnesses (OSHA Form 300) if the employer is required to maintain one.

In addition, incident/near-miss logs, a facility's general event, or daily log and police reports should be reviewed to identify assaults relative to particular:

- Departments/units.
- Work areas.
- Job titles.
- Activities – such as transporting patients between units or facilities and patient intake.
- Time of day.

Job Hazard Analysis

A JHA is an assessment that focuses on job tasks to identify hazards. Details on conducting JHAs can be found in Roughton and Crutchfield (11). It is a common tool used in occupational safety and health. Through review of procedures and operations connected to specific tasks or positions to identify if they contribute to hazards related to workplace violence and/or can be modified to reduce the likelihood of violence occurring, it examines the relationship between the employee, the task, tools, and the work environment. Once again, this process should include worker participation and the TAT team. Priority for conducting JHAs should be given to:

- Jobs with high assault rates due to workplace violence.
- Jobs that are new to an operation or have undergone procedural changes that may increase the potential for workplace violence.
- Jobs that require written instructions, such as procedures for administering medicine and steps required for transferring patients.

After an incident or near miss, the analysis should focus on:

- Analyzing those positions that were affected.
- Identifying if existing procedures and operations were followed and if not, why not (in some instances, not following procedures could result in more effective protections).
- Identifying if staff were adequately qualified and/or trained for the tasks required.
- Developing, if necessary, new procedures and operations to improve staff safety and security.

Employee Surveys

OSHA recommends conducting employee surveys for eliciting information from employees. We have conducted numerous surveys. In Ostrom et al. (12), we detail how to conduct a safety culture survey, and this paper serves as the basis for how safety surveys should be conducted. There is a whole social science surrounding how surveys should be conducted. Please consult the literature as to how to properly develop and conduct an employee survey. Questionnaires and/or surveys are an effective way for employers to identify potential hazards that may lead to violent incidents, identify the types of problems workers face in their daily activities, and assess the effects of changes in work processes, again, if the survey is properly developed and administered. Detailed baseline screening surveys can help pinpoint tasks that put workers at risk. The periodic review process should also include feedback and follow-up.

Survey questions can be open-ended, short answer, multiple choice, or the agreement/disagreement type. The agreement/disagreement type of question is used most in the surveys we have developed. A Likert scale is normally used in these types of questions. The Likert scale can be traditionally a 5-point, 7-point, or 9-point scale. We have used the 7-point the most.

The following are sample questions types:

Examples of open-ended question:

- What daily activities, if any, expose you to the greatest risk of violence?
- What, if any, work activities make you feel unprepared to respond to a violent action?
- Can you recommend any changes or additions to the workplace violence prevention training you received?
- Can you describe how a change in a patient's daily routine affected the precautions you take to address the potential for workplace violence?

Examples of short answer questions:

- Which entrance to the facility do you feel poses the most risk?
- What time of the day do you feel has the most risk for violence?

Example of a multiple-choice question:

- How can management best protect the workforce?
  - (a) Hire guards
  - (b) Use card readers on entrances
  - (c) Have numerous cameras

Example of an agreement/disagreement question using a 5-point Likert scale:
Provide your level of agreement on the statements on the Likert scales provided.
Management listens to my concerns for workplace safety.

Agree	Somewhat agree	Neither agree nor disagree	Somewhat disagree	Disagree
1	2	3	4	5

I feel safe at work.

Agree	Somewhat agree	Neither agree nor disagree	Somewhat disagree	Disagree
1	2	3	4	5

24.4.5 Hazard Prevention and Control

After the systematic worksite analysis is complete, the employer should take the appropriate steps to prevent or control the hazards that were identified. To do this, the employer should (i) identify and evaluate control options for workplace hazards, (ii) select effective and feasible controls to eliminate or reduce hazards, (iii) implement these controls in the workplace, (iv) follow up to confirm that these controls are being used and maintained properly, and (v) evaluate the effectiveness of controls and improve, expand, or update them as needed.

In the field of occupational safety and health, these steps are generally categorized, in order of effectiveness, as (i) substitution, (ii) engineering controls, and (iii) administrative and work practice controls. These principles that are described in more detail below can also be applied to the field of workplace violence. In addition, employers should ensure that, if an incident of workplace violence occurs, post-incident procedures and services are in place and/or immediately made available.

Substitution

The best way to eliminate a hazard is to substitute the practice with a safer one. These substitutions may be difficult. However, substituting a different process might increase efficiency as well.

Engineering Controls and Workplace Adaptations to Minimize Risk

Engineering controls are physical changes that either remove the hazard from the workplace or create a barrier between the worker and the hazard. In facilities where it is appropriate, there are several engineering control measures that can effectively prevent or control workplace hazards. Engineering control strategies include (i) using physical barriers (such as enclosures or guards) or door locks to reduce employee exposure to the hazard, (ii) metal detectors, (iii) panic buttons, (iv) better or additional lighting/automatic lighting, and (v) more accessible exits (where appropriate). The measures taken should be site-specific and based on the hazards identified in the worksite analysis appropriate to the specific therapeutic setting. For example, closed-circuit videos and bulletproof glass may be appropriate in a hospital or other institutional setting, but not in a community care facility. Similarly, it should be noted that services performed in the field (e.g. home health or social services) often occur in private residences where some engineering controls may not be possible or appropriate. In these cases either substation or administrative controls should be used.

If new construction or modifications are planned for a facility, assess any plans to eliminate or reduce security hazards. Listed below are possible engineering controls:

- Readily accessible alarms – in some cases, silent alarms.
- Where possible, each room should have two exits.

- Designate a safe room.
- Arrange furniture so workers have a clear exit route.
- Metal detectors in high-risk locations – trained personnel must be at these stations.
- Monitored cameras.
- Curved mirrors at strategic locations.
- Glass panels in door/walls to observe passage ways.
- Enclosed reception areas with bulletproof glass.
- Deep counters to prevent someone from reaching across.
- Lock all unused doors.
- Where possible, secure furniture that could be used as a weapon.
- Pad or replace sharp edged furniture.
- Reduce or eliminate hallway obstructions like water fountains.
- Install effective lighting in all areas that personnel travel.
- Maintain lights.
- Ensure vehicles are properly maintained.

Administrative and Work Practice Controls

Administrative and work practice controls are appropriate when engineering controls are not feasible or not completely protective. These controls affect the way staff perform jobs or tasks. Changes in work practices and administrative procedures can help prevent violent incidents. As with engineering controls, the practices chosen to abate workplace violence should be appropriate to the type of site and in response to hazards identified.

In addition to the specific measures listed below, training for administrative and treatment staff should include therapeutic procedures that are sensitive to the cause and stimulus of violence. For example, research has shown that trauma-informed care is a treatment technique that has been successfully instituted in inpatient psychiatric units as a way to reduce patient violence and the need for seclusion and restraint. As explained by the Substance Abuse and Mental Health Services Administration, trauma-informed services are based on an understanding of the vulnerabilities or triggers of trauma for survivors and can be more supportive than traditional service delivery approaches, thus avoiding re-traumatization.

The following are possible administrative controls that could apply in different settings:

- There should be log-in and log-out procedures for employees and visitors.
- Supervise visitors in high-risk areas.
- Instruct employees to avoid confrontations.

- Have security personnel available when meeting with high-risk individuals.
- Ensure multiple employees are available when working with high-risk individuals.
- Use a buddy system or have security individuals available when working at night or going out vehicles at night.
- Require identification badges.
- Have employees remove potential weapons from their desks and open areas.
- Ensure there are well-planned and tested emergency procedures.
- Run drills periodically.

24.5 AFTER AN EVENT

As with any work-related incident with the potential for injuries, a series of actions need to be taken. Investigating such incidents of workplace violence thoroughly will provide a roadmap to avoiding fatalities and injuries associated with future incidents. The purpose of the investigation should be to identify the "root cause" of the incident. Root causes, if not corrected, will inevitably recreate the conditions for another incident to occur.

Again, as with all workplace-related incidents, the immediate first steps are to provide first aid and emergency care for the injured worker(s) and to take any measures necessary to prevent others from being injured. All workplace violence programs should provide comprehensive treatment for workers who are victimized personally or may be traumatized by witnessing a workplace violence incident. Injured staff should receive prompt treatment and psychological evaluation whenever an assault takes place, regardless of its severity – free of charge.

Also, injured workers should be provided transportation to medical care, if not available on-site.

Victims of workplace violence could suffer a variety of consequences in addition to their actual physical injuries. These may include:

- Short- and long-term psychological trauma.
- Fear of returning to work.
- Changes in relationships with coworkers and family.
- Feelings of incompetence, guilt, and powerlessness.
- Fear of criticism by supervisors or managers.

Consequently, a strong follow-up program for these workers will not only help them address these problems but also help prepare them to confront or prevent future incidents of violence.

Several types of assistance can be incorporated into the post-incident response. For example, trauma/crisis counseling, critical incident stress debriefing, or employee assistance programs may be provided to assist victims.

Whether the support is trauma-informed or not, certified employee assistance professionals, psychologists, psychiatrists, clinical nurse specialists, or social workers should provide this counseling. Alternatively, the employer may refer staff victims to an outside specialist. In addition, the employer may establish an employee counseling service, peer counseling, or support groups.

Counselors should be well trained and have a good understanding of the issues and consequences of assaults and other aggressive, violent behavior. Appropriate and promptly rendered post-incident debriefings and counseling reduce acute psychological trauma and general stress levels among victims and witnesses. In addition, this type of counseling educates staff about workplace violence and positively influences workplace and organizational cultural norms to reduce trauma associated with future incidents.

24.5.1 Investigation of Incidents

Once these immediate needs are taken care of, the investigation should begin promptly. The basic steps in conducting incident investigations are:

1. *Report as required.* Determine who needs to be notified, both within the organization and outside (e.g. authorities), when there is an incident. Understand what types of incidents must be reported and what information needs to be included. If the incident involves hazardous materials, additional reporting requirements may apply.

2. *Involve workers in the incident investigation.* The employees who work most closely in the area where the event occurred may have special insight into the causes and solutions.

3. *Identify root causes.* Identify the root causes of the incident. Don't stop an investigation at "worker error" or "unpredictable event." Ask "why" the patient or client acted, "why" the worker responded in a certain way, etc.

4. *Collect and review other information.* Depending on the nature of the incident, records related to training, maintenance, inspections, audits, and past incident reports may be relevant to review.

5. *Investigate near misses.* In addition to investigating all incidents resulting in a fatality, injury, or illness, any near miss (a situation that could potentially have resulted in death, injury, or illness) should be promptly investigated as well. Near misses are caused by the same conditions that produce more serious outcomes and signal that some hazards are not being adequately controlled or that previously unidentified hazards exist.

24.5.2 Safety and Health Training

Education and training are key elements of a workplace violence protection program and help ensure that all staff members are aware of potential hazards and how to protect themselves and their coworkers through established policies and procedures. Such training can be part of a broader type of instruction that includes protecting patients and clients (such as training on de-escalation techniques). However, employers should ensure that worker safety is a separate component that is thoroughly addressed.

Training for All Workers

Training can (i) help raise the overall safety and health knowledge across the workforce, (ii) provide employees with the tools needed to identify workplace safety and security hazards, and (iii) address potential problems before they arise and ultimately reduce the likelihood of workers being assaulted. The training program should involve all workers, including contract workers, supervisors, and managers. Workers who may face safety and security hazards should receive formal instruction on any specific or potential hazards associated with the unit or job and the facility. Such training may include information on the types of injuries or problems identified in the facility and the methods to control the specific hazards. It may also include instructions to limit physical interventions in workplace altercations whenever possible.

Every worker should understand the concept of "universal precautions for violence" – that is, that violence should be expected but can be avoided or mitigated through preparation. In addition, workers should understand the importance of a culture of respect, dignity, and active mutual engagement in preventing workplace violence.

New and reassigned workers should receive an initial orientation before being assigned their job duties. All workers should receive required training annually. In high-risk settings and institutions, refresher training may be needed more frequently, perhaps monthly or quarterly, to effectively reach and inform all workers. Visiting staff, such as physicians, should receive the same training as permanent staff and contract workers. Qualified trainers should instruct at the comprehension level appropriate for the staff. Effective training programs should involve role-playing, simulations, and drills.

Training Topics

In general, training should cover the policies and procedures for a facility as well as de-escalation and self-defense techniques. Both de-escalation and self-defense training should include a hands-on component. The following provides a list of possible topics:

- The workplace violence prevention policy.
- Risk factors that cause or contribute to assaults.

- Policies and procedures for documenting patients' or clients' change in behavior.

- The location, operation, and coverage of safety devices such as alarm systems, along with the required maintenance schedules and procedures.

- Early recognition of escalating behavior or recognition of warning signs or situations that may lead to assaults.

- Ways to recognize, prevent, or diffuse volatile situations or aggressive behavior, manage anger, and appropriately use medications.

- Ways to deal with hostile people other than patients and clients, such as relatives and visitors.

- Proper use of safe rooms – areas where staff can find shelter from a violent incident.

- A standard response action plan for violent situations, including the availability of assistance, response to alarm systems, and communication procedures.

- Self-defense procedures where appropriate.

- Progressive behavior control methods and when and how to apply restraints properly and safety when necessary.

- Ways to protect oneself and coworkers, including use of the "buddy system."

- Policies and procedures for reporting and recordkeeping.

- Policies and procedures for obtaining medical care, trauma-informed care, counseling, workers' compensation, or legal assistance after a violent episode or injury.

Training for Supervisors and Managers

Supervisors and managers must be trained to recognize high-risk situations, so they can ensure that workers are not placed in assignments that compromise their safety. Such training should include encouraging workers to report incidents and to seek the appropriate care after experiencing a violent incident.

Supervisors and managers should learn to reduce safety hazards and ensure that workers receive appropriate training. Following training, supervisors and managers should be able to recognize a potentially hazardous situation and make any necessary changes in the physical plant, patient care treatment program and staffing policy, and procedures to reduce or eliminate the hazards.

Training for Security Personnel

Security personnel need specific training from the hospital or clinic, including the psychological components of handling aggressive and abusive clients and ways to handle aggression and defuse hostile situations.

Evaluation of Training

The training program should also include an evaluation. At least annually, the team or coordinator responsible for the program should review its content, methods, and the frequency of training. Program evaluation may involve supervisor and employee interviews, testing, observing, and reviewing reports of behavior of individuals in threatening situations.

24.6 RECORDKEEPING AND PROGRAM EVALUATION

Recordkeeping and evaluation of the violence prevention program are necessary to determine its overall effectiveness and identify any deficiencies or changes that should be made.

Accurate records of injuries, illnesses, incidents, assaults, hazards, corrective actions, patient histories, and training can help employers determine the severity of the problem; identify any developing trends or patterns in particular locations, jobs, or departments; evaluate methods of hazard control; identify training needs; and develop solutions for an effective program. Records can be especially useful to large organizations and for members of a trade association that "pool" data. Key records include:

- *OSHA Log of Work-Related Injuries and Illnesses (OSHA Form 300)*. Covered employers are required to prepare and maintain records of serious occupational injuries and illnesses, using the OSHA 300 Log. As of January 2015, all employers must report (i) all work-related fatalities within 8 hours and (ii) all work-related inpatient hospitalizations, all amputations, and all losses of an eye within 24 hours. Injuries caused by assaults must be entered on the log if they meet the recording criteria.

- *Medical reports of work injury, workers' compensation reports and supervisors' reports for each recorded assault*. These records should describe the type of assault, such as an unprovoked sudden attack or patient-to-patient altercation, who was assaulted, and all other circumstances of the incident. The records should include a description of the environment or location, lost work time that resulted, and the nature of injuries sustained. These medical records are confidential documents and should be kept in a locked location under the direct responsibility of a healthcare professional.

- *Records of incidents of abuse, reports conducted by security personnel, verbal attacks or aggressive behavior that may be threatening*, such as pushing or shouting and acts of aggression toward other clients. This may be kept as part of an assaultive incident report. Ensure that the affected department evaluates these records routinely.

- *Information on patients with a history of past violence, drug abuse or criminal activity recorded on the patient's chart*. Anyone who cares for a potentially aggressive, abusive, or violent client should be aware of the person's background and history, including triggers and de-escalation responses. Log the admission of violent patients to help determine potential risks. Log violent events on patients' charts and flagged charts.

- *Documentation of minutes of safety meetings, records of hazard analyses and corrective actions recommended and taken.*
- *Records of all training programs, attendees, and qualifications of trainers.*

24.7 ELEMENTS OF A PROGRAM EVALUATION

As part of their overall program, employers should evaluate their safety and security measures. Top management should review the program regularly and, with each incident, evaluate its success. Responsible parties (including managers, supervisors, and employees) should reevaluate policies and procedures on a regular basis to identify deficiencies and take corrective action.

Management should share workplace violence prevention evaluation reports with all workers. Any changes in the program should be discussed at regular meetings of the safety committee, union representatives, or other employee groups.

All reports should protect worker and patient confidentiality either by presenting only aggregate data or by removing personal identifiers if individual data are used.

Processes involved in an evaluation include:

- Establishing a uniform violence reporting system and regular review of reports.
- Reviewing reports and minutes from staff meetings on safety and security issues.
- Analyzing trends and rates in illnesses, injuries, or fatalities caused by violence relative to initial or "baseline" rates.
- Measuring improvement based on lowering the frequency and severity of workplace violence.
- Keeping up-to-date records of administrative and work practice changes to prevent workplace violence to evaluate how well they work.
- Surveying workers before and after making job or worksite changes or installing security measures or new systems to determine their effectiveness.
- Tracking recommendations through to completion.
- Keeping abreast of new strategies available to prevent and respond to violence in the healthcare and social service fields as they develop.
- Surveying workers periodically to learn if they experience hostile situations in performing their jobs.
- Complying with OSHA and state requirements for recording and reporting injuries, illnesses, and fatalities.

- Requesting periodic law enforcement or outside consultant review of the work-site for recommendations on improving worker safety.

24.8 WORKPLACE VIOLENCE CHECKLISTS

These checklists can help you or your workplace violence/crime prevention committee evaluate the workplace and job tasks to identify situations that may place workers at risk of assault. It is not designed for a specific industry or occupation and may be used for any workplace. Adapt the checklist to fit your own needs. It is very comprehensive and not every question will apply to your workplace – if the question does not apply, either delete or write "N/A" in the Notes column. Add any other questions that may be relevant to your worksite.

24.8.1 Risk Factors for Workplace Violence

Cal/OSHA and National Institute for Occupational Safety and Health (NIOSH) have identified the following risk factors that may contribute to violence in the workplace (10). If you have one or more of these risk factors in your workplace, there may be a potential for violence.

	Yes	No	Notes/follow-up action
Do employees have contact with the public?			
Do they exchange money with the public?			
Do they work alone?			
Do they work late at night or during early morning hours?			
Is the workplace often understaffed?			
Is the workplace located in an area with a high crime rate?			
Do employees enter areas with a high crime rate?			
Do they have a mobile workplace (patrol vehicle, work van, etc.)?			
Do they deliver passengers or goods?			
Do employees perform jobs that might put them in conflict with others?			

	Yes	No	Notes/follow-up action
Do they ever perform duties that could upset people (deny benefits, confiscate property, terminate child custody, etc.)?			
Do they deal with people known or suspected of having a history of violence?			
Do any employees or supervisors have a history of assault, verbal abuse, harassment, or other threatening behavior?			
Other risk factors (please describe):			

24.8.2 Inspecting Work Areas

- Who is responsible for building security?_____
- Are workers told or can they identify who is responsible for security? Yes No

You or your workplace violence/crime prevention committee should now begin a "walk-around" inspection to identify potential security hazards. This inspection can tell you which hazards are already well controlled and what control measures need to be added. Not all of the following questions may be answered through simple observation. You may also need to talk to workers or investigate in other ways.

	All areas	Some areas	Few areas	No areas	Notes/follow-up action
Are nametags or ID cards required for employees (omitting personal information such as last name and home address)?					
Are workers notified of past violent acts in the workplace?					
Are trained security and counseling personnel accessible to workers in a timely manner?					

	All areas	Some areas	Few areas	No areas	Notes/follow-up action
Do security and counseling personnel have sufficient authority to take all necessary action to ensure worker safety?					
Is there an established liaison with state police and/or local police and counseling agencies?					
Are bullet-resistant windows or similar barriers used when money is exchanged with the public?					
Are areas where money is exchanged visible to others who could help in an emergency? (For example, can you see cash register areas from outside?)					
Is a limited amount of cash kept on hand, with appropriate signs posted?					
Could someone hear a worker who calls for help?					
Can employees observe patients or clients in waiting areas?					
Do areas used for patient or client interviews allow coworkers to observe any problems?					
Are waiting areas and work areas free of objects that could be used as weapons?					

	All areas	Some areas	Few areas	No areas	Notes/follow-up action
Are chairs and furniture secured to prevent their use as weapons?					
Is furniture in waiting areas and work areas arranged to prevent entrapment of workers?					
Are patient or client waiting areas designed to maximize comfort and minimize stress?					
Are patients or clients in waiting areas clearly informed how to use the department's services so they will not become frustrated?					
Are waiting times for patient or client services kept short to prevent frustration?					
Are private, locked restrooms available for employees?					
Is there a secure place for workers to store personal belongings?					

24.8.3 Inspecting Exterior Building Areas

	Yes	No	Notes/follow-up action
Do workers feel safe walking to and from the workplace?			
Are the entrances to the building clearly visible from the street?			
Is the area surrounding the building free of bushes or other hiding places?			

	Yes	No	Notes/follow-up action
Is lighting bright and effective in outside areas?			
Are security personnel provided outside the building?			
Is video surveillance provided outside the building?			
Are remote areas secured during off shifts?			
Is a buddy escort system required to remote areas during off shifts?			
Are all exterior walkways visible to security personnel?			

24.8.4 Inspecting Parking Areas

	Yes	No	Notes/follow-up action
Is there a nearby parking lot reserved for employees only?			
Is the parking lot attended or otherwise secured?			
Is the parking lot free of blind spots and is landscaping trimmed back to prevent hiding places?			
Is there enough lighting to see clearly in the parking lot and when walking to the building?			
Are security escorts available to employees walking to and from the parking lot?			

24.8.5 Security Measures

Does the workplace have:	In place	Should add	Does not apply	Notes/follow-up action
Physical barriers (Plexiglas partitions, bullet-resistant customer window, etc.)?				
Security cameras or closed-circuit TV in high-risk areas?				
Panic buttons?				
Alarm systems?				
Metal detectors?				
Security screening device?				
Door locks?				
Internal telephone system to contact emergency assistance?				
Telephones with an outside line programmed for 911?				
Two-way radios, pagers, or cellular telephones?				
Security mirrors (e.g. convex mirrors)?				
Secured entry (e.g. "buzzers")?				
Personal alarm devices?				
"Drop safes" to limit the amount of cash on hand?				
Broken windows repaired promptly?				
Security systems, locks, etc. tested on a regular basis and repaired promptly when necessary?				

24.8.6 Workplace Violence Prevention Program Assessment Checklist

Use this checklist as part of a regular safety and health inspection or audit to be conducted by the Health and Safety, Crime/Workplace Violence Prevention Coordinator, or joint labor/management committee. If a question does not apply to the workplace, then write "N/A" (not applicable) in the notes column. Add any other questions that may be appropriate.

	Yes	No	Notes
Staffing			
Is there someone responsible for building security?			
Who is it?			
Are workers told who is responsible for security?			
Is adequate and trained staffing available to protect workers who are in potentially dangerous situations?			
Are there trained security personnel accessible to workers in a timely manner?			
Do security personnel have sufficient authority to take all necessary action to ensure worker safety?			
Are security personnel provided outside the building?			
Is the parking lot attended or otherwise secure?			
Are security escorts available to walk employees to and from the parking lot?			
Training			
Are workers trained in the emergency response plan (for example, escape routes, notifying the proper authorities)?			
Are workers trained to report violent incidents or threats?			

	Yes	No	Notes
Are workers trained in how to handle difficult clients or patients?			
Are workers trained in ways to prevent or defuse potentially violent situations?			
Are workers trained in personal safety and self-defense?			
Facility design			
Are there enough exits and adequate routes of escape?			
Can exit doors be opened only from the inside to prevent unauthorized entry?			
Is the lighting adequate to see clearly in indoor areas?			
Are there employee-only work areas that are separate from public areas?			
Is access to work areas only through a reception area?			
Are reception and work areas designed to prevent unauthorized entry?			
Could someone hear a worker call for help?			
Can workers observe patients or clients in waiting areas?			
Do areas used for patient or client interviews allow coworkers to observe any problems?			
Are waiting and work areas free of objects that could be used as weapons?			
Are chairs and furniture secured to prevent their use as weapons?			
Is furniture in waiting and work areas arranged to prevent workers from becoming trapped?			
Are patient or client areas designed to maximize comfort and minimize stress?			
Is a secure place available for workers to store their personal belongings?			
Are private, locked restrooms available for staff?			

	Yes	No	Notes
Security measures – does the workplace have?			
Physical barriers (Plexiglas partitions, elevated counters to prevent people from jumping over them, bullet-resistant customer windows, etc.)?			
Security cameras or closed-circuit TV in high-risk areas?			
Panic buttons (portable or fixed)?			
Alarm systems?			
Metal detectors?			
X-ray machines?			
Door locks?			
Internal phone system to activate emergency assistance?			
Phones with an outside line programmed to call 911?			
Security mirrors (convex mirrors)?			
Secured entry (buzzers)?			
Personal alarm devices?			
Outside the facility			
Do workers feel safe walking to and from the workplace?			
Are the entrances to the building clearly visible from the street?			
Is the area surrounding the building free of bushes or other hiding places?			
Is video surveillance provided outside the building?			
Is there enough lighting to see clearly outside the building?			
Are all exterior walkways visible to security personnel?			

	Yes	No	Notes
Is there a nearby parking lot reserved for employees only?			
Is the parking lot free of bushes or other hiding places?			
Is there enough lighting to see clearly in the parking lot and when walking to the building?			
Have neighboring facilities and businesses experienced violence or crime?			

	Yes	No	Notes
Workplace procedures			
Are employees given maps and clear directions in order to navigate the areas where they will be working?			
Is public access to the building controlled?			
Are floor plans posted showing building entrances, exits, and location of security personnel?			
Are these floor plans visible only to staff and not to outsiders?			
Is other emergency information posted, such as the telephone numbers?			
Are special security measures taken to protect people who work late at night (escorts, locked entrances, etc.)?			
Are visitors or clients escorted to offices for appointments?			
Are authorized visitors to the building required to wear ID badges?			
Are identification tags required for staff (omitting personal information such as the person's last name and social security number)?			
Are workers notified of past violent acts by particular clients, patients, etc.?			

	Yes	No	Notes
Is there an established liaison with local police and counseling agencies?			
Are patients or clients in waiting areas clearly informed how to use the department's services so they will not become frustrated?			
Are waiting times for patient or client services kept short to prevent frustration?			
Are broken windows and locks repaired promptly?			
Are security devices (locks, cameras, alarms, etc.) tested on a regular basis and repaired promptly when necessary?			
Field work – staffing			
Are escorts or "buddies" provided for people who work in potentially dangerous situations?			
Is assistance provided to workers in the field in a timely manner when requested?			
Field work – training			
Are workers briefed about the area in which they will be working (gang colors, neighborhood culture, language, drug activity, etc.)?			
Can workers effectively communicate with people they meet in the field (same language, etc.)?			
Are people who work in the field late at night or early mornings advised about special precautions to take?			
Field work – work environment			
Is there enough lighting to see clearly in all areas where workers must go?			
Are there safe places for workers to eat, use the restroom, store valuables, etc.?			

	Yes	No	Notes
Are there places where workers can go for protection in an emergency?			
Is safe parking readily available for employees in the field?			
Field work – security measures			
Are workers provided two-way radios, pagers, or cellular phones?			
Are workers provided with personal alarm devices or portable panic buttons?			
Are vehicle door and window locks controlled by the driver?			
Are vehicles equipped with physical barriers (Plexiglas partitions, etc.)?			
Field work – work procedures			
Are employees given maps and clear directions for covering the areas where they will be working?			
Are employees given alternative routes to use in neighborhoods with a high crime rate?			
Does a policy exist to allow employees to refuse service to clients or customers (in the home, etc.) in a hazardous situation?			
Has a liaison with the police been established?			
Do workers avoid carrying unnecessary items that someone could use as weapon against them?			
Does the employer provide a safe vehicle or other transportation for use in the field?			
Are vehicles used in the field routinely inspected and kept in good working order?			
Is there always someone who knows where each employee is?			
Are nametags required for workers in the field (omitting personal information such as last name and social security number)?			
Are workers notified of past violent acts by particular clients, patients, etc.?			

	Yes	No	Notes
Field work – are special precautions taken when workers?			
Have to take something away from people (remove children from the home)?			
Have contact with people who behave violently?			
Use vehicles or wear clothing marked with the name of an organization that the public may strongly dislike?			
Perform duties inside people's homes?			
Have contact with dangerous animals (dogs, etc.)?			

24.9 TEAM-BASED THREAT ASSESSMENT APPROACH

Organization should develop a team-based threat assessment approach or TAT (6, 9, 13). This team that we will label as TAT should be an interdisciplinary team of company personnel that can properly assess threats to personnel and property.

The primary goals of the TAT are to assess the conditions, policies, and procedures of the organization in order to prevent or reduce the chances that a potentially violent situation will occur, and in the event of a threat, the TAT is responsible for:

- Ensuring that security is immediately provided to all affected parties.
- Acquiring the consultation and resources necessary for a comprehensive investigation.
- Investigating and assessing the risk posed by the circumstance.
- Planning and implementing a risk abatement action plan.
- Determining the appropriate interventions for both the subject and the target/s.
- Determining those factors that will prevent or mitigate an event in the future.

The members of the team should include personnel who can readily respond when someone there is a threat. The core group typically includes representatives from human resources, security, and employee assistance. Depending on the organization, the team might also include medical personnel, mental health professionals, community resource personnel, union/employee representatives, management, and supervisors. The team should be sensitive to individuals' legal and civil rights, confidentiality issues, and cultural issues and should represent the diversity of the clientele. A human resources or risk management representative is usually the central contact person who coordinates and convenes the TAT meetings, contacts resources, documents proceedings, and provides policy/liability information.

24.9.1 Risk Assessment and Abatement

The TAT should be the team that does the threat risk assessment. This assessment guides the TAT in evaluating the risk posed to the target/s and developing a list of viable options for managing both the immediate and long-term implications of the situation. It is important to consider the breadth of data when assessing the risk level of a situation, because individuals utter threats for many reasons – only some of which truly involve violent intentions. The team should clearly document the specific threat parameters if the risk assessment indicates evidence of conditions and behaviors consistent with the capacity for carrying out a threat. The risk abatement plan specifically details the actions that will be taken to adjust the current conditions and to reduce the potential for future violence. As stated previously in this chapter, threats should be eliminated or mitigated by one of the means stated in Section 24.4.4.

24.10 CONCLUSION

Threat for school and workplace violence is oh so real. The individuals who commit these sorts of crimes can come from all walks of life and lifestyles and commit these sorts of crimes for a variety of reasons. Using a team-based approach can help elucidate potential threats with the goal of eliminating or mitigating them. In the words of Sergeant Phil Esterhaus from Hill Street Blues, "Let's be careful out there."

Self-Check Question

1. Conduct a threat assessment of your school, workplace, or home. Discuss what you found and how mitigated any findings.

REFERENCES

1. CNBC (2018). 17 School shootings in 45 days – Florida massacre is one of many tragedies in 2018. https://www.cnbc.com/2018/02/14/florida-school-shooting-brings-yearly-tally-to-18-in-2018.html (accessed February 2018).
2. CNN (2018). Suspect in quadruple killing at car wash, dies. https://www.cnn.com/2018/01/29/us/pennsylvania-car-wash-shooting/index.html (accessed February 2018).
3. Detroit Free Press (2018). Sterling Heights man charged in workplace shooting in Pontiac. https://www.freep.com/story/news/local/michigan/oakland/2018/02/06/sterling-heights-man-charged-workplace-shooting-pontiac/311491002 (accessed February 2018).

4. The New York Times (2018). 5 People Dead in Shooting at Maryland's Capital Gazette Newsroom. https://www.nytimes.com/2018/06/28/us/capital-gazette-annapolis-shooting.html (accessed July 2018).
5. OSHA (2016). *Guidelines for Preventing Workplace Violence for Healthcare and Social Service Workers*, OSHA 3148-06R. Washington, DC: US Department of Labor, Occupational Safety and Health Administration.
6. National Association of School Psychologists (2018). Threat Assessment for School Administrators & Crisis Teams. https://www.nasponline.org/resources-and-publications/resources/school-safety-and-crisis/threat-assessment-at-school/threat-assessment-for-school-administrators-and-crisis-teams (accessed July 2018).
7. CDC (2018). Violence Prevention. https://www.cdc.gov/violenceprevention/youthviolence/index.html (accessed July 2018).
8. OSHA (2018). Workplace Violence. https://www.osha.gov/SLTC/workplaceviolence (accessed July 2018).
9. Nolan, Esq., J.J., Dinse, K., and McAndrew, P.C. (with contributions from Marisa Randazzo and Gene Deisinger). (2013). Implementing threat assessment and management best practices in the higher education workplace. *NACUA Annual Conference*, Philadelphia (19–22 June 2013).
10. CAL OSHA (2018). Workplace Violence Prevention Guidelines and Model Program. http://www.calhr.ca.gov/documents/model-workplace-violence-and-bullying-prevention-program.pdf (accessed August 2018).
11. Roughton, J. and Crutchfield, N. (2007). *Job Hazard Analysis: A Guide for Voluntary Compliance and Beyond*, 1e. Butterworth-Heinemann.
12. Ostrom, L.T., Wilhelmsen, C.A., and Kaplan, K. (1994). Assessing safety culture. *Journal of Nuclear Safety* 34 (2).
13. US Department of Justice Office of Community Oriented Policing Services (2009). Maximizing Effectiveness of Threat Assessment Teams Training Manual. US Government Report, Washington, DC.

CHAPTER **25**

Project Risk Management

25.1 INTRODUCTION

Project risk management (PRM) is the practice of minimizing the negative impacts of threats to projects and maximizing the upside impact of opportunities. Enterprise risk management (ERM), which will be discussed in Chapter 26, is the process of planning, organizing, leading, and controlling the activities of an organization in order to minimize the effects of **risk** on an organization's capital and earnings. Many similar tools are used for both processes and other types of risk assessment. This chapter will provide an overview of the PRM process and examples of how various tools are used. This chapter will not cover the overall project management process. Also, information for this chapter came from a variety of sources including the Department of Energy (DOE) Risk Management Guide (1), NASA Risk Management Handbook (2), and Project Management Institute's (PMI) website (3). Figure 25.1 shows a general diagram of the PRM process.

To begin the PRM process,

- Senior leadership commitment and participation is required.
- Stakeholder commitment and participation is required.
- Risk management is made a program-wide priority and "enforced" as such throughout the program's life cycle.
- Technical and program management disciplines are represented and engaged. Both program management and engineering specialties need to be communicating risk information and progress toward mitigation.
- Program management needs to identify contracting, funding concerns, and subject matter experts (SMEs) need to engage across the team and identify risks, costs, and potential ramifications, if the risk were to occur, as well as mitigation

Risk Assessment: Tools, Techniques, and Their Applications, Second Edition. Lee T. Ostrom and Cheryl A. Wilhelmsen.
© 2019 John Wiley & Sons, Inc. Published 2019 by John Wiley & Sons, Inc.
Companion website: www.wiley.com/go/Ostrom/RiskAssessment_2e

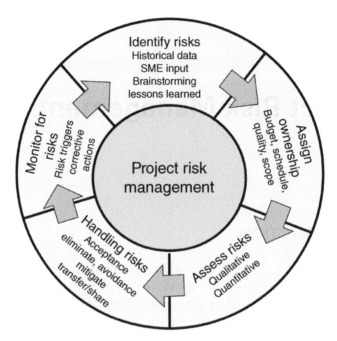

FIGURE 25.1   Project risk management process.

plans (actions to reduce the risk and cost/resources needed to execute successfully).

- Risk management integrated into the program's business processes and systems engineering plans. Examples are risk status included in management meetings and/or program reviews, risk mitigation plan actions tracked in schedules, and cost estimates reflective of risk exposure.

Mitre Corporation lists 21 PRM "musts" (4):

1. Risk management must be a priority for leadership and throughout the program's management levels. Maintain leadership priority and open communication. Teams will not identify risks if they do not perceive an open environment to share risk information (messenger not shot) or management priority on wanting to know risk information (requested at program reviews and meetings), or if they do not feel the information will be used to support management decisions (lip service, information not informative, team members will not waste their time if the information is not used).
2. Risk management must never be delegated to staff that lack authority.
3. A formal and repeatable risk management process must be present in one that is balanced in complexity and data needs, such that meaningful and actionable insights are produced with minimum burden.
4. The management culture must encourage and reward identifying risk by staff at all levels of program contribution.

5. Program leadership must have the ability to regularly and quickly engage SMEs.
6. Risk management must be formally integrated into program management.
7. Participants must be trained in the program's specific risk management practices and procedures.
8. A risk management plan (RMP) must be written with its practices and procedures consistent with process training.
9. Risk management execution must be shared among all stakeholders.
10. Risks must be identified, assessed, and reviewed continuously, not just prior to major reviews.
11. Risk considerations must be a central focus of program reviews.
12. Risk management working groups and review boards must be rescheduled when conflicts arise with other program needs.
13. Risk mitigation plans must be developed, success criteria defined, and their implementation monitored relative to achieving success criteria outcomes.
14. Risks must be assigned only to staff with authority to implement mitigation actions and obligate resources.
15. Risk management must never be outsourced.
16. Risks that extend beyond traditional impact dimensions of cost, schedule, and technical performance must be considered (e.g. programmatic, enterprise, cross-program/cross-portfolio, and social, political, and economic impacts).
17. Technology maturity and its future readiness must be understood.
18. The adaptability of a program's technology to change in operational environments must be understood.
19. Risks must be written clearly using the Condition-If-Then protocol.
20. The nature and needs of the program must drive the design of the risk management process within which a risk management tool/database conforms.
21. Risk management tool/database must be maintained with current risk status information; preferably, employ a tool/database that rapidly produces "dashboard-like" status reports for management.

25.2 PRM AND PROJECT LIFE CYCLE

A project life cycle is traditionally divided into four phases (3). Figure 25.2 shows the relationship of these phases. These are the following:

1. Initiation.
2. Planning.
3. Execution.
4. Closeout.

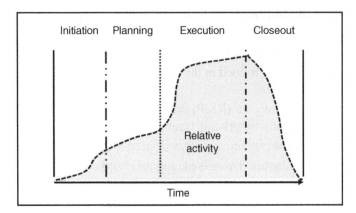

FIGURE 25.2 Project life cycles.

A set of PRM activities occur in each of these phases.

Initiation phase – If the project is, for example, a typical road construction project, then the risk history from other projects can be used to help determine the major risks that might be associated with the project. These could include the following:

- Road design issues.
- Weather.
- Contractor issues.
- Labor issues.
- Safety and health issues.
- Financing.
- Equipment availability.

However, if the project is, for instance, a research and development (R&D) project, then the major risks might be unknown. However, the basic risks will remain the same. Some are listed earlier. Unknown risks could include as follows:

- Available technical personnel.
- Laboratory space.
- Laboratory equipment.
- Graduate students' availability for university-related projects.

Brainstorming or Delphi sessions can help elucidate potential risks for the project (Chapter 9).

Planning phase – During this phase, initial task scheduling, cost estimating, long lead procurements are initiated, permitting, and actual contractor occurs. Each of these activities has risk associated with them. For instance, if the task scheduler or cost estimator is optimistic, then there is the risk of the project going over budget. Also, being optimistic about the weather can also cause schedules to be blown. Purchasing steel well in advance of a project can be risky in the age of tariffs. Permitting in certain areas of the country can have issues as well. During this phase, the risk breakdown structure (RBS) can be developed. The RBS will be discussed in detail later in this chapter.

Project execution phase – Of course, during the execution phase of the project is when the risks are generally realized. This is when weather, labor issues, missing equipment, lack of financing, changes in laws and regulations, and the other risks identified in the first two phases of the project raise their ugly heads. Risk monitoring and communication are very important during this phase. If proper mitigation and contingency plans were put in place, then impact on the project will be held in check. If not, well then, the project will join the hundreds of others that fail on a yearly basis. At the end of this chapter, there is the case study of the Panama Canal that discusses the risk issues that existed and were realized.

Closeout phase – During the closeout phase, the building, dam, aircraft carrier, or garden shed is turned over to the owner/operator. During this phase, it is important to ensure all the necessary paperwork, operating instructions and procedures, and maintenance information are delivered, and, also, to ensure any punch lists are satisfied. At the end of this phase, there should be no loose ends. The prime contractor should no longer have any non-warranty obligations. A colleague of mine was selling her house. After all the paperwork was signed, an inspection was conducted and signed off, and the new buyers took legal ownership. They tried to come back and now claim that things did not work. However, an inspection was performed and signed off. At that point, the previous owner has no legal or moral obligation to the new owners. Buyer, beware.

25.3 AN OVERVIEW OF PRM

As discussed earlier, risk management is performed in all phases of a project life cycle. The key points of PRM are listed in the following:

- Development of a PRM team.
- Risk planning.
- Risk assessment.
- Risk identification.
- Assignment of risk owner.
- Assignment of probabilities and consequences.
- Assignment of risk triggers.
- Risk register.
- Risk handling.

- Risk monitoring.
- RMP.
- Risk communications.

25.4 DEVELOPMENT OF A PRM TEAM

A risk management team (workgroup) is a separate and often independent unit within the project management team headed by the risk manager or the chief risk officer. It helps place a value on the project's activities (such as procuring, communicating, controlling quality, staffing, etc.). A well-organized team needs to be developed to perform PRM as with any other risk assessment effort. Obviously, the team should be scoped with the relative size of the project. For a very small project, it might be that the "team" is composed of only the project manager (PM). For a large project, there might be many people on the PRM team. The types of personnel who should be considered to be part of the PRM team are as follows:

- PM.
- Scheduler/cost estimator.
- Risk management specialist.
- Representatives of the crafts involved in the project.
- SMEs for each of the technical areas.
- Lawyers, if the project is of critical nature.

25.4.1 Risk Planning

The risk planning process should begin as early in the project life cycle as possible, usually in the initiation and planning phases of the project. Planning sets the stage and tone for risk management and involves many critical initial decisions that should be documented and organized for interactive strategy development.

Risk planning is conducted by the PRM team or the PM. Risk planning should establish methods to manage risks, including metrics and other mechanisms or determining and documenting modifications to those metrics and mechanisms. A communication structure should be developed to determine whether a formal risk management communication plan should be written and executed as part of the tailoring decisions to be made in regard to the project. Input to the risk planning process includes the project objectives, assumptions, mission need statement, customer/stakeholder expectations, and site office risk management policies and practices.

The team should also establish what resources, both human and material, would be required for successful risk management on the project. Further, an initial reporting structure and documentation format should also be established for the project.

Overall objectives for risk planning should:

Establish the overall risk nature of the project. Does the project entail hazardous materials or hazardous processes? Is this a project of national security interests or is it a garden shed?

Establish the overall experience and project knowledge desired of the PRM team.

An initial responsibility assignment matrix with roles and responsibilities for various risk management tasks should be developed. Through this responsibility assignment matrix, gaps in expertise should be identified and plans to acquire that expertise should be developed.

The result of the risk planning process is the RMP. The RMP ties together all the components of risk management – i.e. risk identification, analysis, and mitigation – into a functional whole. The plan is an integral part of the project plan that informs all members of the project team and stakeholders how risk will be managed and who will manage them throughout the life of the project. It should be part of the initial project approval package. A companion to the RMP is a risk register, which is updated continuously and used as a day-to-day guide by the project team.

25.4.2 Risk Assessment

In general, the same risk assessment techniques that are described throughout this text are the same that are used for PRM. However, the context is specific for each type of project. The following sections explain this more.

25.4.3 Risk Identification

As with each step in the risk management process, risk identification should be done continuously throughout the project life cycle (1–3). To begin risk identification, break the project elements into an RBS. The RBS is the hierarchical structuring of risks. It is a structured and organized method to present the project risks and to allow for an understanding of those risks in one or more hierarchical manners to demonstrate the most likely source of the risk. The RBS provides an organized list of risks that represents a coherent portrayal of project risk and lends itself to a broader risk analysis. The upper levels of the structure can be set to project, technical, external, and internal risks; the second tier can be set to cost, schedule, and scope. Each tier can be broken down further as it makes sense for the project and lends itself to the next step of risk analysis. To be useful, the RBS should have at least three tiers. Developing the RBS is just one methodology, as the type of project or project organization may dictate. Figure 25.3 shows an example RBS.

The RBS can be used to inform the following:

- Updates of the RMP.

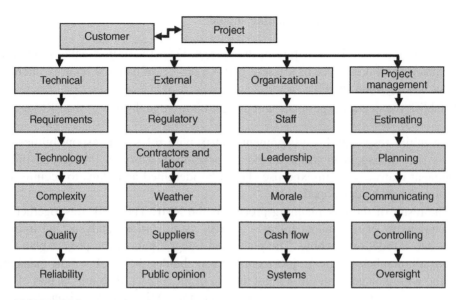

FIGURE 25.3   Example risk breakdown structure.

- Work breakdown structure (WBS).
- Cost estimates.
- Key planning assumptions.
- Preliminary schedules.
- Acquisition strategy documents.
- Technology readiness assessment (TRA) information.
- Project definition rating index (PDRI) analyses.
- Safety-in-design considerations.
- Safety analysis assumptions.
- Environmental considerations such as seismic, wind, and flooding.
- Safeguards and security analysis assumptions.
- Requirements documents or databases.
- SME interviews.
- Stakeholder input.
- Designs or specifications.
- Historical records.
- Lessons learned.
- Any legislative language pertaining to the project. Other similar projects.
- Pertinent published materials.

Various techniques that can be used to elicit risks include brainstorming, interviews, and diagram techniques. Regardless of the technique, the result should not be limiting and should involve the greatest number of knowledgeable participants that can be accommodated within their constraints. In addition, the participants need to address risks that affect the project but are outside of the project ability to control.

As the PRM team identifies risks, it is important that they are aware of biases that may influence the information. Typical biases the facilitator of the risk identification should be aware of include the following (3):

- Confirming evidence bias – information that supports existing points of view are championed while avoiding information that contradicts.

- Anchoring – disproportionate weight is given to the first information provided.

- Sunk cost – tend to make choices in a way that justify past choices, unwillingness to change direction.

When identifying a risk, it should be stated clearly in terms of both the risk event and the consequences to the project. The format for the risk identified should generally be cause/risk/effect.

One may choose to record cause, risk, and effect in separate fields to facilitate grouping of risks into categories based on commonality of these attributes.

Risks should be linked to activities or WBS as much as possible. Consult a project management text concerning the design and uses of the WBS. The linkage is important, especially if the risk owner is different as the risk owners may need to coordinate their efforts on the risk-handling strategies.

The PRM team should capture both opportunities and threats. Opportunities are often shared between and among projects. It should be noted that opportunities for one participant could be detrimental to another; therefore, they should be worked cooperatively. Examples of opportunities include the following:

- Available human resources with flexible scheduling can be shared to the advantage of two or more projects.

- A crane is available at another site at a lower cost than purchasing a new or a used one.

- Duel purposing a piece of equipment.

In addition to identifying a risk in terms of the causal event and consequence, the pertinent assumptions regarding that risk should be captured in the risk register to aid in future reporting of the risk. These assumptions might include items such as, but not limited to, interfaces among and between sites, projects, agencies, and other entities; dependencies on human resources, equipment, facilities, or others; and historically known items that may impact the project either positively or negatively. The assumptions should be kept current and should be validated through various methods including documentation and SMEs.

25.4.4 Assignment of the Risk Owner

Before assigning a qualitative assessment to the dimensions of a risk (probability and consequence), a risk owner should be identified (1–3). The risk owner is the team member responsible for managing a specific risk from risk identification to risk close-out, should ensure that effective handling responses or strategies are developed and implemented, and should file appropriate reports on the risk in a timely fashion. The risk owner should also validate the qualitative and quantitative assessments assigned to their risk. Finally, the risk owner should ensure that risk assumptions are captured in the risk register for future reference and assessment of the risk and to assist possible risk transfer in the future. Any action taken in regard to a risk should be validated with the risk owner before closure on that action can be taken.

25.4.5 Assignment of Probability and Consequence

Techniques discussed in Chapters 7–12 can be used to aid in establishing probabilities and consequences.

25.4.6 Assignment of Risk Trigger Metrics

A risk trigger metric is an event, occurrence, or sequence of events that indicates that a risk may be about to occur, or the pre-step for the risk indicating that the risk will be initiated (1–3). The risk trigger metric is assigned to the risk at the time the risk is identified and entered into the risk register. The trigger metric is then assigned a date that would allow both the risk owner and the PRM team to monitor the trigger. The purpose of monitoring the trigger is to allow adequate preparation for the initiation of the risk-handling strategy and to verify that there is adequate cost and schedule to implement the risk-handling strategy.

25.4.7 Risk Register

The risk register is the information repository for each identified risk. It provides a common, uniform format to present the identified risks (1–3). The level of risk detail may vary depending upon the complexity of the project and the overall risk level presented by the project as determined initially at the initiation phase of the project.

The fields stated here are those that should appear in the risk register, whether the risks presented are a threat or an opportunity:

- Project title and code – denote how the project is captured in the tracking system used by the site office and/or contractor.
- Unique risk identifier – determined by the individual site.
- Risk statement – considers separate subfields to capture cause/risk/effect format to facilitate automated search capabilities on common causes of risks.

- Risk category – project, technical, internal, external, and any subcategory that may be deemed unique to the project such as safety or environment.
- Risk owner.
- Risk assumptions.
- Probability of risk occurrence and basis.
- Consequence of risk occurrence and basis.
- Risk cause/effect.
- Trigger event.

Handling strategy (type and step-wise approach with metrics, which has the action, planned dates, and actual completion dates) includes the probability of success for the risk-handling strategy and considers probabilistic branching to account for the handling strategy failing.

The risk register may also include backup strategies for primary risks, risk-handling strategies for residual and secondary risks, the dates of upcoming or previous risk reviews, and a comment section for historical documentation, lessons learned, and SMEs' input.

25.4.8 Risk Analysis

Risk analysis should begin as early in the project life cycle as possible (1–3). The simplest analysis is a cost and benefit review, a type of qualitative review. The qualitative approach involves listing the presumed overall range of costs over the presumed range of costs for projected benefits. This chapter discusses risk analysis in detail. The following is a high-level view.

Qualitative Risk Analysis

The purpose of qualitative risk analysis is to provide a comprehensive understanding of known risks for prioritization on the project. Qualitative risk assessment calls for several risk characteristics to be estimated:

- Relative probabilities – binning of potential risks.
- Ranking relative consequences.
- Potential initiating events.

These items should be captured in the risk register. The initial qualitative assessment is done without considering any mitigation of the risk, that is, prior to the implementation of a handling strategy.

Qualitative analysis, or assessment as it is sometimes referred, is the attempt to adequately characterize risk in words to enable the development of an appropriate risk-handling strategy. Additionally, qualitative analysis assigns a risk rating to each risk, which allows for a risk grouping process to occur. This grouping of risks may identify patterns of risk on the project. The patterns are indicative of the areas of risk exposure on the project. The qualitative analysis may be the foundation for initiating the quantitative risk analysis, if required.

25.5 QUALITATIVE MATRICES ANALYSIS

One of the tools used to assign risk ratings is a qualitative risk analysis matrix, also referred to as a probability impact diagram or matrix. Risk ratings are also often referred to as risk impact scores. The matrix shown in Table 25.1 is an example of the tool and could be modified by site and contractor, or any other category as required. Similar matrices have been discussed in other chapters. The matrix combines the probability and consequence of a risk to identify a risk rating for each individual risk. Each of these risk ratings represents a judgment as to the relative risk to the project and categorizes at a minimum, each risk as low, moderate, or high. Based on these risk ratings, key risks, risk-handling strategies, and risk communication strategies can be identified.

Risk ratings should be assigned via a matrix to the risk, threat, or opportunity, based upon the risk classification. Typical risk classifications are low, moderate, or high. Another option could be to use numerical values for ratings. The numerical value could be tailored to the project or standardized for a program (1). Risks that have a determinative impact upon project cost or schedule will generally rate toward the higher end of the qualitative scale. However, there may be little or no correlation between a risk's determinative impact and the qualitative risk rating, so caution with the lowest rated risks in the qualitative analysis. Care should be taken when comparing project risk scores of different projects as the project risk scores are a result of a subjective process and are prepared by different project teams. Qualitative risk analysis could also be performed on residual risks and secondary risks, but only after the handling strategy has been determined for the primary risk. Again, the risk owner should validate and accept the risk rating.

As the information is gathered and finalized, the data should be analyzed for bias and perception errors. While the data will not be systematically used for a quantitative analysis, it should still be analyzed and perceptions scrutinized.

Relative risk rankings can be included on a project network diagram. Though, once again, the specific network diagraming techniques are not described in this text. Figure 25.4 shows an example of a network diagram with risk information included and a critical path.

Risk is mapped to the WBS element that would be impacted if it occurred. The pattern that emerges allows one to either use the assigned expected value score or to count the number of risks associated with the element. This method allows attention to be focused on specific areas of risks.

TABLE 25.1
Relative Risk Matrix

		Relative consequence				
		Negligible	Marginal	Significant	Critical	Crisis
	Cost	Minimal or no consequence. No impact to project cost.	Small increase in meeting objectives. Marginally increases costs.	Significant degradation in meeting objectives significantly increases cost; fee is at risk.	Goals and objectives are not achievable. Additional funding may be required; loss of fee and/or fines and penalties imposed.	Project stopped. Funding withdrawal; withdrawal of scope, or severe contractor cost performance issues.
	Schedule	Minimal or no consequence. No impact to project schedule.	Small increase in meeting objectives. Marginally impacts schedule.	Significant degradation in meeting objectives, significantly impacts schedule.	Goals and objectives are not achievable. Additional time may need to be allocated. Missed incentivized and/or regulatory milestones.	Project stopped. Withdrawal of scope or severe contractor schedule performance issues.
Relative Probability	Very High >90%	Low	Moderate	High	High	High
	High 75% – 90%	Low	Moderate	Moderate	High	High
	Moderate 26% – 74%	Low	Low	Moderate	Moderate	High
	Low 10% – 25%	Low	Low	Low	Moderate	Moderate
	Very Low <10%	Low	Low	Low	Low	Moderate

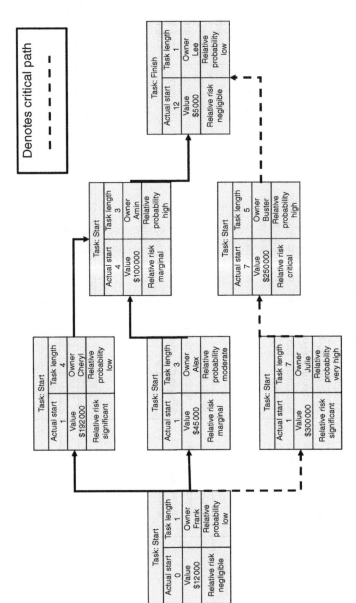

FIGURE 25.4 Example network diagram with relative risk rankings.

25.6 QUANTITATIVE RISK ANALYSIS

Chapters 7–10 discuss quantitative risk analysis in detail.

25.7 CONTINGENCY FUNDING

Numerous tools exist to analyze the adequacy of the contingency valuation that has resulted from the qualitative and/or quantitative analysis of the risks (1). Various cost estimating guidance documents have been compiled by industry and are available in texts and journals (e.g. the Association for the Advancement of Cost Engineering International) and are updated on a regular basis. These references provide percent ranges of the base that a contingency should represent in order to be considered adequate. Further, the contingency value should be commensurate with the maturity and type of the project, project size, and risks, including technical and technology uncertainties.

If a quantitative risk analysis will not be conducted, estimates for cost contingency and schedule contingency should be provided. As a general rule, the project should use various inputs to determine those values: interviewing staff, crafts, retirees, and others familiar with similar work efforts at the site or other sites and technical records such as safety analysis documents including the risk and opportunity assessment, quality assessments, and environmental assessments. As the information is gathered and finalized, the data should be analyzed for bias and perception errors. While the data will not be systematically used for a quantitative analysis, it should still be analyzed and perceptions scrutinized.

25.7.1 Risk Handling

Risk handling covers a number of risk strategies, including acceptance, avoidance, mitigation, and transfer (1–3). When weighing these approaches, the following should be taken into account:

- The feasibility of the risk-handling strategy.
- The expected effectiveness of the risk-handling strategy based upon the tools used.
- The results of a cost/benefit analysis, i.e. how do the costs of the handling strategy compare to the benefits derived from not realizing the risk event?
- The impacts of the strategy on other technical portions of the project.
- Any other analysis deemed relevant to the decision process.

Many parameters of the project can change over time that can impact the risk-handling strategies (e.g. scope of the project, available resources, internal and external environments, technical advancements, etc.). Thus, risk handling should be an iterative process. One or more of these items can change a step in a risk-handling strategy, or even the complete strategy, which then changes the cost and/or the schedule for implementation of the risk-handling strategy.

Risk-handling strategies should consider the probability and consequence of the risk and, if deemed necessary by the risk owner, should allow for a backup risk-handling strategy that is documented in the risk register. If backup risk-handling strategies are documented in the risk register, they should be documented at the same level of detail as the primary risk-handling strategy. Documentation at the same level as the primary strategy will ease implementation if the primary risk-handling strategy is deemed unsuitable or inadequate. Further, the cost and necessary schedule for the backup risk-handling strategy should be calculated and noted in the risk register.

The cost for the risk-handling strategy for the primary risk should typically be included in the baseline as direct project costs if the handling action will be performed (see further discussion in the following paragraph). The process includes identifying the scope, cost, and schedule associated with implementing the risk-handling strategy, and assigning a unique WBS number and activity to the strategy so that it can be tracked and monitored. The project team should develop the risk-handling implementation plans with the appropriate level of detail. The project activities should include the detailed work plans (for whichever phase the project is then in) with the associated budget and schedule identified in the project WBS.

Some project teams make the mistake of thinking that all handling costs should be part of the project contingency. If the handling actions will be performed, then do not include the costs of these handling actions in the risk contingency. These are known, identified project work activities and need to be planned accordingly, and included as part of the direct project costs.

However, if the handling action will not occur until some event that may or may not occur, e.g. a risk trigger event, then it is appropriate to assign those costs to project. If the triggering event occurs, then the project would process a change using the project change control system, to take cost/schedule from contingency and assign it to the project handling activities. This latter approach is more the exception than the rule.

There may be occasions when a primary risk is not added to the baseline until a change control action occurs, such as when it is predicted during a monthly project review or a review of lessons learned.

Risk-handling strategies should be regularly reviewed throughout the project life cycle for their affordability, achievability, effectiveness, and resource availability as described in the reporting requirements of the RMP.

The common risk handing strategies are as follows:

- Acceptance of the risk.
- Risk voidance.
- Mitigate the risk.
- Transfer/share the risk.

25.7.2 Acceptance of the Risk

Acceptance as a risk-handling strategy should be a deliberate decision and documented in the risk register. Acceptance of the risk does not mean that the risk is ignored. The risk should be included in the cost and schedule contingency impact analysis.

25.7.3 Risk Avoidance

Avoidance, as a risk-handling strategy, is done by planning the project activities in such a way as to eliminate the potential threat. Avoidance should be considered the most desirable risk-handling strategy. However, avoidance should be analyzed for its cost/benefit to the project within the current funded boundaries of the project. The cost/benefit analysis should also take into consideration the impact on the overall project and the available funding for handling the other identified risks. The decision processes used to determine whether or not to pursue the avoidance risk-handling strategy for risks on the project should be documented.

25.7.4 Mitigate the Risk

Mitigation is a risk-handling strategy that is taken to reduce the likelihood of occurrence and/or impact of an identified negative risk or threat. Enhancement is a risk-handling strategy used to increase the likelihood of occurrence and/or benefit of an identified positive risk or opportunity. The goal of a mitigation risk-handling strategy is to reduce the risk to an acceptable level.

25.7.5 Transfer/Share the Risk

In this case, the risk is passed to the insurance company or to another entity. The other party accepts the risk for a fee. Risk transference indicates a transfer of ownership, and therefore written acceptance of the risk should be obtained before transfer is complete. When risk has been transferred, the transfer of the risk should be reviewed to ensure it did not create other risks and that it does not impact the project mission and objectives.

25.7.6 Residual Risk

Residual risk (post-mitigated risk) is the risk that remains after the risk-handling strategy (accept, avoid, mitigate, or transfer) has been performed to the original primary risk to which they had been assigned in the risk register. A residual risk may end up being the same risk as the original risk (pre-mitigated risk) if the risk-handling strategy does not reduce or mitigate the risk or the risk is one that recurs. The fact that residual risk remains does not mean that the risk handling was not effective, only that it did not completely avoid a risk remaining. It is up to the risk owner to decide whether the residual risk will be moved to a primary risk position.

This remaining or residual risk should be qualitatively analyzed. Through this process, a decision should be made as to when the risk planning process should stop. Those residual risks for which no risk strategies are planned are accepted and should be clearly communicated to the team and management.

25.7.7 Risk Monitoring

Risk monitoring involves the systematic, continuous tracking and evaluation of the effectiveness and appropriateness of the risk-handling strategy, techniques, and actions established within the RMP. Monitoring is performed for individual risks per the risk metrics and overall project risk status. The risk monitoring process should provide both qualitative and quantitative information to decision makers regarding the progress of the risks and risk-handling actions being tracked and evaluated (1).

Risk monitoring may also provide information that can assist in identifying new risks or changes in the assumptions for risks captured previously on the risk register. These results should be used to initiate another risk identification process. The risk monitoring process should be tailored to the project and described in the RMP. The risk monitoring process should be more than a risk tracking documentation process.

Risk owner monitoring – the risk owner has a significant role in risk monitoring. As part of the risk monitoring process, the risk owner should update information in the risk register through an agreed upon process as stated in the RMP. Any changes that a risk owner makes to the risk register should be discussed at the risk meetings to ensure that changes in the conditions of one risk do not impact another risk or create another potential risk. It may be necessary to conduct an analysis study depending upon the extent of the impact of the change to the risk register.

Integrated risk monitoring – integrated risk monitoring occurs when risk management metric monitoring is integrated with other standard project metrics such as earned value or safety metrics. The determination as to the root cause of any negative or positive impact upon a metric should include a determination as to whether it involved a risk including whether it involved the positive benefit risk known as an opportunity. The output of the reporting process can be the input to the risk management process for further risk identification, analysis of consequence and impact ratings, and the analysis of the handling strategy as planned or as being implemented.

25.7.8 Risk Management Plan

The RMP is the governing document for the risk management process on a project (1–3). The RMP includes by reference the risk register, risk analysis, and other risk data and risk database information that is updated more frequently but is not reissued whenever such data is changed or updated. Results from the risk analyses are recommended for inclusion in monthly progress reports if the analyses are updated more frequently than annually.

The RMP should include the following sections:

- Project summary.
- Responsibility assignment matrix.
- Key definitions.
- Key requirements documents and regulatory drivers.
- Assumptions and constraints.
- Risk and opportunity management process.
- Risk planning.
- Risk assessment.
- Risk identification.
- Risk analysis.
- Risk and opportunity handling.
- Risk monitoring.
- Risk feedback.
- Risk documentation and communication.

Risk Management Communication Plan

To ensure project success, the RMP should address how information related to risk, and risk status is communicated to the project team and stakeholders. This communication information could be addressed in either the project execution plan or a communication plan or could be included in the RMP. A separate risk management communication plan could also be developed as part of the tailoring decisions. The risk management communication plan should also specifically address the integration points with the DOE enterprise-wide lessons learned systems.

25.8 SUMMARY

Risk management is used to help ensure project success. A real designed risk management system and plan will ensure the projects stays on schedule, within budget and quality, and help ensure the health and safety of people and to protect the environment by controlling or mitigating risk. The RMP evaluates and treats identified risks, evaluates controls and limits proposed by the applicant, and considers general risk management measures.

Case Study

Read the following case study and discuss the risk management issues.

The Panama Canal Project
Introduction

The Panama Canal has been in operation over 100 years. It was officially opened on 15 August 1914. However, the idea of a canal across the Isthmus of Panama was since 1513, when Vasco Núñez de Balboa first crossed the isthmus (5). The history of the canal's construction is long and fraught with disease and death. The following discusses the project and presents a project risk analysis.

History

The following history of the canal is primarily drawn from a book by Saxon Mills (5). In 1869, President Ulysses S. Grant establishes the Interoceanic Canal Commission (IOCC) and sends out an expedition to investigate possible routes for a canal to provide faster shipping route between the Atlantic and Pacific oceans. A US-led investigatory team surveys Panama, at that time a part of the Republic of Colombia, for a feasible canal route. The team determines a Panama Canal will be too expensive and propose a canal in Nicaragua. The idea of a canal across Nicaragua is not new either. There were a number of failed attempts to build a Nicaragua canal connecting the Caribbean Sea/Atlantic Ocean with the Pacific Ocean. These attempts go back at least to 1825 when the Federal Republic of Central America hired surveyors to study a route via Lake Nicaragua. This canal would have been 32.7 m (107 ft) above sea level. Many other proposals have followed. On 26 September 2012, the Nicaraguan Government and the newly formed Hong Kong Nicaragua Canal Development Group (HKND) signed a memorandum of understanding that committed HKND to financing and building the "Nicaraguan Canal and Development Project" (6, 7).

However, in 1879, 10 years later, the French government approves Ferdinand de Lesseps' plan for a sea-level canal (5). Figure 25.5 shows a map of the Isthmus. The cost is estimated at F1.2 billion ($240 million). De Lesseps had exclusive rights from the Columbian Government. He arrived in the Panama region and begins construction on his sea level canal plan. The plan included constructing a 40-m high dam at Gamboa and a 24-m wide path through the Culebra Cut. The dam was to hold back the Chagres River. Plagued by flooding and landslides, the French team finds itself behind schedule; only 660 000 m^3 of earth have been excavated, though de Lesseps had promised over 5 million cubic meters completed by spring 1883. Figure 25.6 shows aftermath of one of the slides.

FIGURE 25.5   Map of Panama. *Source*: Courtesy of Project Gutenberg (5).

FIGURE 25.6   Photo of a mudslide. *Source*: Courtesy of Project Gutenberg (5).

Early in 1884, an epidemic of yellow fever panics the workers. An estimated 400 workers have died of disease, compared with only 126 the year before. More than 300 French engineers ask to return home and are denied. To further complicate efforts, an outbreak of dysentery cripples the already largely weakened workforce, affecting about 30% of workers. In total, more than 25 000 workers die during the French construction attempt (5).

Racial tensions between native Panamanians and Jamaican workers lead to large numbers of Jamaicans to return home, and the canal loses much of its primary source of manual laborers. By the summer of 1885, only 8 million cubic meters of earth was excavated, out of the 120 required for the canal. After four years of excavations, only a few feet have been removed from the top of Culebra Cut out of the hundreds necessary to reach sea level.

De Lesseps runs out of money after nine-month fundraising campaign that included borrowing F30 million from friends and selling lottery tickets. His company collapses, ruining the fortunes of 800 000 private investors. In 1892, Ferdinand de Lesseps and his son Charles are found guilty of fraud and maladministration of the Canal project. De Lesseps will die within 2 years, at the age of 89.

President Ulysses S. Grant establishes the IOCC and sends out an expedition to investigate possible routes for a canal. The idea of a canal will continue to be a priority for Grant throughout his presidency as he seeks a faster shipping route between the Atlantic and Pacific oceans (5).

Theodore Roosevelt is a major proponent for a US-led effort to build a canal to connect the Atlantic and the Pacific. Once sworn into office in 1901, he declares the need to build a canal in Central America. The United States formally took control of the French-canal-related property on 4 May 1904. The newly created Panama Canal Zone Control came under the control of the Isthmian Canal Commission during canal construction. The plan is to build a series of locks on both ends of the canal to raise and lower ships from Lake Gatun. The length of the proposed Panama Canal is approximately 51 miles. The canal route from its Atlantic entrance would take you through a 7-mile dredged channel in Limón Bay. The canal then proceeds for a distance of 11.5 miles to the locks. There are a series of three locks that raise ships 26 m to Gatun Lake. The canal then continues south through a channel in Gatun Lake for 32 miles to Gamboa, where the Culebra Cut begins. This channel through the cut will be 8-miles long and 150-m wide. At the end of this cut are the locks at Pedro Miguel. The Pedro Miguel locks lower ships 9.4 m to a lake that then takes you to the locks, which will lower ships 16 m to sea level into the bay of Panama.

The value of the canal to the United States at the time was that the military could move ships, soldiers, and material to either coast much quicker via the Panama Canal, than by going around the tip of South America. The completed canal helped make the United States a world power.

Risk Management of the Project

Think of this project in modern risk management terms. Risk management is an integral part of project management. Proactive risk management completed in a comprehensive manner will assess and document risks and uncertainty. The PM working with the project team and project sponsors needs to ensure that risks are actively identified, analyzed, and managed throughout the life of the project. Risks need to be identified as early as possible in the project so as to minimize their impact. The steps for accomplishing this were discussed in the main body of the chapter. The PM or other designee serve as the risk manager for this project. During the time the French were attempting the canal, the construction technology was not as advanced and the nature of disease was not as well known. By the early 1900s, construction technology had advanced and diseases were being better controlled.

Risk Identification

As discussed in the main body of the chapter, risk identification is the first activity in the risk management process. The PRM team needs to look at historical data and to ask "What can go wrong?" This step involves examining the technical aspects of a program to determine risk events that may have negative cost, schedule, and performance impacts. What risks did the French face?

- Disease.
- Technology.
- Financing.
- Labor issues.
- Poor planning/management.
- Political issues.
- Weather/mudslides.

How might the American effort benefit from these historical data?

Risk identification involves the PRM team and appropriate stakeholders and should include an evaluation of environmental factors, organizational culture, and the project management plan including the project scope. The canal's size and duration of construction leads to a great number of risks. Based on cost, the most severe were loss of human life due to diseases, mudslides, and the constant use of explosives. Another great challenge is the technical challenges of constructing the massive locks.

Since the United States took over the project, diseases became less of a problem, but health concerns remain. In 1904, the Isthmian Canal Commission, accompanied by Col. W. C. Gorgas, Medical Corps, US Army; John W. Ross, Medical Director, US Navy; Capt. C. E. Gillette, Corps of Engineers, US Army; and Maj. Louis A. LaGarde, Medical Corps, US Army, as experts on sanitation, inspected the potential site of construction. These experts prepared a plan for the sanitation of the Canal Zone and the cities of Panama and Colon. On 30 June 1904, the Sanitary Department was formed with Colonel Gorgas as its head (8). Many workers and managers have stated tropical disease among their reasons for resigning from the project. The massive death tolls from the French attempt have made this exposure a top priority. The Sanitary Department ordered the following tasks to be performed to reduce the risk of disease:

- Drainage: All pools within 200 yards of all villages and 100 yards of all individual houses were drained. Subsoil drainage was preferred followed by concrete ditches. Lastly, open ditches were constructed. Paid inspectors made sure ditches remained free of obstructions.

- Brush and grass cutting: All brush and grass was cut and maintained at less than 1 ft high within 200 yards of villages and 100 yards of individual houses. The rationale was that mosquitoes would not cross open areas over 100 yards.

- Oiling: When drainage was not possible along the grassy edges of ponds and swamps, oil was added to kill mosquito larvae.

- Larviciding: When oiling was not sufficient, larvaciding was done. At the time, there were no commercial insecticides. Joseph Augustin LePrince, Chief Sanitary Inspector for the Canal Zone, developed a larvacide mixture of carbolic acid, resin, and caustic soda that was spread in great quantity.

- Prophylactic quinine: Quinine was provided freely to all workers along the construction line at 21 dispensaries. In addition, quinine dispensers were on all hotel and mess tables. On average, half of the work force took a prophylactic dose of quinine each day.

- Screening: Following the great success in Havana, all governmental buildings and quarters were screened against mosquitoes.

- Killing adult mosquitoes: Because the mosquitoes usually stayed in the tent or the house after feeding, collectors were hired to gather the adult mosquitoes that remained in the houses during the daytime. This proved to be very effective. Mosquitoes that were collected in tents were examined by Dr. Samuel T. Darling, Chief of the Board of Health Laboratory. Cost of adult mosquito killing was $3.50/per capita/per year for whole population of the strip.

Mudslides were still common in the early American effort. The unstable soil was made worse by relatively frequent earthquakes. Whenever the sides of the cut collapsed, there was the danger to the work crews and potential serious damage to the digging equipment and the railroad. In addition, it was discouraging to continuously repair and rework the canal after each slide. Additionally, repeatedly reworking the same location multiplies the cost of construction. This risk was having a very high impact to both schedule and budget. Despite precautions, major setbacks were frequent.

Another issue was explosives were in use everywhere. In the Culebra Cut, massive boulders were a problem and workers must set off dynamite charges to reduce them to moveable pieces. The planned path for ships through the man-made lakes is a rain forest filled with large trees that also needed to be removed with explosives. The dynamite at the time was unstable. The probability of premature detonation was high, and the risk to human life was extreme.

The largest technical challenge on the project was the locks. They are proposed to be gigantic mechanisms, among the largest and most complex construction ever attempted. Locks had never before been constructed for large ocean-going ships. The doors for the locks were to be huge and very heavy. The volume of water that must be held by the locks when filled is so great that the pressure on the doors is immense, and the precision required for the seams where the doors close to hold in the water must be precise. The locks design was enormous boxes with sides and bottoms formed of concrete. Probably the biggest technological hurdle is the requirement that all operations be electric. This new technology of electric power and the hydroelectric installations required to supply enough electricity is a poorly understood technology and had never be attempted. Without the locks and the electricity, the canal would be useless, and the risks associated with resolving all of these technical problems were large.

The following is a preliminary list of potential risks:

- Inefficient organizational structure.
- Channel disruption.
- Environmental concerns/planning.
- Poor communications.
- Execution.
- Project completion delay.
- Changes in cost projections/overruns.
- Change in project scope, design, or definition.
- Recruitment and retention of skilled labor.
- Quality and quantity of skilled labor.
- Employee safety.
- Inaccurate revenue projections.
- Improperly trained program manager or team.
- Lack of controls.
- Inefficient planning.
- Inefficient contracting process.
- General inflation.
- Referendum delays.
- Extreme bad weather.
- Technology lag.
- Insufficient revenues.
- Inadequate claims administration.

- Lack of skilled and local labor.
- Material, equipment, and labor cost.
- Disease – yellow fever and malaria.
- Poorly understood technology.
- Lack of proper equipment.

Besides the other tools discussed in this text, one of the tools that can be used to analyze risks associated with such a project is a fishbone diagram. The fishbone diagram can be used to link a potential problem with many possible causes for a problem. The following is an example of a fishbone diagram for examining the risks of the Panama Canal construction (Figure 25.7).

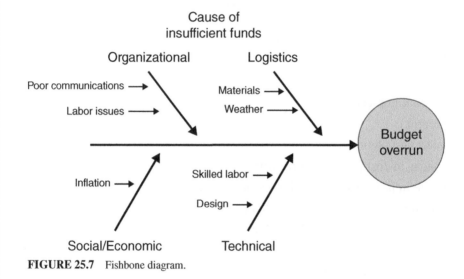

FIGURE 25.7   Fishbone diagram.

The risk management process is a living document. As time progresses, additional risk will be identified and current risks will either be mitigated or factors will change that will increase the risk.

Handling the Risks

After each risk is identified, it must be assigned to a project team member for monitoring purposes to ensure that the risk will not "fall through the cracks." For each major risk, one of the following approaches will be selected to address it:

- Avoid – eliminate the threat by eliminating the cause.

- Mitigate – identify ways to reduce the probability or the impact of the risk.
- Accept – nothing will be done.
- Transfer – make another party responsible for the risk (buy insurance, outsourcing, etc.).

Summary

This case study has provided information relating to the construction of the Panama Canal. It is a fact that the canal was constructed successfully by the United States. Using the material provided, your imagination and the list of references develop the RMP.

REFERENCES

1. Department of Energy (2015). Risk Management Guide. DOE G 413.3-7A Chg 1.
2. NASA (2011). Risk Management Handbook. NASA/SP-2011-3422, Version 1.0.
3. Project Management Institute (2017). *PMBOK® Guide*, 6e. Project Management Institute.
4. The Mitre Corporation. 21 Project Risk Management Musts. https://www.mitre.org/publications/systems-engineering-guide/acquisition-systems-engineering/risk-management/risk-management-approach-and-plan (accessed August 2018).
5. Mills, S. (2010). *The Project Gutenberg eBook of The Panama Canal*. London: Thomas Nelson and Sons.
6. Passary, Sumit (2015). Scientists Wary about Environmental Effects of Canal-Building Project in Nicaragua. *Tech Times* (5 March 2015).
7. Nicaragua taps China for canal project. *Nicaragua dispatch*. 2012.
8. Center for Disease Control (2018). Malaria. https://www.cdc.gov/malaria/about/history/panama_canal.html (accessed August 2018).

OTHER RESOURCES

US Department of State, Office of the Historian. Milestones. https://history.state.gov/milestones/1899-1913/panama-canal (accessed August 2018).

Hanily, A., Alvarado, P., and Ungo, R. (2005). Development and implementation of a risk model and contingency estimation for the Panama Canal Expansion Program, ACP. Final Report. https://docs.micanaldepanama.com/plan-maestro/Study_Plan/Financial_and_Economic/Development_and_implementation_of_a_risk_model/0302-01.pdf (accessed August 2018).

Kendrick, T. (2016). A Tale of Two Projects: The Panama Canal and the Birth of Project and Risk Management. http://failureproofprojects.com/Panama2006.pdf (accessed August 2018).

Nix, E. (2014). Fascinating Facts about the Panama Canal, 15 August 2014. http://www.history.com/news/7-fascinating-facts-about-the-panama-canal (accessed August 2018).

Defense for Systems Engineering (2014). *Department of Defense Risk Management Guide for Defense Acquisition Programs*, 7e(Interim Release). Washington, DC: Office of the Deputy Assistant Secretary of Defense for Systems http://acqnotes.com/wp-content/uploads/2014/09/DoD-Risk-Mgt-Guide-v7-interim-Dec2014.pdf (accessed August 2018).

CHAPTER 26

Enterprise Risk Management Overview

26.1 INTRODUCTION

Chapter 25 discussed project risk management (PRM). This chapter is focused on enterprise risk management (ERM). Though similar, ERM is not only concerned with risks but opportunities as well. Risks in this regard can be the traditional risks that this book has discussed up to this point, but also missed opportunities, loss of intellectual property, fluctuations in the economy, and political risks. The terms used in ERM are essentially the same as in PRM or risk assessment in general. However, they tend to be used in a broader sense. Other less common terms will be explained in the various sections of this chapter.

The ERM model, once again, is similar to PRM, but is in a different context. Figure 26.1 shows one version of the ERM model.

According to the Committee of Sponsoring Organizations of the Treadway Commission (COSO), ERM encompasses the following attributes (1):

- Aligning risk appetite and strategy – Management considers the entity's risk appetite in evaluating strategic alternatives, setting related objectives, and developing mechanisms to manage-related risks.

- Enhancing risk response decisions – ERM provides the rigor to identify and select among alternative risk responses – risk avoidance, reduction, sharing, and acceptance.

- Reducing operational surprises and losses – Entities gain enhanced capability to identify potential events and establish responses, reducing surprises and associated costs or losses.

Risk Assessment: Tools, Techniques, and Their Applications, Second Edition. Lee T. Ostrom and Cheryl A. Wilhelmsen.
© 2019 John Wiley & Sons, Inc. Published 2019 by John Wiley & Sons, Inc.
Companion website: www.wiley.com/go/Ostrom/RiskAssessment_2e

FIGURE 26.1   Enterprise risk model.

- Identifying and managing multiple and cross-enterprise risks – Every enterprise faces a myriad of risks affecting different parts of the organization, and ERM facilitates effective response to the interrelated impacts and integrated responses to multiple risks.

- Seizing opportunities – By considering a full range of potential events, management is positioned to identify and proactively realize opportunities.

- Improving deployment of capital – Obtaining robust risk information allows management to effectively assess overall capital needs and enhance capital allocation.

The ERM Committee discusses the following aspects of ERM (2):

- Establishing context: This includes an understanding of the current conditions in which the organization operates on an internal, external, and risk management context.

- Identifying risks: This includes the documentation of the material threats to the organization's achievement of its objectives and the representation of areas that the organization may exploit for competitive advantage.

- Analyzing/quantifying risks: This includes the calibration and, if possible, creation of probability distributions of outcomes for each material risk.

- Integrating risks: This includes the aggregation of all risk distributions, reflecting correlations and portfolio effects, and the formulation of the results in terms of impact on the organization's key performance metrics.
- Assessing/prioritizing risks: This includes the determination of the contribution of each risk to the aggregate risk profile and appropriate prioritization.
- Treating/exploiting risks: This includes the development of strategies for controlling and exploiting the various risks.
- Monitoring and reviewing: This includes the continual measurement and monitoring of the risk environment and the performance of the risk management strategies.

These ideas are then tailored to specific organizations. For instance, Stanford University's ERM purpose and objectives are (3) as follows:

The purpose of ERM activities at Stanford University is to provide a comprehensive program to proactively manage the portfolio of what leadership collectively believes are the most critical risks to the achievement of the entity's mission and objectives.

ERM promotes an ongoing, risk-aware culture across the University to enable decision makers to perform a risk–reward analysis of choices and make decisions with an understanding of implications of such actions while pursuing the mission and goals of Stanford University. It is not intended to be a one-time process or a prescriptive method for managing individual risks, but instead a tool for leadership to use in managing existing and emerging risks within their portfolio of activities.

- Identifying and assessing a broad array of risks that could negatively impact the achievement of institutional goals and objectives.
- Ensuring appropriate ownership and accountability of risks.
- Developing and implementing appropriate risk mitigation and monitoring plans by risk owners.
- Establishing a program structure that engages functional leaders across the campus to identify and prioritize risks.
- Providing senior leadership with key information to make risk-informed decisions and to effectively allocate resources.

However, a bank would have a different risk management profile. In fact, take Wells Fargo, for instance. Before the financial crisis of 2008, Wells Fargo took few risks. They developed a very risky strategy around 2008 where the bank sought to incentivize tellers and branch bankers to have customers open numerous new accounts (4). What this resulted in was that these employees began to fraudulently open numerous accounts in customers' names without their consent. This began to catch up to the bank several years later. This was not the only fraudulent process the bank was doing. The timeline for the problems at Wells Fargo is as follows:

- 2009–2016 – Wells Fargo perpetrates a massive cross-selling scandal in which millions of accounts were created without consumers' consent.
- September 2016 – The Consumer Fraud Protection Bureau (CFPB) levies a $185 million fine, the highest in their operational history.
- August 2017 – The bank accidentally leaks the personal information for over 50 000 accounts.
- August 2017 – Wells Fargo charges 800 000 customers for insurance they did not need.
- October 2017 – The bank wrongly charges homebuyers with fees to lock in mortgage rates.
- March 2017 – The Federal Reserve imposes unprecedented sanctions on Wells Fargo prohibiting them from growing beyond their holdings in 2017.
- April 2018 – Wells Fargo nears $1 billion settlement with its federal regulators.

Currently in 2018, the bank is running an advertisement campaign aimed at restoring Wells Fargo's dignity. However, at some point, the bank made these risky decisions with the goal of increasing profits. What happened was the bank lost reputation, as well as having to pay huge fines. Was this a wise ERM strategy? Time will only tell.

26.2 ENTERPRISE RISK MANAGEMENT VERSUS PROJECT RISK MANAGEMENT

As previously stated, ERM and PRM are similar in that both processes involve identifying and handling risks. The methods in which risks are identified in ERM are essentially the same as for PRM. These methods are, once again, discussed throughout this text. However, where ERM and PRM differ is more in the areas ERM touches. These areas include:

- Strategic planning – Identifies external threats and competitive opportunities, along with strategic initiatives to address them.
- Marketing – Understands the target customer to ensure product/service alignment with customer requirements.
- Compliance and ethics – Monitors compliance with code of conduct and directs fraud investigations.
- Accounting/financial compliance – Directs the Sarbanes–Oxley Section 302 and 404 assessment, which identifies financial reporting risks.
- Law department – Manages litigation and analyzes emerging legal trends that may impact the organization.
- Insurance – Ensures the proper insurance coverage for the organization.

- Treasury – Ensures cash is sufficient to meet business needs while managing risk related to commodity pricing or foreign exchange.
- Operational quality assurance – Verifies operational output is within tolerances.
- Operations management – Ensures that the business runs day to day and that related barriers are surfaced for resolution.
- Credit – Ensures any credit provided to customers is appropriate to their ability to pay.
- Customer service – Ensures customer complaints are handled promptly and root causes are reported to operations for resolution.
- Internal audit – Evaluates the effectiveness of each of the above risk functions and recommends improvements.

One major difference is that ERM really seeks to also identify opportunities and the risk can be the possibility of missing those opportunities. For instance, take a small investment firm that wants to invest in something that will turn a good profit.

- First choice – The company could invest in a relatively safe venture, but the returns would be low.
- Second choice – The company could invest in a riskier venture and receive a reasonably good return but a moderate probability of loss.
- Third choice – The company could invest in a risk venture and possibly have a great return or lose a chunk of change.

The second choice is probably the most logical. This maximizes profits while minimizing risks.

The curve in Figure 26.2 gives a graphical representation of this idea.

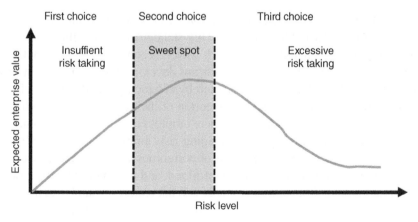

FIGURE 26.2   Risk versus reward.

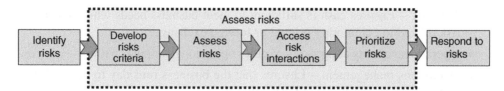

FIGURE 26.3 Risk assessment flow for ERM.

Decision analysis techniques are discussed in Chapter 16. These techniques can be used to help determine what would be the best choice.

Though the same tools can be used to assess the risks for ERM, the assessment flow is slightly different. Figure 26.3 shows a model assessment flow for ERM (5).

COSO discusses a risk impact scale, which is analogous to qualitative risk scales used in other areas. However, in this context, it more ties to financial impact (5). The five levels they present are as follows:

- Incidental.
- Minor.
- Moderate.
- Major.
- Extreme.

Each organization would need to determine what these levels mean to them.

26.3 INTERNATIONAL ENTERPRISE RISK MANAGEMENT STANDARD: ISO 31000

International Organization for Standardization (ISO) developed an ERM standard. This is ISO 31000 – Risk Management. This standard is copyrighted, so it is not possible to provide a detailed description. However, according to the ISO 31000 marketing brochure, "ISO 31000 helps organizations develop a risk management strategy to effectively identify and mitigate risks, thereby enhancing the likelihood of achieving their objectives and increasing the protection of their assets. Its overarching goal is to develop a risk management culture where employees and stakeholders are aware of the importance of monitoring and managing risk. Implementing ISO 31000 also helps organizations see both the positive opportunities and negative consequences associated with risk, and allows for more informed, and thus more effective, decision making, namely in the allocation of resources. What's more, it can be an active component in improving an organization's governance and, ultimately, its performance."

Please visit the ISO website for more information (6).

26.4 SUMMARY

ERM is a process designed to identify potential events that may affect an institution, and manage risk to be within an acceptable level, to provide reasonable assurance regarding the achievement of institutional objectives. ERM:

- Is a tool to enhance management decision making, corporate governance, and accountability.
- Facilitates effective management of the uncertainty and associated risks and opportunities facing an organization.
- Helps an organization "get to where it wants to go, and avoid pitfalls and surprises along the way."
- Is a systematic approach to a historically intuitive exercise.

Activity   Review the following case study written by Justin Walters and discuss the risks and opportunities.

26.5 INTRODUCTION

The following case study is based on information found in Refs. (7–9).

In the early 1990s Pepsi was losing the soda battle to Coca-Cola in the Philippines. The Philippines at the time was the twelfth largest soda-consuming country in the world and held a large amount of profit for Pepsi to make if they could overtake Coca-Cola as the leading selling soda company. In 1989 Pepsi released Pepsi A.M. in the country to try to appeal to a morning crowd who did not drink coffee for religious reasons. This attempt failed dramatically and sent Pepsi back to the drawing board to begin a new effort to overtake Coca-Cola. In 1992 marketing executives came up with a lottery-style idea of printing numbers on the bottom of the bottle caps of their best-selling soda brands in that country – Pepsi, Mountain Dew, and 7 Up. Monetary prizes would be given to those buyers who had specific numbers printed on their caps. The values would vary between 100 pesos (5 USD) and 1 000 000 pesos (40 000 USD). Most prizes would be the 100 pesos, but two lucky winners would get the 1 000 000 pesos prize. The game was a success, and Pepsi was able to increase sales by 40% and capture 26% of the country's market share. Two months after starting the promotion, the winning number for the 1 000 000 pesos prize was announced as 349. Once multiple buyers began coming forward with the winning bottle cap, Pepsi executives realized that 800 000 bottle caps had been printed with the number 349. Once the incident had been discovered, buyers were being turned away as they came to collect their prize money, which created unrest in the region. Riots ensued, and many Pepsi executives were sent death threats daily. Once the high level of emotions started to subside, many lawsuits were filed against Pepsi alleging that the money was still owed to all of the bottle cap holders.

26.6 ANALYSIS

26.6.1 Management Systems

In all articles, it is stated that a subcontractor, D.G. Consultores, was responsible for picking the 60 winning number combinations. D.G. Consultores is a marketing firm based in Mexico, and they were using a computer system to determine what the winning number combinations would be random. Once those numbers were chosen, they were sealed in an envelope and secured in a safe deposit box in Manila, which is the capital of the Philippines. Pepsi had reportedly sent the marketing firm a list of numbers they were not supposed to consider, and the number 349 was on that list. Several problems could have arisen by using this method. The first problem being that Pepsi could have kept the responsibility of picking the winning numbers with their in-house marketing department. This would have ensured that the employees selecting numbers were in the direct chain of management under the executives who came up with the idea and likely would have been supervised differently than an outside company. Management at Pepsi should also be held accountable for not pursuing a response to the memo sent to D.G. Consultores that contained the numbers they were not allowed to select as winning numbers. There is also evidence that a project plan had not been thoroughly developed to select and verify winning numbers or on programming printers and whose responsibility it was to ensure the product coming off the line was correct. A security plan was also lacking to provide that someone at the marketing company, as well as Pepsi, was able to check and verify numbers before products were shipped while still ensuring that nobody would leak that information to the public. There was also a safety concern once the rioting started that should have had management accountability. With the rioting and destructive nature, the citizens of Manila were exhibiting it is commendable that Pepsi paid for bodyguards as well as armed passengers in delivery trucks; however, with some forethought into the planning process, those security measures would not have been necessary. Upon realization of the incident, executives did offer to pay 20 USD to all buyers who held a supposedly winning cap. This was a good faith effort, and although it did not carry weight with the public, it was a step in the right direction to set things right.

26.6.2 Equipment and Employee Systems

While computers have simplified many aspects of everyday life, they can still have glitches, and in 1992 they were no different. While the computer program that was used to select the winning numbers randomly is unknown, it is highly possible that an error occurred in the program and the numbers it was programmed not to select were ignored or overwritten. This could be amplified by the security measures placed after the numbers were selected and sealed in an envelope and sent to a safe deposit box in Manila. If the employees were not allowed to verify that the selected numbers were different from the blacklisted numbers, the computer error would not be caught.

If the memo was received by the marketing firm, an employee should be held accountable for the lack of information being shared, or if management was aware of the blacklisted numbers, they should be held responsible for not incorporating that information into their computer system. One aspect of the equipment used that did give grounds to turn buyers away for a warranted reason was that the cap not only had to have the correct number but also needed to have the corresponding security code. Upon investigation, only two caps were printed with the winning number and the correct security code. This potentially could mean that Pepsi employees programmed their printing equipment twice with the same number, but used a different security code both times. Again with the security requirements that were in effect with these numbers, there is a chance that employees were not aware that the two caps that were printed had the same number as caps that had been printed 800 000 times.

26.6.3 Environment Systems

Many companies complete work internationally for many different industries. In this case, being located continents away could have resulted in the incident taking place. The memo having to travel either through the physical mail system or electronically could have been misrouted and never received by the marketing firm. The distance the memo went could also have taken more time to arrive than anticipated or could have been processed at the marketing firm, which delayed the information from being incorporated by the marketing firm. At the time the Philippines was an impoverished region, and the promise of two citizens being able to win 1 000 000 pesos caused emotions to run high and drastic actions to be taken when the promotion was pulled.

26.6.4 Monitoring

As in all incident analysis, hindsight is 20/20, and it is far easier to spot faults after the fact. For all of these countermeasures to be effective, Pepsi executives would have to be accountable or would have to appoint specific employees to be responsible for ensuring the promotion went smoothly. Involving a network specialist would ensure the computer programs work as efficiently as possible with a limited amount of glitches.

26.7 SUMMARY

This case study demonstrates a lack of ERM. Discuss how the risks could have been identified and handled better.

REFERENCES

1. Committee of Sponsoring Organizations of the Treadway Commission (COSO) (2004). *Enterprise Risk Management: Integrated Framework*, White Paper. Jersey City, NJ: COSO.

2. Enterprise Risk Management Committee (2003). Overview of Enterprise Risk Management. Casualty Actuarial Society: 11–13. (accessed September 2018)
3. Stanford University (2018). Enterprise Risk Management Objectives. https://acrp.stanford.edu/erm/purpose (accessed September 2018).
4. Emily Flitter and Glenn Thrush (2018). Wells Fargo Said to Be Target of $1 Billion U.S. Fine. *The New York Times* (19 April 2018).
5. Committee of Sponsoring Organizations of the Treadway Commission (COSO) (2012). *Risk Assessment in Practice*. Deloitte & Touche LLP.
6. International Organization for Standardization, ISO 31000 (2018). Risk Management. https://www.iso.org/iso-31000-risk-management.html (accessed September 2018).
7. Rossen, J. (2018). The Computer Error That Led to a Country Declaring War on Pepsi. http://mentalfloss.com/article/558202/pepsi-number-fever-promotion-failure-philippines (accessed September 2018).
8. Srivastava, A. (1992). Marketing disaster: Pepsi-Cola's "Number Fever" Fiasco, Feb–May 1992. https://marketinglessons.in/marketing-disaster-pepsi-colas-number-fever-fiasco-feb-may-1992 (accessed September 2018).
9. Marketing Schmarketing (2015). Pepsi launched a campaign, proclaimed the "wrong lucky number," caused huge riots. http://marketingshmarketing.net/post/119207342851/pepsi-number-fever-marketing-fail (accessed September 2018).

CHAPTER 27

Process Safety Management and Hazard and Operability Assessment

27.1 INTRODUCTION

This chapter discusses process safety management (PSM), process hazard analysis (PHZA), and hazard and operability analysis (HAZOP). Process hazard analysis is abbreviated as PHZA to distinguish between this analysis method and preliminary hazard analysis (PHA). This chapter is adapted from the Occupational Safety and Health Administration's (OSHA) Process Safety Management Guidelines (1) and other sources.

27.2 PURPOSE

The major objective of PSM of highly hazardous chemicals is to prevent unwanted releases of hazardous chemicals especially into locations that could expose employees and others to serious hazards. An effective PSM program requires a systematic approach to evaluating the whole chemical process. Using this approach, the process design, process technology, process changes, operational and maintenance activities and procedures, nonroutine activities and procedures, emergency preparedness plans and procedures, training programs, and other elements that affect the process are all considered in the evaluation.

Risk Assessment: Tools, Techniques, and Their Applications, Second Edition. Lee T. Ostrom and Cheryl A. Wilhelmsen.
© 2019 John Wiley & Sons, Inc. Published 2019 by John Wiley & Sons, Inc.
Companion website: www.wiley.com/go/Ostrom/RiskAssessment_2e

27.2.1 Application

The various lines of defense that have been incorporated into the design and operation of the process to prevent or mitigate the release of hazardous chemicals need to be evaluated and strengthened to ensure their effectiveness at each level. PSM is the proactive identification, evaluation, and mitigation or prevention of chemical releases that could occur because of failures in processes, procedures, or equipment.

The PSM standard targets highly hazardous chemicals that have the potential to cause a catastrophic incident. The purpose of the standard as a whole is to aid employers in their efforts to prevent or mitigate episodic chemical releases that could lead to a catastrophe in the workplace and possibly in the surrounding community.

OSHA believed PSM would have a positive effect on the safety of employees and will offer other potential benefits to employers, such as increased productivity, and smaller businesses that may have limited resources to them at this time might consider alternative avenues of decreasing the risks associated with highly hazardous chemicals at their workplaces. One method that might be considered is reducing inventory of the highly hazardous chemical. This reduction in inventory will result in reducing the risk or potential for a catastrophic incident. Also, employers, including small employers, may establish more efficient inventory control by reducing, to below the established threshold, the quantities of highly hazardous chemicals on-site. This reduction can be accomplished by ordering smaller shipments and maintaining the minimum inventory necessary for efficient and safe operation. When reduced inventory is not feasible, the employer might consider dispersing inventory to several locations on-site. Dispersing storage into locations so that a release in one location will not cause a release in another location is also a practical way to reduce the risk or potential for catastrophic incidents.

OSHA, through the Voluntary Protection Program (VPP) process, determined businesses were collecting lagging and leading indicators of performance (2). These include the following.

Lagging Metrics

- **Injury and/or Incident Reports Related to Process Safety**: Incident reports are created after an incident investigation has been completed. Incident reports typically describe the causes of an incident that were identified by the investigation and the corrective measures that should be taken to address those causes. VPP sites have used metrics to track a number of process safety incidents and injuries including:
  - Near-miss incidents reported that did or could have led to a loss of containment.
  - Recordable injuries and first aid incidents due to loss of primary containment.
  - Number of incidents versus number of incidents with formal reports.
  - Status of incident investigations.

- **Loss of Containment**: A loss of containment is an unplanned or uncontrolled release of materials. For incidents related to loss of containment, VPP facilities have tracked:
  - The number of incidents.
  - Whether there was primary or secondary containment.
  - The cause and location of the incident.

Leading Metrics

- **Management of Change (MOC)**: A management of change (MOC) is a system that identifies, reviews, and approves all modifications to equipment, procedures, raw materials, and processing conditions other than "replacements in kind," prior to implementation. There are various types of changes that occur in the workplace where a facility may want to track MOC to reduce the likelihood of system failures or catastrophic events. For MOC, VPP facilities have tracked:
  - Overdue MOCs.
  - Approved MOCs.
  - Open MOCs.
  - MOCs performed each month.
- **Preventive Maintenance (PM)**: Preventive maintenance is maintenance that is regularly performed on a piece of equipment to decrease the likelihood of it failing. In their maintenance efforts, VPP facilities have tracked:
  - Completion rates.
  - Open items.
  - Overdue safety critical PMs.
  - Number of inspections.
- **PHZA**: VPP facilities have monitored the PHZA process by tracking:
  - PHZA actions open.
  - PHZAs overdue.
  - PHZAs completed.
  - Scheduled versus completed PHZAs.
  - Status of PHZA/incident recommendations.
  - Status of scheduled PHZA revalidations.

27.2.2 Hazards of the Chemicals Used in the Process

Complete and accurate written information concerning process chemicals, process technology, and process equipment is essential to an effective PSM program and to a PHZA. The compiled information will be a necessary resource to a variety of users including the team performing the PHZA as required by PSM, those developing the training programs and the operating procedures, contractors whose employees will be working with the process, those conducting the pre-startup reviews, local emergency preparedness planners, and insurance and enforcement officials.

The information to be compiled about the chemicals, including process intermediates, needs to be comprehensive enough for an accurate assessment of the fire and explosion characteristics, reactivity hazards, the safety and health hazards to workers, and the corrosion and erosion effects on the process equipment and monitoring tools. Current material safety data sheet (MSDS) information can be used to help meet this but must be supplemented with process chemistry information, including runaway reaction and overpressure hazards, if applicable.

27.2.3 Technology of the Process

Process technology information will be a part of the process safety information package and should include employer-established criteria for maximum inventory levels for process chemicals, limits beyond which would be considered upset conditions, and a qualitative estimate of the consequences or results of deviation that could occur if operating beyond the established process limits. Employers are encouraged to use diagrams that will help users understand the process.

A block flow diagram is used to show the major process equipment and interconnecting process flow lines and flow rates, stream composition, temperatures, and pressures when necessary for clarity. The block flow diagram is a simplified diagram. The Critical Incident Technique (CIT) (Chapter 13) discusses how block flow diagrams can be created.

Process flow diagrams are more complex and show all main flow streams including valves to enhance the understanding of the process as well as pressures and temperatures on all feed and product lines within all major vessels and in and out of headers and heat exchangers and points of pressure and temperature control. Figure 27.1 is a simplified block flow diagram created using symbols from Chapter 13. Also, information on construction materials, pump capacities and pressure heads, compressor horsepower, and vessel design pressures and temperatures are shown when necessary for clarity. In addition, process flow diagrams usually show major components of control loops along with key utilities.

27.2.4 Equipment in the Process

Piping and instrument diagrams (P&IDs) may be the more appropriate type diagrams to show some of the above details as well as display the information for the piping designer and engineering staff. The P&IDs are to be used to describe the relationships

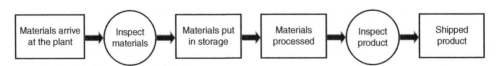

FIGURE 27.1 Simplified block flow diagram.

between equipment and instrumentation as well as other relevant information that will enhance clarity. Computer software programs that do P&IDs or other diagrams useful to the information package may be used to help meet this requirement. Figure 27.2 shows a P&ID for an energy conversion loop (ECL) (3). Figure 27.3 is a photograph of the ECL and Figure 27.4 is a Solidworks drawing of the ECL.

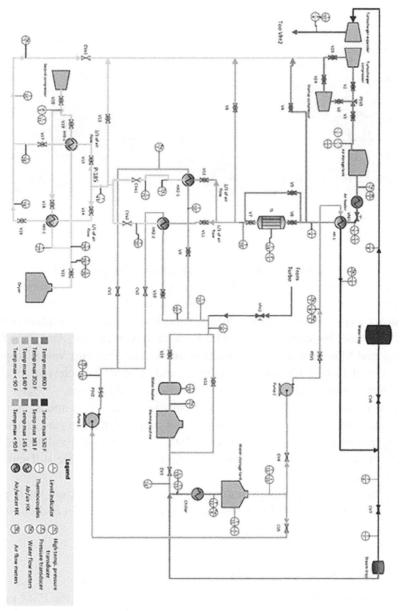

FIGURE 27.2   Piping and instrument diagrams (P&ID) for an energy conversion loop.

FIGURE 27.3   Photo of the energy conversion loop.

The information pertaining to process equipment design must be documented. In other words, what codes and standards were relied on to establish good engineering practice? These codes and standards are published by such organizations as the American Society of Mechanical Engineers, American Petroleum Institute, American National Standards Institute, National Fire Protection Association, American Society for Testing and Materials, National Board of Boiler and Pressure Vessel Inspectors, National Association of Corrosion Engineers, and American Society of Exchanger Manufacturers Association, as well as model building code groups.

For existing equipment designed and constructed many years ago in accordance with the codes and standards available at that time and no longer in general use today, the employer must document which codes and standards were used and that the design and construction along with the testing, inspection, and operation are still suitable for the intended use.

Where the process technology requires a design that departs from the applicable codes and standards, the employer must document that the design and construction are suitable for the intended purpose.

27.2.5 Process Hazard Analysis

A PHZA, or evaluation, is one of the most important elements of the PSM program. It is an organized and systematic effort to identify and analyze the significance of potential hazards associated with the processing or handling of highly hazardous

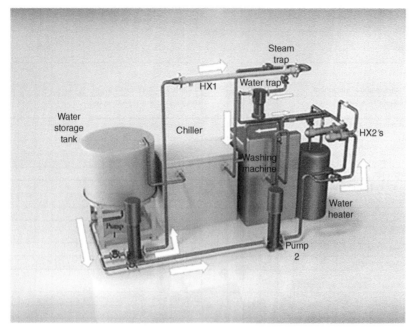

FIGURE 27.4   Solidworks drawing of the energy conversion loop.

chemicals. It also provides information that will assist employers and employees in making decisions for improving safety and reducing the consequences of unwanted or unplanned releases of hazardous chemicals.

A PHZA analyzes potential causes and consequences of fires, explosions, releases of toxic or flammable chemicals, and major spills of hazardous chemicals. The PHZA focuses on equipment, instrumentation, utilities, human actions (routine and nonroutine), and external factors that might affect the process.

The selection of a PHZA methodology or technique is influenced by factors including how much is known about the process. Is it a process that has been operated for a long period of time with little or no innovation and extensive experience has been generated with its use? Or is it a new process or one that has been changed frequently by the inclusion of innovation features? Also, the size and complexity of the process will influence the decision as to the appropriate PHZA methodology to use. All PHZA methodologies are subject to certain limitations. For example, the checklist methodology works well when the process is very stable and no changes are made, but it is not as effective when the process has undergone extensive change. The checklist may miss the most recent changes and consequently they would not be evaluated. Figure 27.5 is a checklist from NASA Procedural Guide (4). Another limitation to be considered concerns the assumptions made by the team or analyst. The PHZA is dependent on good judgment, and the assumptions made during the study need to be documented and understood by the team and reviewer and kept for a future PHZA.

Event Name:					Civil Servant Sponsor Phone:				ORG Code:	
Event Building/Location:									Date:		
Item	Yes	No	N/A	Date Fixed	Item	Yes	No	N/A	Fixed Date		
Pre-Event Planning and Management					**Material Storage/Handling**						
Safety Notification Checklist (Form 847) Completed					Materials properly stored/stacked						
Emergency numbers/contacts posted & distributed to staff and volunteers					Dust protection adequate						
Event Safety Plan Prepared					Loads lifted correctly						
Safety Briefing Prepared					**Poster Boards or Displays**						
Housekeeping/sanitation					Secured from falling						
Road closure and traffic control plan prepared					Placed on a level surface						
Size and type of crowd analyzed					**Compressed Gases**						
Hand washing/toilet facilities					Cylinders secured						
Clean eating/dining area ADA					Valve cap in place when not in use						
Fire Prevention					MSDS posted at use location						
Fire extinguishers available					**Ladders**						
Correct extinguishers for the type of fire anticipated (ie: paper, chemical, electrical, etc.)					Ladders in good condition						
No smoking posted and enforced					Side rails extend 36" above landing						
Stages/Speakers Platforms					Proper for job & secure						
Level					Inspected prior to use						
Free of trip hazards					**Scaffolding**						
Tents					Equipment in good condition						
Support posts erected safely					Scaffold is tied to structure						
Posts tied down and secured by water barrels. Note: ground stakes are not permitted					Guardrails, top, mid, toe boards in place						
Pole support lines do not present a hazard					Connections sound & secure						
Hand & Power Tools					Planking cleats in place						
Hand tools in good working condition					Worker protected from falling objects						
Cords in good condition					**Welding & Cutting**						
All mechanical safeguards in place					Screen & shields in place						
Proper tools utilized for each job					Electrical equipment grounded						
Tools grounded or double insulated					Compressed gas cylinders secure/upright						
Heavy Equipment					Proper personnel protection utilized						
Certified/Trained Operator					Fire extinguishers immediately available						
Brakes, lights, signals & alarms operable					Welding cables in good condition						
Wheels chocked when necessary					**Personal Protective Equipment**						
Seat belt equipped and used					Hardhats worn						
Pre-use inspection performed					Gloves available & used						
Barricades & Fencing					Steel toe footwear						
Site fenced					Eye protection utilized						
Roadways & sidewalks fenced					Ear protection utilized						
Floor openings planked or barricaded					Safety belts & lanyards utilized						
Access/traffic controlled					Respirators & masks utilized						

FIGURE 27.5 NASA Ames Procedural Guide example checklist.

The team conducting the PHZA needs to understand the methodology that is going to be used. A PHZA team can vary in size from two people to a number of people with varied operational and technical backgrounds. Some team members may be part of the team for only a limited time. The team leader needs to be fully knowledgeable in the proper implementation of the PHZA methodology to be used and should be impartial in the evaluation. The other full- or part-time team members need to provide the team with expertise in areas such as process technology; process design; operating procedures and practices; alarms; emergency procedures; instrumentation; maintenance procedures, both routine and nonroutine tasks, including how the tasks are authorized; procurement of parts and supplies; safety and health; and any other relevant subjects. At least one team member must be familiar with the process.

The ideal team will have an intimate knowledge of the standards, codes, specifications, and regulations applicable to the process being studied. The selected team members need to be compatible and the team leader needs to be able to manage the team and the PHZA study.

The team needs to be able to work together while benefiting from the expertise of others on the team or outside the team to resolve issues and to forge a consensus on the findings of the study and recommendations.

The application of a PHZA to a process may involve the use of different methodologies for various parts of the process. For example, a process involving a series of unit operations of varying sizes, complexities, and ages may use different methodologies and team members for each operation. Then the conclusions can be integrated into one final study and evaluation.

A more specific example is the use of a PHZA checklist for a standard boiler or heat exchanger and the use of a HAZOP PHZA for the overall process. Also, for batch-type processes like custom batch operations, a generic PHZA of a representative batch may be used where there are only small changes of monomer or other ingredient ratio and the chemistry is documented for the full range and ratio of batch ingredients. Another process where the employer might consider using a generic type of PHZA is a gas plant. Often these plants are simply moved from site to site, and therefore, a generic PHZA may be used for these moveable plants. Also, when an employer has several similar size gas plants and no sour gas is being processed at the site, a generic PHZA is feasible as long as the variations of the individual sites are accounted for in the PHZA.

Finally, when an employer has a large continuous process with several control rooms for different portions of the process, such as for a distillation tower and a blending operation, the employer may wish to do each segment separately and then integrate the final results.

Small businesses covered by this rule often will have processes that have less storage volume and less capacity and may be less complicated than processes at a large facility. Therefore, OSHA would anticipate that the less complex methodologies would be used to meet the PHZA criteria in the standard. These PHZAs can be done in less time and with fewer people being involved. A less complex process generally means that less data, P&IDs, and process information are needed to perform a PHZA.

Many small businesses have processes that are not unique, such as refrigerated warehouses or cold storage lockers or water treatment facilities. Where employer associations have a number of members with such facilities, a generic PHZA, evolved from a checklist or what-if questions, could be developed and effectively used by employers to reflect their particular process; this would simplify compliance for them.

When the employer has a number of processes that require a PHZA, the employer must set up a priority system to determine which PHZAs to conduct first. A PHA may be useful in setting priorities for the processes that the employer has determined are subject to coverage by the PSM standard. Consideration should be given first to those processes with the potential of adversely affecting the largest number of employees. This priority setting also should consider the potential severity of a chemical release, the number of potentially affected employees, the operating history of the process such as the frequency of chemical releases, the age of the process, and any other relevant factors. Together, these factors would suggest a ranking order using either a weighting factor system or a systematic ranking method. The use of a PHA will assist an employer in determining which process should be of the highest priority for hazard analysis resulting in the greatest improvement in safety at the facility occurring first.

Detailed guidance on the content and application of PHZA methodologies is available from the American Institute of Chemical Engineers' Center for Chemical Process Safety, 345 E. 47th Street, New York, New York 10017, (212)705-7319.

27.2.6 Operating Procedures

Operating procedures describe tasks to be performed, data to be recorded, operating conditions to be maintained, samples to be collected, and safety and health precautions to be taken. The procedures need to be technically accurate, understandable to employees, and revised periodically to ensure that they reflect current operations. The process safety information package helps to ensure that the operating procedures and practices are consistent with the known hazards of the chemicals in the process and that the operating parameters are correct. Operating procedures should be reviewed by engineering staff and operating personnel to ensure their accuracy and that they provide practical instructions on how to actually carry out job duties safely. Also, the employer must certify annually that the operating procedures are current and accurate.

Operating procedures provide specific instructions or details on what steps are to be taken or followed in carrying out the stated procedures. The specific instructions should include the applicable safety precautions and appropriate information on safety implications. For example, the operating procedures addressing operating parameters will contain operating instructions about pressure limits, temperature ranges, flow rates, what to do when an upset condition occurs, what alarms and instruments are pertinent if an upset condition occurs, and other subjects. Another example of using operating instructions to properly implement operating procedures is in starting up or shutting down the process. In these cases, different parameters will be required from those of normal operation. These operating instructions need to clearly indicate the distinctions between startup and normal operations, such as the appropriate allowances for heating up a unit to reach the normal operating parameters. Also, the operating instructions need to describe the proper method for increasing the temperature of the unit until the normal operating temperatures are reached.

Computerized process control systems add complexity to operating instructions. These operating instructions need to describe the logic of the software as well as the relationship between the equipment and the control system; otherwise, it may not be apparent to the operator.

Operating procedures and instructions are important for training operating personnel. The operating procedures are often viewed as the standard operating practices (SOPs) for operations. Control room personnel and operating staff, in general, need to have a full understanding of operating procedures. If workers are not fluent in English, then procedures and instructions need to be prepared in a second language understood by the workers. In addition, operating procedures need to be changed when there is a change in the process. The consequences of operating procedure changes need to be fully evaluated and the information conveyed to the personnel. For example, mechanical changes to the process made by the maintenance department (like changing a valve from steel to brass or other subtle changes) need to be evaluated to determine whether operating procedures and practices also need to be changed. All MOC actions must be coordinated and integrated with current operating procedures, and operating personnel must be alerted to the changes in procedures before the change is made. When the process is shut down to make a change, then the operating procedures must be updated before restarting the process.

Training must include instruction on how to handle upset conditions as well as what operating personnel are to do in emergencies such as pump seal failures or pipeline ruptures. Communication among operating personnel and workers within the process area performing nonroutine tasks also must be maintained. The hazards of the tasks are to be conveyed to operating personnel in accordance with established procedures and to those performing the actual tasks. When the work is completed, operating personnel should be informed to provide closure on the job.

27.2.7 Employee Training

All employees, including maintenance and contractor employees involved with highly hazardous chemicals, need to fully understand the safety and health hazards of the chemicals and processes they work with so they can protect themselves, their fellow employees, and the citizens of nearby communities. Training conducted in compliance with the OSHA Hazard Communication standard (*Title 29 Code of Federal Regulations* [CFR] Part 1910.1200) will inform employees about the chemicals they work with and familiarize them with reading and understanding MSDSs. However, additional training in subjects such as operating procedures and safe work practices, emergency evacuation and response, safety procedures, routine and nonroutine work authorization activities, and other areas pertinent to process safety and health need to be covered by the employer's training program.

In establishing their training programs, employers must clearly identify the employees to be trained, the subjects to be covered, and the goals and objectives they wish to achieve. The learning goals or objectives should be written in clear measurable terms before the training begins. These goals and objectives need to be tailored to each of the specific training modules or segments. Employers should describe the important actions and conditions under which the employee will demonstrate competence or knowledge as well as what is acceptable performance.

Hands-on training, where employees actually apply lessons learned in simulated or real situations, will enhance learning. For example, operating personnel, who will work in a control room or at control panels, would benefit by being trained at a simulated control panel. Upset conditions of various types could be displayed on the simulator, and then the employee could go through the proper operating procedures to bring the simulator panel back to the normal operating parameters. A training environment could be created to help the trainee feel the full reality of the situation but under controlled conditions. This type of realistic training can be very effective in teaching employees correct procedures while allowing them also to see the consequences of what might happen if they do not follow established operating procedures. Other training techniques using videos or training also can be very effective for teaching other job tasks, duties, or imparting other important information. An effective training program will allow employees to fully participate in the training process and to practice their skills or knowledge.

Employers need to evaluate periodically their training programs to see if the necessary skills, knowledge, and routines are being properly understood and implemented by their trained employees. The methods for evaluating the training should be developed along with the training program goals and objectives. Training program evaluation will help employers to determine the amount of training their employees understood and whether the desired results were obtained. If, after the evaluation, it appears that the trained employees are not at the level of knowledge and skill that was expected, the employer should revise the training program, provide retraining, or provide more frequent refresher training sessions until the deficiency is resolved. Those who conducted the training and those who received the training also should be consulted as to how best to improve the training process. If there is a language barrier, the language known to the trainees should be used to reinforce the training messages and information.

Careful consideration must be given to ensure that employees, including maintenance and contract employees, receive current and updated training. For example, if changes are made to a process, affected employees must be trained in the changes and understand the effects of the changes on their job tasks. Additionally, as already discussed, the evaluation of the employee's absorption of training will certainly determine the need for further training.

27.2.8 Contractors

Employers who use contractors to perform work in and around processes that involve highly hazardous chemicals have to establish a screening process so that they hire and use only contractors who accomplish the desired job tasks without compromising the safety and health of any employees at a facility. For contractors whose safety performance on the job is not known to the hiring employer, the employer must obtain information on injury and illness rates and experience and should obtain contractor references. In addition, the employer must ensure that the contractor has the appropriate job skills, knowledge, and certifications (e.g. for pressure vessel welders). Contractor work methods and experience should be evaluated. For example, does the contractor conducting demolition work swing loads over operating processes, or does the contractor avoid such hazards?

Maintaining a site injury and illness log for contractors is another method employers must use to track and maintain current knowledge of activities involving contract employees working on or adjacent to processes covered by PSM. Injury and illness logs of both the employer's employees and contract employees allow the employer to have full knowledge of process injury and illness experience. This log contains information useful to those auditing PSM compliance and those involved in incident investigations.

Contract employees must perform their work safely. Considering that contractors often perform very specialized and potentially hazardous tasks, such as confined space entry activities and nonroutine repair activities, their work must be controlled while they are on or near a process covered by PSM. A permit system or work authorization system for these activities is helpful for all affected employers. The use of a work authorization system keeps an employer informed of contract employee activities. Thus, the employer has better coordination and more management control over the work being performed in the process area. A well-run and well-maintained process, where employee safety is fully recognized, benefits all of those who work in the facility whether they are employees of the employer or the contractor.

27.2.9 Pre-startup Safety Review

For new processes, the employer will find a PHZA helpful in improving the design and construction of the process from a reliability and quality point of view. The safe operation of the new process is enhanced by making use of the PHZA recommendations before final installations are completed. P&IDs should be completed, the operating procedures put in place, and the operating staff trained to run the process, before startup. The initial startup procedures and normal operating procedures must be fully evaluated as part of the pre-startup review to ensure a safe transfer into the normal operating mode.

For existing processes that have been shut down for turnaround or modification, the employer must ensure that any changes other than "replacement in kind" made to the process during shutdown go through the MOC procedures. P&IDs will need to be updated, as necessary, as well as operating procedures and instructions. If the changes made to the process during shutdown are significant and affect the training program, then operating personnel as well as employees engaged in routine and nonroutine work in the process area may need some refresher or additional training. Any incident investigation recommendations, compliance audits, or PHZA recommendations need to be reviewed to see what affect they may have on the process before beginning the startup.

27.2.10 Mechanical Integrity of Equipment

Employers must review their maintenance programs and schedules to see if there are areas where "breakdown" is used rather than the more preferable ongoing mechanical integrity program. Equipment used to process, store, or handle highly hazardous chemicals has to be designed, constructed, installed, and maintained to minimize the risk of releases of such chemicals. This requires that a mechanical integrity program be in place to ensure the continued integrity of process equipment.

Elements of a mechanical integrity program include identifying and categorizing equipment and instrumentation, inspections, and tests and their frequency; maintenance procedures; training of maintenance personnel; criteria for acceptable test results; documentation of test and inspection results; and documentation of manufacturer recommendations for equipment and instrumentation.

27.2.11 Process Defenses

The first line of defense an employer has is to operate and maintain the process as designed and to contain the chemicals. This is backed up by the second line of defense, which is to control the released chemicals through venting to scrubbers or flares, or to surge or overflow tanks designed to receive such chemicals. This also would include fixed fire protection systems like sprinklers, water spray, or deluge systems, monitor guns, dikes, designed drainage systems, and other systems to control or mitigate hazardous chemicals once an unwanted release occurs.

27.2.12 Written Procedures

The first step of an effective mechanical integrity program is to compile and categorize a list of process equipment and instrumentation to include in the program. This list includes pressure vessels, storage tanks, process piping, relief and vent systems, fire protection system components, emergency shutdown systems and alarms, and interlocks and pumps. For the categorization of instrumentation and the listed equipment, the employer should set priorities for which pieces of equipment require closer scrutiny than others.

27.2.13 Inspection and Testing

The mean time to failure of various instrumentation and equipment parts would be known from the manufacturer's data or the employer's experience with the parts, which then influence inspection and testing frequency and associated procedures. Also, applicable codes and standards – such as the National Board inspection Code, or those from the American Society for Testing and Materials, American Petroleum Institute, National Fire Protection Association, American National Standards institute, American Society of Mechanical Engineers, and other groups – provide information to help establish an effective testing and inspection frequency, as well as appropriate methodologies.

The applicable codes and standards provide criteria for external inspections for such items as foundation and supports, anchor bolts, concrete or steel supports, guy wires, nozzles and sprinklers, pipe hangers, grounding connections protective coatings and insulation, and external metal surfaces of piping and vessels. These codes and standards also provide information on methodologies for internal inspection and a frequency formula based on the corrosion rate of the materials of construction. Also, internal and external erosion must be considered along with corrosion effects for piping and valves. Where the corrosion rate is not known, a maximum inspection frequency is recommended (methods of developing the corrosion rate are available in the codes). Internal inspections need to cover items such as the vessel shell, bottom and head; metallic linings; nonmetallic linings; thickness measurements for vessels and piping; inspection for erosion, corrosion, cracking, and bulges; and internal equipment like trays, baffles, sensors ad screens for erosion, corrosion, or cracking and other deficiencies. Some of these inspections may be performed by state or local government inspectors under state and local statutes. However, each employer must develop procedures to ensure that tests and inspections are conducted properly and that consistency is maintained even where different employees may be involved. Appropriate training must be provided to maintenance personnel to ensure that they understand the PM program procedures, safe practices, and the proper use and application of special equipment or unique tools that may be required. This training is part of the overall training program called for in the standard.

27.2.14 Quality Assurance

A quality assurance system helps ensure the use of proper materials of construction, the proper fabrication and inspection procedures, and appropriate installation procedures that recognize field installation concerns. The quality assurance program is an essential part of the mechanical integrity program and will help maintain the primary and secondary lines of defense designed into the process to prevent unwanted chemical releases or to control or mitigate a release. "As built" drawings, together with certifications of coded vessels and other equipment and of construction, must be verified and retained in the quality assurance documentation.

Equipment installation jobs need to be properly inspected in the field for use of proper materials and procedures and to ensure that qualified craft workers do the job. The use of appropriate gaskets, packing, bolts, valves, lubricants, and welding rods needs to be verified in the field. Also, procedures for installing safety devices need to be verified, such as the torque on the bolts on rupture disk installations, uniform torque on flange bolts, and proper installation of pump seals. If the quality of parts is a problem, it may be appropriate for the employer to conduct audits of the equipment supplier's facilities to better ensure proper purchases of required equipment suitable for intended service. Any changes in equipment that may become necessary will need to be reviewed for MOC procedures.

27.2.15 Nonroutine Work Authorizations

Nonroutine work conducted in process areas must be controlled by the employer in a consistent manner. The hazards identified involving the work to be accomplished must be communicated to those doing the work and to those operating personnel whose work could affect the safety of the process. A work authorization notice or permit must follow a procedure that describes the steps the maintenance supervisor, contractor representative, or other person needs to follow to obtain the necessary clearance to start the job. The work authorization procedures must reference and coordinate, as applicable, lockout/tagout procedures, line breaking procedures, confined space entry procedures, and hot work authorizations. This procedure also must provide clear steps to follow once the job is completed to provide closure for those that need to know the job is now completed and that equipment can be returned to normal.

27.2.16 Managing Change

To properly manage changes to process chemicals, technology, equipment, and facilities, one must define what is meant by change. In the PSM standard, change includes all modifications to equipment, procedures, raw materials, and processing conditions other than "replacement in kind." These changes must be properly managed by identifying and reviewing them prior to implementing them. For example, the operating procedures contain the operating parameters (pressure limits, temperature ranges, flow rates, etc.) and the importance of operating within these limits. While the operator must have the flexibility to maintain safe operation within the established parameters, any operation outside of these parameters requires review and approval by a written MOC procedure. MOC also covers changes in process technology and changes to equipment and instrumentation. Changes in process technology can result from changes in production rates, raw materials, experimentation, equipment unavailability, new equipment, new product development, change in catalysts, and changes in operating conditions to improve yield or quality. Equipment changes can be in materials of construction, equipment specifications, piping prearrangements, experimental equipment, computer program revisions, and alarms and interlocks. Employers must establish means and methods to detect both technical and mechanical changes.

Temporary changes have caused a number of catastrophes over the years, and employers must establish ways to detect both temporary and permanent changes. It is important that a time limit for temporary changes be established and monitored since otherwise, without control, these changes may tend to become permanent. Temporary changes are subject to the MOC provisions. In addition, the MOC procedures are used to ensure that the equipment and procedures are returned to their original or designed conditions at the end of the temporary change. Proper documentation and review of these changes are invaluable in ensuring that safety and health considerations are incorporated into operating procedures and processes. Employers may wish to develop a form or clearance sheet to facilitate the processing of changes through the MOC procedures. A typical change form may include a description and the purpose of the change, the technical basis for the change, safety and health considerations, documentation of changes for the operating procedures, maintenance procedures, inspection and testing, P&IDs, electrical classification, training and communications, pre-startup inspection, duration (if a temporary change), approvals, and authorization. Where the impact of the change is minor and well understood, a checklist reviewed by an authorized person, with proper communication to others who are affected, may suffice (See Figure 27.2 for a sample request for change form that can be helpful in guiding this procedure.).

For a more complex or significant design change, however, a hazard evaluation procedure with approvals by operations, maintenance, and safety departments may be appropriate. Changes in documents such as P&IDs, raw materials, operating procedures, mechanical integrity programs, and electrical classifications should be noted so that these revisions can be made permanent when the drawings and procedure manuals are updated. Copies of process changes must be kept in an accessible location to ensure that design changes are available to operating personnel as well as to PHZA team members when a PHZA is being prepared or being updated.

27.2.17 Incident Investigation

Incident investigation is the process of identifying the underlying causes of incidents and implementing steps to prevent similar events from occurring. The intent of an incident investigation is for employers to learn from past experiences and thus avoid repeating past mistakes. The incidents OSHA expects employers to recognize and to investigate are the types of events that resulted in or could reasonably have resulted in a catastrophic release. These events are sometimes referred to as "near misses," meaning that a serious consequence did not occur, but could have.

Employers must develop in-house capability to investigate incidents that occur in their facilities. A team should be assembled by the employer and trained in the techniques of investigation including how to conduct interviews of witnesses, assemble needed documentation, and write reports. A multidisciplinary team is better able to gather the facts of the event and to analyze them and develop plausible scenarios as to what happened and why. Team members should be selected on the basis of their training, knowledge, and ability to contribute to a team effort to fully investigate the incident.

Employees in the process area where the incident occurred should be consulted, interviewed, or made a member of the team. Their knowledge of the events represents a significant set of facts about the incident that occurred. The report, its findings, and recommendations should be shared with those who can benefit from the information. The cooperation of employees is essential to an effective incident investigation. The focus of the investigation should be to obtain facts and not to place blame. The team and the investigative process should clearly deal with all involved individuals in a fair, open, and consistent manner.

27.2.18 Emergency Preparedness

Each employer must address what actions employees are to take when there is an unwanted release of highly hazardous chemicals. Emergency preparedness is the employer's third line of defense that will be relied on along with the second line of defense, which is to control the release of chemical. Control releases and emergency preparedness will take place when the first line of defense to operate and maintain the process and contain the chemicals fails to stop the release. In preparing for an emergency chemical release, employers will need to decide the following:

- Whether they want employees to handle and stop small or minor incidental releases.
- Whether they wish to mobilize the available resources at the plant and have them brought to bear on a more significant release.
- Whether employers want their employees to evacuate the danger area and promptly escape to a preplanned safe zone area and then allow the local community emergency response organizations to handle the release.
- Whether the employer wants to use some combination of these actions.

Employers will need to select how many different emergency preparedness or third lines of defense they plan to have, develop the necessary emergency plans and procedures, appropriately train employees in their emergency duties and responsibilities, and then implement these lines of defense.

Employers, at a minimum, must have an emergency action plan that will facilitate the prompt evacuation of employees when there is an unwanted release of a highly hazardous chemical. This means that the employer's plan will be activated by an alarm system to alert employees when to evacuate and that employees who are physically impaired will have the necessary support and assistance to get them to a safe zone. The intent of these requirements is to alert and move employees quickly to a safe zone. Delaying alarms or confusing alarms are to be avoided. The use of process control centers or buildings as safe areas is discouraged. Recent catastrophes indicate that lives are lost in these structures because of their location and because they are not necessarily designed to withstand overpressures from shock waves resulting from explosions in the process area.

When there are unwanted incidental releases of highly hazardous chemicals in the process area, the employer must inform employees of the actions/procedures to take. If the employer wants employees to evacuate the area, then the emergency action plan will be activated. For outdoor processes, where wind direction is important for selecting the safe route to a refuge area, the employers should place a wind direction indicator, such as a wind sock or pennant, at the highest point visible throughout the process area. Employees can move upwind of the release to gain safe access to a refuge area by knowing the wind direction.

If the employer wants specific employees in the release area to control or stop the minor emergency or incidental release, these actions must be planned in advance and procedures developed and implemented. Handling incidental releases for minor emergencies in the process area must include preplanning, providing appropriate equipment for the hazards, and conducting training for those employees who will perform the emergency work before they respond to handle an actual release. The employer's training program, including the Hazard Communication standard training, is to address, identify, and meet the training needs for employees who are expected to handle incidental or minor releases.

Preplanning for more serious releases is an important element in the employer's line of defense. When a serious release of a highly hazardous chemical occurs, the employer, through preplanning, will have determined in advance what actions employees are to take. The evacuation of the immediate release area and other areas, as necessary, would be accomplished under the emergency action plan. If the employer wishes to use plant personnel, such as a fire brigade, spill control team, a hazardous materials team, or employees to render aid to those in the immediate release area and to control or mitigate the incident, refer to OSHA's Hazardous Waste Operations and Emergency Response (HAZWOPER) standard (Title 79CFR Part 1910.1 20). If outside assistance is necessary, such as through mutual aid agreements between employers and local government emergency response organizations, these emergency responders are also covered by HAZWOPER. The safety and health protection required for emergency responders is the responsibility of their employers and of the on-scene incident commander.

Responders may be working under very hazardous conditions; therefore, the objective is to have them competently led by an on-scene incident commander and the commander's staff, properly equipped to do their assigned work safely, and fully trained to carry out their duties safely before they respond to an emergency. Drills, training exercises, or simulations with the local community emergency response planners and responder organizations are one means to obtain better preparedness. This close cooperation and coordination between plant and local community emergency preparedness managers also will aid the employer in complying with the Environmental Protection Agency's Risk Management Plan criteria.

An effective way for medium to large facilities to enhance coordination and communication during emergencies within the plant and with local community organizations is by establishing and equipping an emergency control center. The emergency control center should be located in a safe zone so that it could be occupied throughout the duration of an emergency. The center should serve as the major communication link between the on-scene incident commander and plant or corporate management as well as with local community officials. The communication equipment in the emergency control center should include a network to receive and transmit information by telephone, radio, or other means. It is important to have a backup communication network in case of power failure or if one communication means fails. The center also should be equipped with the plant layout; community maps; utility drawings, including water for fire extinguishing; emergency lighting; appropriate reference materials such as a government agency notification list, company personnel phone list, Superfund Amendments and Reauthorization Act (SARA) Title III reports and MSDSs, emergency plans and procedures manual; a listing the location of emergency response equipment and mutual aid information; and access to meteorological data and any dispersion modeling data.

27.2.19 Compliance Audits

An audit is a technique used to gather sufficient facts and information, including statistical information, to verify compliance with standards. Employers must select a trained individual or assemble a trained team to audit the PSM system and program. A small process or plant may need only one knowledgeable person to conduct an audit. The audit includes an evaluation of the design and effectiveness of the PSM system and a field inspection of the safety and health conditions and practices to verify that the employer's systems are effectively implemented. The audit should be conducted or led by a person knowledgeable in audit techniques who is impartial toward the facility or area being audited. The essential elements of an audit program include planning, staffing, conducting the audit, evaluating hazards and deficiencies and taking corrective action, performing a follow-up, and documenting actions taken.

27.2.20 Planning

Planning is essential to the success of the auditing process. During planning, auditors should select a sufficient number of processes to give a high degree of confidence that the audit reflects the overall level of compliance with the standard. Each employer must establish the format, staffing, scheduling, and verification methods before conducting the audit. The format should be designed to provide the lead auditor with a procedure or checklist that details the requirements of each section of the standard. The names of the audit team members should be listed as part of the format as well.

The checklist, if properly designed, could serve as the verification sheet that provides the auditor with the necessary information to expedite the review of the program and ensure that all requirements of the standard are met. This verification sheet format could also identify those elements that will require an evaluation or a response to correct deficiencies. This sheet also could be used for developing the follow-up and documentation requirements.

27.2.21 Staffing

The selection of effective audit team members is critical to the success of the program. Team members should be chosen for their experience, knowledge, and training and should be familiar with the processes and auditing techniques, practices, and procedures. The size of the team will vary depending on the size and complexity of the process under consideration. For a large, complex, highly instrumented plant, it may be desirable to have team members with expertise in process engineering and design, process chemistry, instrumentation and computer controls, electrical hazards and classifications, safety and health disciplines, maintenance, emergency preparedness, warehousing or shipping, and process safety auditing. The team may use part-time members to provide the expertise required and to compare what is actually done or followed with the written PSM program.

27.2.22 Conducting the Audit

An effective audit includes a review of the relevant documentation and process safety information, inspection of the physical facilities, and interviews with all levels of plant personnel. Utilizing the audit procedure and checklist developed in the preplanning stage, the audit team can systematically analyze compliance with the provisions of the standard and any other corporate policies that are relevant. For example, the audit team will review all aspects of the training program as part of the overall audit. The team will review the written training program for adequacy of content, frequency of training, and effectiveness of training in terms of its goals and objectives as well as to how it fits into meeting the standard's requirements. Through interviews, the team can determine employees' knowledge and awareness of the safety procedures, duties, rules, and emergency response assignments. During the inspection, the team can observe actual practices such as safety and health policies, procedures, and work authorization practices. This approach enables the team to identify deficiencies and determine where corrective actions or improvements are necessary.

27.2.23 Evaluation and Corrective Action

The audit team, through its systematic analysis, should document areas that require corrective action as well as where the PSM system is effective. This provides a record of the audit procedures and findings and serves as a baseline of operation data for future audits. It will assist in determining changes or trends in future audits.

Corrective action is one of the most important parts of the audit and includes identifying deficiencies and planning, following up, and documenting the corrections. The corrective action process normally begins with a management review of the audit findings. The purpose of this review is to determine what actions are appropriate and to establish priorities, timetables, resource allocations and requirements, and responsibilities. In some cases, corrective action may involve a simple change in procedures or a minor maintenance effort to remedy the problem. MOC procedures need to be used, as appropriate, even for a seemingly minor change. Many of the deficiencies can be acted on promptly, while some may require engineering studies or more detailed review of actual procedures and practices. There may be instances where no action is necessary; this is a valid response to an audit finding. All actions taken, including an explanation when no action is taken on a finding, need to be documented.

The employer must assure that each deficiency identified is addressed, the corrective action to be taken is noted, and the responsible audit person or team is properly documented. To control the corrective action process, the employer should consider the use of a tracking system. This tracking system might include periodic status reports shared with affected levels of management, specific reports such as completion of an engineering study, and a final implementation report to provide closure for audit findings that have been through MOC, if appropriate, and then shared with affected employees and management. This type of tracking system provides the employer with the status of the corrective action. It also provides the documentation required to verify that appropriate corrective actions were taken on deficiencies identified in the audit.

27.2.24 Conclusion

OSHA believes the preceding discussion of PSM should help small employers to comply more easily with the new requirements the standard imposes. The end result can only be safer more healthful workplace for all employees – a goal we all share.

27.3 HAZARD AND OPERABILITY STUDY (HAZOP)

HAZOP is a formally structured method of systematically investigating each element of a system for all of the ways in which important parameters can deviate from the intended design conditions to create hazards and operability problems. This was originally defined by Skelton (5) as "a formal, systematic, critical, rigorous examination to the process and engineering intentions of new and existing facilities to assess the hazard potential of mal-operation or mal-function of individual items of equipment and the consequential effects." The HAZOP problems are typically determined by a study of the P&IDs (or plant model) by a team of personnel who critically analyze effects of potential problems arising in each pipeline and each vessel of the operation. A more recent book on the subject is HAZOP: Guide to Best Practice (6).

Pertinent parameters are selected, for example, flow, temperature, pressure, and time. Then the effect of deviations from design conditions of each parameter is examined. A list of key words, for example, "more of," "less of," "part of," are selected for use in describing each potential deviation.

The system is evaluated as designed and with deviations noted. All causes of failure are identified. Existing safeguards and protection are identified. An assessment is made weighing the consequences, causes, and protection requirements involved.

A HAZOP analysis generates a list of identified problems, usually with some suggestions for improvements of the system. It improves safety, reliability, and quality by making people more aware of potential problems. It also helps to identify loopholes and inconsistencies in procedures and force plant personnel to update instructions.

The basic philosophy of HAZOP is that if a process operates within its intended design philosophy, then undesired hazardous events should not occur. The objective of a HAZOP is mainly to identify how process deviations can be prevented or mitigated to minimize process hazards.

The premise of HAZOP is to stimulate the imagination of a review team, including designers and operators, in a systematic way so that they can identify potential hazards in a plant design. Also, to brainstorm, in a controlled fashion, all the possible ways that process and operational failures can occur.

The outcomes of HAZOP analysis are recommended necessary changes to a system to meet company risk guidelines and also to recommend procedures or changes for eliminating or reducing the probability of operating deviations.

The terms traditionally used in HAZOP are:

- Design intent – The way in which the plant is intended to operate.
- Deviation – Any perceived deviations in operation from the design intent.
- Cause – The causes of the perceived deviations.
- Consequence – The consequences of the perceived deviations.
- Study node – Specific process or operation.
- Safeguards – Existing provisions to mitigate the likelihood or consequences of the perceived deviations and to inform operators of their occurrence.
- Actions – The recommendations or requests for information made by the study team to improve the safety and/or operability of the plant.
- Guide words – Simple words used to qualify the intent and hence discover deviations.
- Parameters – Basic process requirements such as "flow," "temperature," "pressure," and so on.

The HAZOP team normally consists of between four and eight members, each of whom can provide knowledge and experience appropriate to the project to be studied. The team needs to be small enough to be efficient and allow each member to make a contribution. The team members should be considered subject matter experts (SMEs) in the topics being analyzed. Two types of person are required in a HAZOP team: those with detailed technical knowledge of the process and those with knowledge and experience of the HAZOP technique and the ability to chair and report upon technical meetings. The makeup of the HAZOP team includes a chairman or team leader, secretary/recorder, process design engineer, control engineer, operations specialist, and project engineer. Other specialists may be consulted or be available for specific points. Chairman or team leader is selected for his or her ability to effectively lead the study. He/she should have sufficient seniority to give the study recommendations the proper level of authority and has a knowledge and experience of the HAZOP technique. The secretary/recorder should have a technical appreciation of the project and be familiar with the HAZOP technique. The technical members are usually part of the project design team.

27.4 HAZOP PROCEDURE

The overall HAZOP process is shown in Figure 27.6.

1. Develop a detailed flow diagram of the entire process.
2. Divide the flow diagram into a number of process units. For example, the reactor area might be one unit, and the storage tank farm another. Select a particular unit for study.
3. Choose a study node (vessel, line, operating instruction).
4. Describe the design intent of the study node. For example, Reactor P-1 is designed is to react carbon monoxide with chlorine to produce raw phosgene.
5. Pick a process parameter (flow, level, temperature, pressure, concentration, pH, viscosity, power, or inert).
6. Apply a guide word to the process parameter to suggest possible deviations. (NO, MORE, LESS, REVERSE).
7. If the deviation is applicable, determine possible causes and note any protective systems.
8. Evaluate the consequences of the deviation (if any).
9. Recommend action: what, by whom, and by when.
10. Record all information.
11. Repeat steps 5 through 9 until all applicable guide words have been applied to the chosen process parameter.
12. Repeat steps 4 through 10 until all applicable process parameters have been considered for the given study node.
13. Repeat steps 2 through 11 until all study nodes have been considered for the given section and proceed to the next section on the flow sheet.

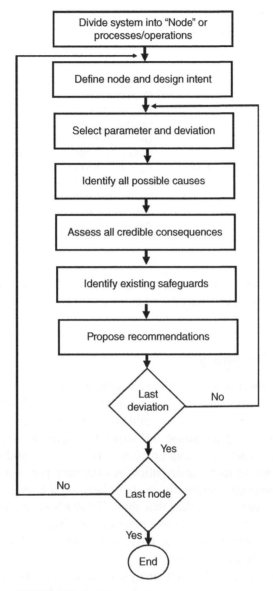

FIGURE 27.6   HAZOP flow.

There is a set of standard HAZOP guide words. Example guide words are contained in Table 27.1. An example is the best way to show how HAZOP is performed. Table 27.2 is an example worksheet. The various sources have different versions of a HAZOP worksheet, and this is just one example. The hot water tank (Figure 27.7) discussed in Chapter 15 will be used to provide a partial example of this technique. In a way, it is very similar to a failure mode and effects analysis (FMEA)

TABLE 27.1
HAZOP Guide Words

Guide word	Meaning
NO OR NOT	Complete negation of the design intent
MORE	Quantitative increase
LESS	Quantitative decrease
AS WELL AS	Qualitative modification/increase
PART OF	Qualitative modification/decrease
REVERSE	Logical opposite of the design intent
OTHER THAN/INSTEAD	Complete substitution
EARLY	Relative to the clock time
LATE	Relative to the clock time
BEFORE	Relating to order or sequence
AFTER	Relating to order or sequence

as described in Chapter 11. However, the technique is from a process perspective, rather than a component level analysis. Also, as with FMEA, HAZOP is a structured "what-if" technique.

27.4.1 HAZOP Summary

As Gordon McKay discusses, HAZOP is a technique that does work if it is conducted correctly. It is very methodical and systematic. It identifies the chain of cause to consequence of a hazard (7). This facilitates risk assessment and the design of preventative and mitigating measures. Subsequent designs are improved by a more systematic approach. Also, operating problems are identified and communicated to the entire design team. In turn, participation by operating personnel promotes training and design acceptance. Despite the costs of keeping a team together for several days, there are many detailed studies that show HAZOPs save a lot of money from the costs of accidents and more effective plant operability. Reports on several major accidents have shown that a well-done HAZOP would have identified the potential hazards that initiated the major disasters.

However, the downside of HAZOP analyses is that sometimes they are viewed as solving all problems, when in fact they are just another risk assessment tool. Process hazards are identified but hazards due to layout or the failure of components in detail are often missed. They are only as good as the information that is used and the people who use it. If the HAZOP is carried out on an early version of the design, it will be necessary to do another study. Deviations from the proposal are considered rather than alternatives. The conventional HAZOP is normally carried out late in design. It brings hazards and operating problems to light at a time when they can be put right with a drawing change rather than a mechanical engineering operation, but at a time when it is too late to make fundamental changes in design.

TABLE 27.2
HAZOP Worksheet

Number	Guide word	Element	Deviation	Possible causes	Consequences	Safeguards	Comments	Actions	Responsible party
Assign a unique identifier	Provide a deviation guide word	Describe what the guide word pertains to	Describe the deviation	Describe how the deviation may occur	Describe what may happen if the deviation occurs	List controls (preventive or reactive) that reduce deviation likelihood or severity	Provide important information	Identify any hazard mitigation or control actions required	Provide a responsible person and/or organization
1	No flow	Hot water outlet	No hot water is flowing from tank	Inlet is blocked. Inlet valve is closed. Outlet valve is closed	Hot water service is not available	Ensure inlet is clear. Ensure outlet valve is open	The hot water tank valves need to be designed so that there is clear indication that the valves are open or closed. Inlet needs to be designed to ensure it can be cleaned out	Design team to redesign valves	Valve designers
2	More temperature	Tank thermostat	High-temperature water	Thermostat failed. Thermostat incorrectly adjusted	Potential scalding from shower	Adjust thermostat or replace	Ensure quality thermostat is installed	Provide instructions on what to do if water temperature is too high	Procedure writers

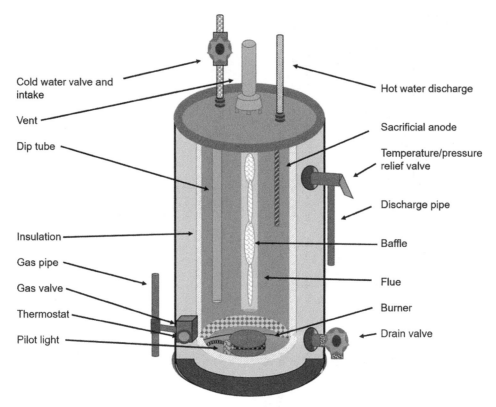

FIGURE 27.7   Hot water tank.

Self-Check Questions

1. What differentiates PSM and HAZOP from other types of risk assessment?

2. A new process plant is becoming operational. What types of steps should be performed?

3. Figure 27.2 is a P&ID of an experimental piece of equipment. What hazards can you identify by looking at the figure?

4. A process involving high-temperature/pressure steam is hazardous. A process involving a chemical like ethylene oxide is also hazardous. Compare and contrast these two hazard types.

5. Events like the BP and Phillips chemical plant explosions discussed in Chapter 29 are examples of events where PSM and HAZOP were not applied. How might applying PSM and HAZOP prevented these incidents?

REFERENCES

1. US Department of Labor Occupational Safety and Health Administration (2000). Process Safety Management Guidelines, OSHA 3132.
2. US Department of Labor Occupational Safety and Health Administration (2016). Process Safety Management: The Use of Metrics in Process Safety Management (PSM) Facilities Fact Sheet, OSHA FS 3896.
3. Ostrom, L.T. and Verner, K. (2017). Utilizing waste heat from nuclear power plants. *Presented at the 2017 Ecology and Safety Conference*, Etienne/Sunny Beach Bulgaria, June 2017.
4. NASA Ames Procedural Guide (2017). NASA Research Park Design Review Program.
5. Skelton, B. (1997). *Process Safety Analysis*, 1e. Newnes.
6. Crawley, F. and Tyler, B. (2015). *HAZOP: Guide to Best Practice*, 3e. Elsevier.
7. McKay, G. (2008). Process Safety Management and Risk Hazard Analysis. Retrieved August 2018. https://cbe.ust.hk/safetycourse/download/09.1HAZOPStudyTrainingCourse.pdf

CHAPTER 28

Emerging Risks

28.1 INTRODUCTION

It is harder to define an emerging risk than a known risk. There are a few definitions of an emerging risk, depending on the industry or facility. Characteristics of emerging risks are common. There is usually a high level of uncertainty, in which frequency and potential impact of risks are difficult to assess. An emerging risk is typically characterized by a low frequency or in other words not likely to happen and has a high impact. It is difficult to assess an emerging risk because it may be something that has never happened before or rarely occurs. A small, seemingly low risk could be a cascading failure where a small risk or event creates other problems until there is a major crisis. The International Risk Governance Council (IRGC) defines emerging risks as "new risks or familiar risks that become apparent in new or unfamiliar conditions" (1).

Emerging risks can start as a trend or even a slight shift that can impact years down the road. A cyber risk was considered an emerging risk over five years ago. We hear about it more today, but it is still an emerging risk due to the uncertainty, unpredictability, and changes occurring in the technology industry. We will discuss cyber risks later in this chapter.

28.2 POSSIBLE EMERGING RISKS

It is important to be aware of potential emerging risks. The following is a small list of potential areas for emerging risks:

- Housing bubble.
- Corruption and fraud.
- Market crash.

Risk Assessment: Tools, Techniques, and Their Applications, Second Edition. Lee T. Ostrom and Cheryl A. Wilhelmsen.
© 2019 John Wiley & Sons, Inc. Published 2019 by John Wiley & Sons, Inc.
Companion website: www.wiley.com/go/Ostrom/RiskAssessment_2e

- Default of US national debt.
- United States entering a new war.
- Climate changes.
- Natural catastrophes.
- Food production's impact on freshwater and food supply.
- Fragile electric power grid.
- Pandemic.
- Impact of government regulation.
- Civil unrest.
- Social network misuse.
- Terrorism.
- Cyberterrorism.
- Hacker attacks.
- Cybercrime (2).

This chapter will address a few of these risks and others.

What emerging risks do you see impacting your organization in the next year, five years, or even longer? When you look past the risks of today and focus on the emerging trends, you can start to develop strategies. Strategic planning around these risks is an opportunity for your risk management functions to be proactive and add tremendous value to your company's future. Technology-based change is happening continuously, and most organizations struggle to see the change in advance. An assessment of the risks should be the first line of defense.

An assessment of emerging risk is characterized by the early detection of facts related to that risk. This assessment can be from research or from monitoring programs. The assessment must be flexible to accommodate changes in the conditions that affect the risks. The monitoring programs should be reliable, sensitive, and quantifiable and should provide the information on the nature of the hazard and the source of the risk. These monitoring programs should identify indicators that point to a specific emerging risk in different ways, either directly or indirectly. An example may be identifying the lack of safety conditions that could possibly cause a release of a toxic or radioactive agent from a chemical or nuclear power plant. Identifying a detection of significant levels of toxic or radioactive chemicals on or in vegetables grown near the releasing plant provides a direct indication of the nature and extent of the emerging risk.

Intelligent automation, such as machine learning, and robotic process automation are changing the business world dramatically. This fairly new technology complements and augments human skills that can increase speed, precision, quality, and operational efficiency.

Smart machines perform activities and make decisions that humans controlled. The industries where humans perform repeated manual tasks over and over will soon be replaced by machines. The companies replacing the human by machines must be cognizant of the risks related to intelligent automated tools and tasks. The risks need to be identified, evaluated, and mitigated. Some of the risks include business disruption, skill gaps, insufficient cybersecurity, and lack of effective controls.

A survey has been conducted with over a thousand employees from all over the globe, as well as a network of scientists. The survey targets experts in diverse backgrounds, skills, and cultures to compile a picture of emerging risks. The survey is entitled "Emerging Risk Survey," and a few of the results are found in the following (3).

Internet of Things (IoT), which is artificial intelligence and robotization, is a multifaceted risk category that touches on advances in artificial intelligence with the development of deep learning and the development of robotics, as well as the rising number of connected objects, IoT. Two risks evolve from this development: first, cybersecurity of data and second, cybersecurity of devices. The development of deep learning is a software that mimics the activity in the brain where thinking occurs. The software learns to recognize patterns in digital representations of sounds, images, and other data (4).

The main concerns with this technology are the fear of state actors, terrorists, stalkers, or teenage vandals that may want to use automated systems to cause physical destruction, data destruction, and network disruption. The risks are changing shape as the research advances, and many have raised questions regarding the reliability of these systems (3).

Another emerging risk from the survey is natural resources management. There is a stress on our natural resources as the global population continues to grow, resulting in an increase in the demand of food, water, and energy.

Water is a crucial resource for life on Earth, but it is also a central component of all economic development (food production, energy, and industry). A water supply crisis due to lack of resources or overwhelming demand would paralyze entire regions and place the lives of many in peril (3).

Other emerging risks from the survey target health risks that are things such as pandemics, resistance to antibiotics, and new medical practices. Surveys aid in identifying and monitoring of emerging risk dynamics and trends, proving a worthwhile asset for establishing adaptable risk anticipation strategies (3).

As mentioned earlier in the chapter, emerging risks are sometimes present at low impact levels with the potential to grow very fast, which results in a more significant level of impact. So, what are the drivers of an emerging risk event? Because an emerging risk is relatively new, unknown, and/or changing in some new way, there is no set consensus on the impacts or the root causes of these types of risks, which, in turn, makes understanding and managing the emerging risk very challenging.

As quoted in the Survey of Emerging Risks published by the Society of Actuaries, "assessment of emerging risks 'requires managers and modelers to think outside their comfort zone. Often there is no incentive for firms to contemplate risks that others are ignoring'" (5).

28.2.1 Communication

How difficult is it to develop understanding about an emerging risk? We have discussed earlier on the definition of an emerging risk as so unlikely to occur that it does not warrant attention. Remember "it can't happen here" syndrome. This makes communication difficult.

Example: Prior to the attack on 9/11, there were few resources allocated to terrorism preparedness. However, since 9/11, terrorism has been forefront in the planning standards, and departments have been organized at government levels with funding. Of course, terrorism was known before 11 September 2001, but the syndrome of it can't happen here was deeply rooted, so it did not have the allocated time and funding or even resources to prepare or mitigate that big of an incident. This was an emerging risk that was ignored until it actually happened. So how do you communicate the importance of the risks that had not yet been experienced?

Risk management resources tend to be focused on current operational, financial, and compliance risks. The well-known phrase "failing to plan is planning to fail" comes into play here.

28.2.2 What Can Be Done?

- A formal, documented process for identifying, assessing and periodically reviewing emerging risks should be established. The review process should incorporate features that allow for immediate communication of new information about risk as it is discovered.

- Include emerging risk reviews in the strategic planning process. Conduct risk reviews in your strategic planning process.

- Challenge the conventional thought processes and expectations. Perform the "what if" scenarios and how it will impact you and your organizations. Emerging risks differ from the conventional expectation (5).

More Examples of Emerging Risks

Gibson describes a few examples of actively emerging risks today on his blog that is Gibson Strength Against Risk (6):

Brand reputation in social media – A simple tweet or post can be viewed as "news" and go viral before a company has any chance to react. A recent example is Newell Brands, parent company to Crock-Pot®, and the portrayal of a fire caused by a faulty slow cooker. Social media reaction and commentary caused a

24% drop in stock prices. This forced the parent company to open social media accounts to defend their product. All this purely due to a fictional TV event and viral reaction. In an ever-shrinking global community, any type of brand reputation management through social media will continue to be a threat or an opportunity!

Sexual Harassment in the Workplace – Recent allegations and media attention on the topic are bringing a new level of risk and awareness to the forefront. The list of high-profile household names under scrutiny continues to grow. *Time* magazine even selected "The Silence Breakers" as the 2017 Person of the Year.

This movement is bringing the conversation to the surface. Equal Employment Opportunity Commission (EEOC)-related accusations are growing regardless of industry segment. Turning a blind eye to poor actions is not acceptable for a board or corporate governing entity. It is being suggested that, in addition to the individual accused, companies and their individual directors' and officers' should also be held liable for allowing such activity to occur. Allegations of misconduct can no longer be kept under wraps. All companies, public and private, need to review their programs and policies around such matters.

Cryptocurrency – This digital or virtual currency is designed to work as a medium of exchange. In 2017 new cryptocurrencies raised more than $3.5 billion in initial coin offerings, a type of fundraising like an initial public offering. The number of companies that now accept an alternate currency is growing. How will this impact future trade of your product or service? As popularity continues to grow, what happens to the traditional dollar and commerce system we know today? How does a company protect the currency in a digital world of hackers and cybercrime? (6)

28.3 A VERY BRIEF INTRODUCTION TO CYBERSECURITY

What is security? We know it's important. What is it?
Security is defined as follows:

- A part of resiliency.
- Risk management.
- Compliance, regulation, contracts, liability.
- Maintain confidentiality, integrity, availability.
- Privacy.
- Trust.
- Logical and physical; computers, machines, and humans.
- Offense versus defense; white hat versus black hat.

Below is a list of core principles of security:

- Least privilege.
- Complete mediation.
- Fail-safe defaults.
- Open design.
- Economy of mechanism.
- Layers of defense.
- Psychological acceptability.

The following is a list of some of the laws and regulations in place:

- Computer Fraud and Abuse Act (1984, 1996).
- Computer Security Act (1987).
- Digital Millennium Copyright Act (1998).
- Gramm-Leach-Bliley Act (1999).
- Sarbanes-Oxley Act (2002).
- HIPAA (1996, 2006).
- FISMA (2002) – and DITSCAP, DIACAP.
- USA PATRIOT Act (2001).
- PCI Data Security Standard.
- NERC CIP.

There are many types and categories of attacks within this realm. Below is a small list of some of the attacks:

- Denial of service (DOS) and distributed denial of service (DDoS).
- Spoofing.
- Sniffing.
- Password cracking and brute force.
- Buffer overflows.
- Code injections.
- Cross-site scripting (XSS) and cross-site request forgery (CSRF).
- Man-in-the-middle (MITM).
- Misconfiguration and default settings.

What is cybercrime?

- Vandalism.

- Theft – identity theft.
- Fraud.
- Extortion and data ransom.
- Phishing, spear phishing.
- Doxing motivations: notoriety, money, activism, war.

What is cyberterrorism and cyberespionage?

- Nation-state conducted or sponsored.
- China, United States, Iran, Russia, United Kingdom, Syria, Brazil, Germany, North Korea, anyone with military.
- Advanced persistent threats.
- Rules of engagement for information warfare.
- Cyber suspicion and cyber blame in cyber arms race.

What could go wrong?

- Embarrassment, loss of brand trust.
- Higher costs.
- Loss of availability of power – an hour? a day? a month?
- People die, panic ensues, mob rule, martial law, etc.

Critical infrastructure is critical. So, what can we do?

- Defense in depth.
- Least privilege.
- Passwords.
- Firewalls and antivirus.
- Air gapping.
- Configuration management – start with asset identification and inventory, whitelisting, and patching.
- Supply chain verification.
- What about insider threats? Trusted business partners.

A good defense is to be prepared. There are policies, procedures, guidelines, training and education, and Cyber Incident exercises, tabletop events, and fire drills. We need better detection, better logs, and better audits, and we better pay attention! The key is to reduce time between an exploit and the discovery (minutes, not months).

28.4 CYBER ATTACKS

A very important issue in cybersecurity is the protection of the national infrastructure from foreign or domestic interference. There must be a focus on the systems that have highest impact on the nation, critical systems that must remain functioning, and then prioritizing the protections around those systems. A concerted effort must be made to prioritize those systems at greatest risk and to prepare and implement appropriate protective interoperable security capabilities that are in place to secure the operation of critical functions for the nation.

There are potential threats to the "Smart" grid, which is an innovation of the current electric grid to provide power across our nation and the world. This new grid is exciting and will enhance technology in many ways. Unfortunately, this innovation requires hardware and software components that make it a threat to cyberattacks. It is vital that cyberattacks are mitigated for these systems. Nuclear laboratories are conducting research through Supervisory Control and Data Acquisition (SCADA) testbeds that create testbeds to analyze real-life threats in a real-world environment. It is encouraging to see our nation take a proactive approach and address risk, research, and development efforts in cybersecurity as this system continues to develop.

Cyber breaches seem to be growing in scope and frequency due to rapid shifts in technology. New capabilities and techniques are constantly being developed by very sophisticated and well-funded hackers that include insiders, organized crime, and others. They target companies both directly and indirectly through phishing scams.

Companies need to stay on top of emerging threats and remain vigilant. Too many times a company is breached, and the personal data of thousands is exposed, and people become vulnerable to identity theft.

A top-down risk assessment should be performed around the company's cybersecurity processes.

28.5 BULK ENERGY SYSTEM (BES): A SMART GRID SCENARIO

Before tackling the threat of cybersecurity within the Smart grid, it is important to first understand what it is. The Smart grid is a new innovation of the current electric grid to provide power across our nation and the world. The original grid was developed in the 1890s and consisted of transmission lines, substations, transformers, and other methods of power distribution to bring electricity from a power plant to homes and businesses.

This system is an engineering marvel that has supplied power around the world for more than 100 years. However, it still has limitations, such as the threat of large blackouts. This can have a domino effect on banking, communications, security, and heating and cooling, among other drastic threats. The Smart grid transition includes efficient transmission of electricity, reduced length of power outages, opportunity for lower utility costs, increased integration of large-scale renewable energy systems, and improved security. The Smart grid will allow for better preparedness and encourage innovation to prevent or reduce the risk of failure (7).

As the Smart grid innovation continues to evolve, there is a lot to be excited about. However, there remain many oppositions and concerns. Moving these systems to more of an online approach increases the vulnerability to cyberattacks. Although there are many efforts to mandate and enforce cyber and physical security standards, oftentimes this is not enough to completely neutralize these threats. Smart grid components are built around microprocessors and other hardware devices whose basic functions are controlled by software programing. These systems must be linked to the Internet, and therefore this same innovation that allows for such an advanced system of power also increases the potential for cyberattacks (8).

The Smart grid also contains sensors that gather information and send them back and forth between the control centers. Hackers have the ability to interrupt these systems. This is called a denial of service or DoS attack. Because the Smart grid will need to contain a large number of access points, it will be difficult to defend all potential weak points. The ability to control, interrupt, or manipulate power systems across the nation and world presents a giant bull's-eye for cyberattack and terrorist attack. Although the government has listed these cyberattacks as an imminent danger, many of the protection standards are treated more as suggestions rather than mandates. This is because many of the electricity facilities and equipment are owned in the private sector, thus making it difficult to create a widespread standardization of protocols (9).

The grid has already experienced serval types of cyber intrusions over the last several years. In October 2014 an announcement was provided indicating that a malware program called BlackEnergy infected several control systems with a Trojan horse approach. This malware has the capability of moving through several network files onto removable storage media. This tactic increases the chance for lateral movement within the affected environment (8).

An older malware system HAVEX has targeted the energy sector since August 2012. HAVEX is typically distributed via spam email or spear-phishing attacks. These phishing attacks trick individuals to click or download malicious links that allows the distribution of the malware (8). Additionally, in December 2015 a cyberattack successfully shut down the power grid in Ukraine. The attack left more than 230 000 residents without power, and the hackers also disabled backup power supplies in many of the power centers. The power was only out for six hours or so, but for more than two months after the attack, the control centers were not fully operational (10).

With the constantly changing and growing technology in the world, the shift to the Smart grid is inevitable and necessary. It is vital that we develop methods to mitigate or stop the large amount of cyberattacks that threaten this system. The US Department of Energy has prioritized research and development to combat the vulnerability of cyberattacks. There has been a strong focus on risk analysis and risk mitigation in these areas. With the help of funding sources from the US Department of Energy and the Department of Homeland security, many research and development dollars have been distributed across the nation to solve these issues (11).

Additionally, many research testbeds have been created to analyze real-life threats in a real-world environment. The SCADA testbeds are hosted at an Idaho laboratory. This work has resulted in the discovery of multiple cyber vulnerabilities. Additional work at another laboratory has utilized research in both physical and virtual components (9).

Training and understanding of potential threats are also a key to mitigate cyber risks. For personnel working in the control centers, it is important to be aware and trained to spot phishing or other threats that can enter through software systems. Cyber teams send out internal phishing emails to let the employees know when they click on a link that could be a true threat. They provide detailed information including what to look for and advise to never click on links or attachments that seem out of the ordinary. This training and others will help prevent future threats (9).

The shift to the Smart grid will not happen overnight, but is continually evolving piece by piece, and will continue to do so over the next decade or two. As these technologies evolve, so will the opportunity for cyberattacks. It is encouraging to see our nation take a proactive approach and address risk, research, and development efforts as this system continues to develop.

Many efforts are in place to encourage children at a young age to study science, engineering, technology, and math (STEM) courses, particularly in computer science and cybersecurity.

These educational efforts will bring forth future talent and innovative ways to mitigate cyberattacks on our Smart grids. These attacks will likely never go away, but it has allowed for future generations to become more involved in solving this threat (9).

28.6 RECENT CYBERATTACKS

Global shipping ports in San Diego, California, and in Barcelona, Spain, have been the target of recent cyberattacks. They were hit with ransomware, and it halted service in the loading and unloading of boats for a few days as the technology groups tried to fix the problem. Business was interrupted in both major ports. Integrated risk management (IRM) is a recommended approach to helping to understand the strategic risks that can result in business failure (12).

28.7 BRITISH AIRWAYS (BA) DATA BREACH EXAMPLE

The recent announcement of the British Airways (BA) data breach brought into focus the General Data Protection Regulation (GDPR).

BA's data breach involved roughly 380,000 customers who transacted on BA's websites from late August through the first week of September. Given the requirement by GDPR to disclose data breaches within 72 hours of discovery, BA had little time to understand the potential impacts from the breach. So, their disclosure was vague and left many customers guessing what to do in the wake of the breach. What is most concerning now is the potential for a GDPR fine which can range up to 4% of annual revenues or, in BA's case, £500 million. It's no wonder that CEOs consider risk management as one of their top priorities in 2018.

(13)

The GDRP does not allow organizations to take as much time as they want to disclose information about a data breach. The information to those involved needs to be accurate. An IRM allows a quick response that is quickly coordinated among the stakeholders. The IRM coordinates with policies, the enterprise risk management, operational risk management, IT risk management, and any standards or metrics involved in risk assessment. The consensus at a London Summit is that IRM is the solution and should be a priority for organizations involved in the new GDPR (13).

Self-Check Questions

1. What is the definition of an emerging risk?
2. Name some of the emerging risks today and in the future.
3. What is cybersecurity?
4. What are some of the laws and regulations in place for cybersecurity today?
5. Describe what the Smart grid is and how it can be compromised.
6. What is IRM and how can it help in cyberattacks?
7. Describe what GDPR is.

REFERENCES

1. IRGC (2015). *Guidelines for Emerging Risk Governance*. Lausanne: International Risk Governance Council (IRGC).
2. Weymann, M. and Egloff, R. (2017). Identifying Emerging Risks-Early anticipation of the future risk landscape. https://theactuarymagazine.org/identifying-emerging-risks/#enref-4118-2 (accessed October 2018).
3. AXA (2017). Emerging Risks: what are the main risks for 2025?. https://group.axa.com/en/about-us/emerging-risks-survey (accessed November 2018).
4. MIT Technology Review. Deep Learning: with massive amounts of computational power, machines can now recognize objects and translate speech in real time. Artificial intelligence is finally getting smart. https://www.technologyreview.com/s/513696/deep-learning (accessed November 2018).
5. RIMS RIMS Executive Report: The Risk Perspective. https://www.rims.org/resources/ERM/Documents/EmergingRisk_ERMweb.pdf (accessed November 2018).

6. Gibson (2018). Emerging Risks. https://www.gibsonins.com/blog/emerging-risks (accessed November 2018).

7. Wikipedia (n.d.). Smart grid. https://en.wikipedia.org/wiki/Smart_grid#Oppositions_and_concerns (accessed November 2018).

8. Campbell, R.J. (2015). Cybersecurity issues for the bulk power system. https://web.archive.org/web/20150628192830/http://www.fas.org/sgp/crs/misc/R43989.pdf (accessed November 2018).

9. Stridar, S., Hahn, A., and Govindarasu, M. (2011). Cyber-physical system security for the electric power grid. *Proceedings of the IEEE* 100 (1): 210–224.

10. Zetter, K. (2016). Inside the cunning, unprecedented hack of Ukraine's power grid. https://www.wired.com/2016/03/inside-cunning-unprecedented-hackukraines-power-grid (accessed November 2018).

11. US Department of Energy (n.d.). What is the smart grid? https://www.Smartgrid.gov/the_Smart_grid/Smart_grid.html (accessed November 2018).

12. Wheeler, J.A. (2018). IRM is critical for ERM success. https://blogs.gartner.com/john-wheeler/recent-cyber-attacks-demonstrate-why-irm-is-critical-for-erm-success (accessed November 2018).

13. Wheeler, J.A. (2018). GDPR requires IRM for fast and effective response. https://blogs.gartner.com/john-wheeler/gdpr-requires-irm-for-fast-and-effective-response (accessed November 2018).

CHAPTER 29

Process Plant Risk Assessment Example

29.1 INTRODUCTION

A class of industrial facilities that has experienced catastrophic events is process plants, specifically oil and chemical process plants. Some of the most spectacular recent events were:

- Phillips Houston Chemical Complex Incidents.
- British Petroleum (BP) Texas City Refinery (BPTCR) Fire and Explosion.

The series of events that led up to the events are discussed below.

29.2 PHILLIPS HOUSTON CHEMICAL COMPLEX

The Phillips Houston Chemical Complex had endured several accidents with fatalities spanning approximately a decade. Figure 29.1 shows a diagram of the plant (1). Of the three accidents with fatalities, one was due to an explosion of released flammable gases and two were explosions from highly reactive chemicals. The result of the accidents brought on a great amount of pressure from federal agencies, local communities, and unions for the management of Phillips to implement a more comprehensive management of safety.

29.2.1 1989 Explosion

On 23 October 1989, while routine maintenance was being conducted, approximately 85 000 lb of highly flammable gases escaped from process equipment within a very brief period of time. This highly flammable vapor cloud was ignited by an unknown

Risk Assessment: Tools, Techniques, and Their Applications, Second Edition. Lee T. Ostrom and Cheryl A. Wilhelmsen.
© 2019 John Wiley & Sons, Inc. Published 2019 by John Wiley & Sons, Inc.
Companion website: www.wiley.com/go/Ostrom/RiskAssessment_2e

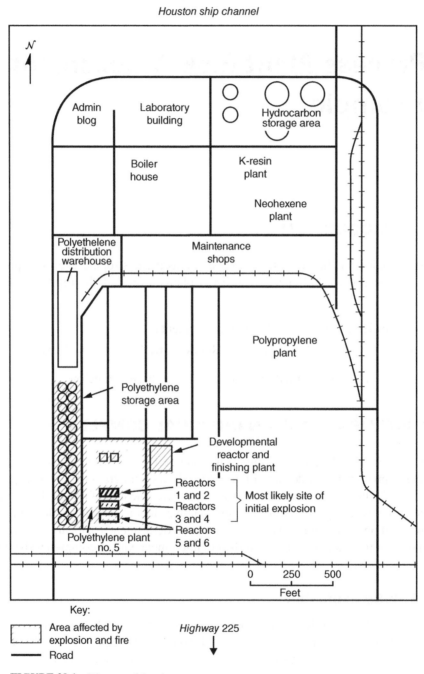

FIGURE 29.1   Diagram of the plant.

ignition source. The resulting explosion and fireball caused the death of 23 people and injured another 314 people. Figure 29.2 shows the affected areas. The damage to the plant from the explosion and fires was estimated as being over $700 million and through the disruption of production and additional $700 million was lost (2). Two of the known gases in the explosive mix were ethylene and isobutane.

A material safety data sheet (MSDS) from Nova Chemicals states that ethylene is an extremely flammable liquefied gas (3). Ethylene has a flammable limit of 2.3–36%, meaning that the material will burn in air from a concentration of 2.3% to a concentration of 36%. On large spills, the initial evacuation is stated to be 2640 ft (800 m) downwind.

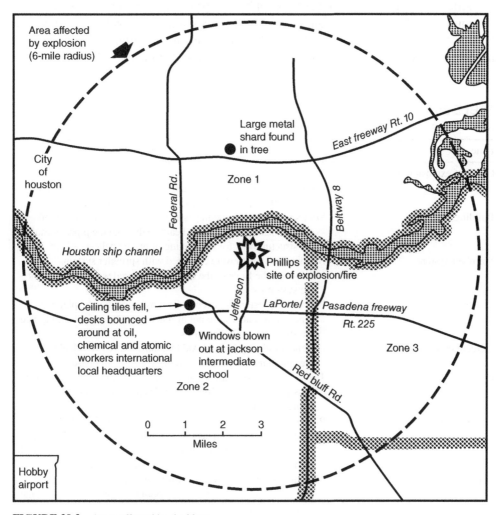

FIGURE 29.2   Areas affected by the blast.

Isobutane is also a very flammable material. The lower flammable limit for this material is 1.8%, and the upper flammable limit is 8.4% (4). Isobutane is also listed as a simple asphyxiant that targets the central nervous system.

According to US Fire Administration, the workers of the plant observed the leak and vapor cloud but only had 60–90 seconds to evacuate as the gases soon found an ignition source (5). The report also states that there were plenty of sources of ignition in the immediate area that included ventilation fans, electrical switches, and gas burn-off flames. The actual ignition source will never be known. The explosion within the complex measured at an estimated 500 ft by 750 ft and produced 3.5 on the Richter scale 25 miles away, having an estimate of 4.0 at the epicenter. Structures at a quarter of a mile away had considerable damage. The US Fire Administration listed several key issues (5). These are contained in Table 29.1.

After a lengthy investigation, the use of backup protection was not used at the plant as required by Phillips own regulations. Due to this failure Occupational Safety and Health Administration (OSHA) charged Phillips Company in failure to train maintenance personnel on "how to work safely with hazardous chemicals." The results were

TABLE 29.1

Key Issues in 1989 Phillips Explosion

Issue	Comments
Cause of explosion/fire	Unknown whether human or mechanical failure. A 10-in high-pressure line carrying ethylene and/or isobutene emitted a flammable vapor cloud into a hostile environment
Building structure	Multistory structure encompassing several acres, primarily of metal construction. Material carried through the pipes within the structure was for the making of plastic pellets
Sprinkler system	Building/structures were equipped with sprinkler systems; however, the force of the explosion severed water supplies for the systems
Evacuation	Because explosion/fire occurred almost immediately, planned evacuation routes were not of much help. Personnel simply fled the structure in all directions
Incident command	Chemical Industry Mutual Aid Organization (CIMA) had a pre-fire plan that worked extremely well in all phases
Casualties	Twenty-three people are known to have been killed. One was still missing at the time of this report. Well in excess of 100 were injured in varying degrees
Fire department's right to know	Responding departments were all members of CIMA. CIMA strictly adheres to Superfund Amendments and Reauthorization Act (SARA) Title III and their own strict guidelines that state all member companies must have an open door policy
Enclave	Although this plant's location was in an unincorporated area, it still fell under the ruling law for both the State of Texas and the City of Pasadena

a $4-million fine and the loss of $1.4 billion due to the accident, not to mention the loss of 23 lives (5).

The positive aspect of the accident was the reaction of the first responders. According to the US Fire Administration, the use of a pre-emergency plan by the Chemical Industry Mutual Aid Organization (CIMA) that had produced its own handbook was invaluable (5). Though initially very tragic, the pre-emergency plan proved successful in response. The use of dual command with the company official in the field command post and the CIMA commanding the central command post provided ease of operation that would have been hindered by a single command being overwhelmed. Also noted was the accessibility to the plant prior to enhance firefighter's knowledge of hazards and layout of the plant and the availability of experts on chemical fires. The use of a separate staging area from the command post prevented bottlenecking and provided traffic control. During the response operations there were no additional deaths or injuries.

29.2.2 Incidents in 1999 and 2000 with Butadiene

The following two accidents involved Butadiene. Butadiene is a colorless, noncorrosive liquefied gas with a mild aromatic or gasoline-like odor (6). "Butadiene is both explosive and flammable because of its low flashpoint." Butadiene is used in several areas of the chemical industry in the production of synthetic rubbers and resins in which the annual growth of its use is estimated at 2% (7). The lower explosive limit is 1.1% and the upper explosive limit is 16.3%. Of special note is the flashpoint, which is $-76\,°C$ (5). Therefore, butadiene is both a fire and explosion hazard".

29.2.3 1999 Explosion

The first accident with butadiene was on 23 June 1999 in the K-Resin facility. A flash fire occurred in the reactor vessel resulting in two deaths. Data on the accident show that a thermal runaway was responsible for the fire and explosion. Findings rule that the following cause the accident: "inadequate hazard evaluation during management of change, inadequate procedures/training, inadequate process hazards analysis, and inadequate emergency relief design" (7).

29.2.4 2000 Explosion

The second accident involving butadiene in the same facility occurred on 27 March 2000. Butadiene in a storage tank was involved. The storage tank was out of service and being cleaned. The absence of any temperature and pressure gauges prevented personnel to have any knowledge of any chemicals, especially the highly volatile butadiene was still present. The result was one person killed and 71 injured (7).

29.2.5 Butadiene: A Reactive Chemical

These two accidents at the Houston Chemical Complex were due to the use of butadiene, which is essential in manufacturing through reaction. "The hazards associated with reactivity are related to process-specific factors, such as operating temperatures, pressures, quantities handled, concentrations, the presence of other substances, and impurities with catalytic effects (8). Since Bhopal, federal agencies have been improving regulations on manufacturing processes in the chemical industry. The US Chemical Safety Board (CSB) conducted an investigation on reactive chemicals by looking into 167 accidents over approximately two decades. With the results, recommendations were given to OSHA and Environmental Protection Agency (EPA) for regulatory changes (9).

The findings on uncontrolled chemical reactivity from 1989 to 2001 were:

1. Of the 167 accidents, 48 resulted in death averaging about 5 deaths a year.
2. Almost one-third affected the public.
3. 42% of the incidents were fire and explosions.
4. Causes and lessons learned were only reported in 20% of the incidents.
5. 50% involved inadequate procedures for storage, handling, or processing of chemicals.

29.3 BP TEXAS CITY REFINERY

Texas City was the site of a large industrial fire and explosion on 23 March 2005, and when this explosion occurred is what considered the worse workplace accident in the United States for the period since 1989 (10, 11). The explosion was at the Texas City Refinery that was owned and operated by BP. Fifteen workers were killed and over 170 others were injured in the blast. Prior to the explosion the Texas City Refinery covered a vast amount of space that was almost 2 miles^2 in size and was considered the third largest refinery within the United States. The official report of the accident cites the following as the causes of the fire and subsequent explosion (10, 11):

- The working environment had eroded over the years to one characterized by resistance to change and lacking of trust, motivation, and a sense of purpose. Coupled with unclear expectations around supervisory and management behaviors, this meant that rules were not consistently followed, rigor was lacking, and individuals felt disempowered from suggesting or initiating improvements.
- Process safety, operations performance, and systematic risk reduction priorities had not been set and consistently reinforced by management.
- Many changes in a complex organization had led to the lack of clear accountabilities and poor communication, which together resulted in confusion in the workforce over roles and responsibilities.

- A poor level of hazard awareness and understanding of process safety on the site resulted in people accepting levels of risk that are considerably higher than comparable installations. One consequence was that temporary office trailers were placed within 150 ft of a blowdown stack, which vented heavier than air hydrocarbons to the atmosphere without questioning the established industry practice.

- Given the poor vertical communication and performance management process, there was neither adequate early warning system of problems nor any independent means of understanding the deteriorating standards in the plant (Chemical Safety).

Table 29.2 contains a partial timeline of the events leading up to the explosion and events subsequent to the explosion (10–12).

It is quite clear from the final investigative reports that there were several errors that not only contributed to the explosion but also caused more people to be in harm's way than needed or should have been (10, 11).

The errors began long before the explosion when mobile work trailers were placed on the site without submitting a change to the siting plan and getting that change approved. Most of the people that died in the explosion were in the mobile work trailers that should not have been located where they were as they were on an approved siting plan as dictated by the refinery's own procedures.

The next error occurred while they were starting to get the isomerization unit running again, and he had some issues during his shift that he was able to correct; however, he did not stay at the site in order to brief the incoming shift supervisor or operator. By doing this there was a lack of communication among the two shifts. During the course of the day, several other factors occurred that aided to the lack of attention to detail and also communication problems among the staff. There were several phone calls that were made during the shift at critical moments that had they not occurred, the operators and/or their supervisors would have potentially caught and rectified the problems early on and ultimately averted the disaster. Also, the report cites when the supervisor and the crew went to lunch and again critical events occurred during these times that if they had been caught soon enough, the problems could have been noted and corrected action taken place (11).

The worst part of this disaster is that it could have been avoided. By enforcing the refinery's own standards and ensuring that workers were properly trained and supervised, the explosion would have never occurred. In addition, if the mobile work trailers had not been moved into the site without first getting an approved change to the siting plan, the majority of the people who lost their lives would not have been in harm's way (11).

This disaster like most ultimately comes down to human error (10, 11). In this case there were many that contributed to the cause of the explosion. Had better controls been in place, this may not have occurred at all. In this case, the US government fined BP with record-setting fines due to them not following rules and regulations,

TABLE 29.2

Timeline of Events Leading Up To The Texas City Explosion

Date	Event/activity
Date event September 2004	British Petroleum (BP) sites the double-wide trailer between the naphtha desulfurization unit (NDU) and isomerization (ISOM) units to house contractor employees for turnaround work in the nearby ultracracker unit
October 2004	The Texas City site leader meets with the R&M Chief Executive and Senior Executive Team to discuss the 2004 incidents; management discusses how these incidents are the result of casual compliance and personal risk tolerance despite two of the three incidents being directly process safety related
2004	The 2004 Process Safety Management (PSM) audit reveals poor PSM performance of the Texas City Refinery, especially in mechanical integrity, training, process safety information, and management of change (MOC)
November 2004	Plant leadership meets with all site supervisors for a "Safety Reality" presentation that declares that Texas City is not a safe place to work
Late 2004	BP Group refining leadership gives the Texas City Refinery business unit leader a 25% budget cut "challenge" for 2005; the business unit leader asks for more funds due to the conditions of the refinery, but less than half of the 25% cuts are restored
Late 2004	The Telos survey is conducted to assess safety culture at the refinery and finds serious safety issues
2004	The refinery-wide Odisha Central Audit Management (OCAM) audit finds that only 25% of ISOM unit operators are given performance appraisals annually and that no individual operator development plans are being developed for unit operators; the audit also finds that the budget allows for no training beyond initial new employee and OSHA-required refresher information
January–February 2005	Nine additional trailers are placed in the area between the NDU and ISOM units
January 2005	The Telos report is issued with recommendations to improve the significantly deficient organizational and cultural conditions of the Texas City Refinery
February 2005	The BP Group VP and the North American VP for Refining meet with refinery managers in Houston, where they are presented with information on the Telos report findings, the deteriorating conditions of the refinery, budget cuts, inadequate training, pressures of production overshadowing safety, and the 2004 fatality incidents
2005	The 2005 Texas City Health, Safety, Security and Environment (HSSE) Business Plan warns management that that refinery will likely "kill someone in the next 12–18 months"
March 2005	The Texas City Process Safety Manager tells management that PSM action item closure is still a significant concern and this metric is finally added to the site's 1000 Day Goals

(continued)

TABLE 29.2
(*Continued*)

Date	Event/activity
23 March 2005	Explosion and fire at the Texas City Refinery results in 15 fatalities and 180+ injuries
April 2005	Due to 23 deaths at the Texas City Refinery in 30 years, OSHA puts BP onto its list of "Enhanced Enforcement Program for Employers Who are Indifferent to Their Obligations"
July 2005	An incident in the RHU results in a shelter in place of the community and $30 million in damage at the refinery
August 2005	A release in the CFHU results in a shelter in place and $2 million in damage at the refinery
September 2005	OSHA fines BP $21 million for 301 egregious willful violations
13 December 2005	During unit startup, a distillation tower at the BP Whiting Refinery in Indiana is overfilled, resulting in fire and damage
June 2006	Settlement Agreement's independent auditor study and recommendations. Included in the study are recommendations to BPTCR to implement the ISA S84.00.01 standard for safety instrumented systems
July 2006	An employee of a contractor was fatally injured when he was crushed between a scissor lift and a pipe rack at BPTCR
January 2008	The top head blew off a pressure vessel, resulting in the death of a BP employee. BP was issued four serious citations related to PSM

had verification measures been enacted a lot of death and destruction could have been averted (13).

29.4 EXAMPLE ANALYSIS

This example provides the basic steps for performing a thorough risk assessment of a process plant. For this example the fictitious chemical reactor from Chapter 15 will be used as the case. The attributes of the reactor are:

- It is a 5500-gal-capacity batch reactor.
- Three chemicals are combined in the reactor to produce Chemical D. These chemicals are Chemicals A, B, and C.
- The following amounts of chemicals are stored on-site:
  - 20 000 gal Chemical A
  - 40 000 gal Chemical B
  - 60 000 gal Chemical C
  - 80 000 gal Chemical D

- The ratio of the three chemicals are:
  - ○ 10% Chemical A
  - ○ 30% Chemical B
  - ○ 60% Chemical C
- Chemicals have to be mixed in the proper ratio for 30 minutes to ensure a successful batch.
- The reaction is exothermic. For each degree over 300 °F the reactor reaches, the quality of the product is reduced. Chemical E is the contaminant produced. The batch becomes 1% Chemical E for degree over the 300° level.
- Increased temperature can cause a spike in reactor pressure. If the pressure reaches 310 °F, the reactor pressure will near the safety factor limits of the reactor. At this point a rupture disk will break and the gases produced will be directed to a scrubber column.
- The product must have less than 2% Chemical E to be successful.

The fictitious chemicals used in this process and their physical and hazardous properties are contained in Table 29.3.

Figure 29.3 shows the diagram of the plant. Table 29.4 contains a list of the symbols that were used for the components of this fictitious plant.

Table 29.5 contains failure/reliability data for the components of the plant that will be used in the analyses.

There can be a large number of risk analyses performed even with this relatively simple plant diagram and limited process information. Therefore, only a few of the many possible example analyses will be presented. The examples risk analyses that will be presented are:

- Preliminary hazard analysis (PHA).
- Failure mode and effect analysis (FMEA).
- Event tree analysis.

TABLE 29.3

Chemical Properties

Chemical	Boiling point	Vapor pressure	Explosive limits (%)	Lethal dose (LD_{50}) or lethal concentration (LC_{50})	Flashpoint
A	173 °F (78.5 °C)	5.7 kPa (20 °C)	3.3–19	3600 mg kg^{-1} oral mouse	54 °F (12 °C)
B	180 °F (82 °C)	4.1 kPa (20 °C)	2.4–8	2745 mg kg^{-1} oral rat	52 °F (11 °C)
C	47 °F (8.5 °C)	132 kPa (20 °C)	NA	5 ppm/1 h inhalation rat	NA
D	214 °F (102 °C)	2 kPa (20 °C)	3–9	1 gm kg^{-1} oral mouse	110 °F
E	240 °F (115 °C)	1 kPa (20 °C)	NA	15 mg kg^{-1} oral rat	160 °F (71 °C)

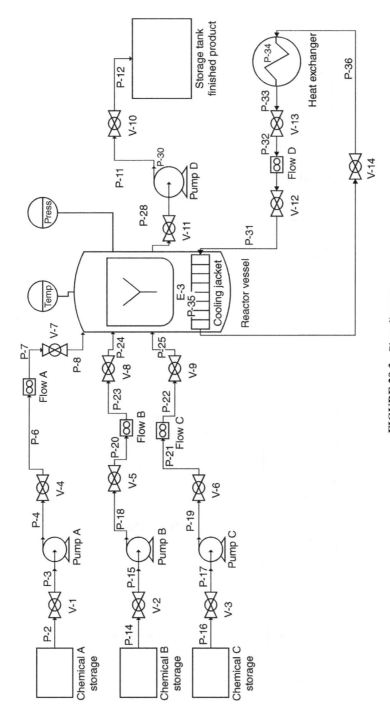

FIGURE 29.3 Plant diagram.

TABLE 29.4

Plant Symbols

Symbol	Description
Chemical A storage	Storage tank
V-1	Valve
Pump A	Pump
Flow B	Flow controller
Temp	Instrument
E-3	Mixer
Heat exchanger	Heat exchanger
Cooling jacket	Cooling jacket

TABLE 29.5
Plant Component Reliability Data

Component	Failure type	Probability of failure
Chemical valves V-1 to V-11	Valve stem seal/shaft failure	1/10 000 valve operations
Cooling water valve V-12, V-13, and V-14	Valve stem seal/shaft failure	1/10 000 valve operations
Chemical pumps, pumps A–C	Seal failure	1/500 demands
	Motor fails and overheats	1/1 000 demands
Cooling jacket	Weld failure	1/1 000 operating hours
Heat exchanger	Weld failure	1/1 000 operating hours
Flow controller – chemical	Inaccurate flow rate	Device needs calibrated every 5 000 operating hours to ensure proper operation
	Catastrophic failure and leaks	1/2 000 demands
Flow controller – cooling water	Inaccurate flow rate	Device needs calibrated every 5 000 operating hours to ensure proper operation
	Catastrophic failure	1/20 000 operating hours
Chemical storage tanks	Weld failure	1/30 000 operating hours
Operator	Fails to monitor	1/1 000
	Omits steps in a procedure	1/500

Data provided for example use only.

- Fault tree analysis (FTA).
- Simplified probabilistic risk assessment (PRA).

29.4.1 Preliminary Hazard Analysis

Table 29.6 shows a sample of the items that would be included in a PHA for the data presented.

29.4.2 Sample Failure Mode and Effect Analysis

The sample FMEA will present an analysis of operational aspects of the process plant since the sample PHA presented an analysis of the hardware aspects. Table 29.7 shows a partial FMEA for the process plant.

29.4.3 Sample Event Trees

As discussed in prior chapters, event trees link events together in a logical sequence. In full PRAs event trees are used to show the progression of events, and for each bifurcation of the event tree there might be a single event, a fault tree, or a human

TABLE 29.6
Sample Preliminary Hazard Analysis for Process Plant

Hazard	Accident	Probably cause	Contingencies, preventative measures	Probability of failure	Severity	Comments
Chemical A	Spill	Seal leaks develop in any of the following valves: V-1, V-4, or V-7	Perform preventative maintenance on valve seals	1/10 000 valve operations based on vendor information on valves. This is for each valve	Toxicity of Chemical A is relatively low, so severity is moderate	Periodic maintenance plans are in place to inspect and repair valve seals on a 12-month basis
		Pump A seal failure	Perform preventative maintenance on pump seals	1/500 operating hours based on vendor information on pumps	Toxicity of Chemical A is relatively low, so severity is moderate	Periodic maintenance plans are in place to inspect and repair pump seals on an 18-month basis
	Fire	Chemical A leaks from pump or valves and ignites from static spark	Perform preventative maintenance on pump and valve seals. Ground process equipment	1/5 years based on operating experience	Chemical A is considered a flammable liquid, and a fire could spread to the storage tanks and process vessel. The consequences could be very severe that could affect on-site assets and the public at large	Periodic maintenance plans are in place to inspect and repair pump seals on an 18-month basis. Periodic maintenance plans are in place to inspect and repair valve seals on a 12-month basis. Process equipment grounding will be tested every 12 months

Chemical	Event	Cause	Action	Frequency	Severity	Comments
Chemical C	Spill	Seal leaks develop in any of the following valves: V-3, V-6, or V-9	Perform preventative maintenance on valve seals. Chemical detectors will be installed to determine if any of these valves develop leaks	1/10 000 valve operations based on vendor information on valves. This is for each valve	Chemical C is highly toxic and a spill would have severe on- and off-site consequences	Periodic maintenance plans are in place to inspect and repair valve seals on a six-month basis. A replacement valve is being sought that will have a higher level of reliability
		Pump C seal failure	Perform preventative maintenance on pump seals	1/500 operating hours based on vendor information on pumps	Chemical C is highly toxic and a spill would have severe on- and off-site consequences	Periodic maintenance plans are in place to inspect and repair pump seals on an 18-month basis. A replacement valve is being sought that will have a higher level of reliability
Chemical D	Spill	Seal leaks develop in any of the following valves: V-10 or V-11	Perform preventative maintenance on valve seals	1/10 000 valve operations based on vendor information on valves. This is for each valve	Low severity	Chemical D has a very low possibility of toxicity

TABLE 29.7
Sample Failure Mode and Effect Analysis for Process Plant

Item, function, activity	Potential failure mode	Potential effects of failure	Severity	Potential cause(s)	Probability	Current controls	Criticality	Risk priority number (RPN) 1–10 (10 highest)	Recommended actions
Reactor cooling	Cooling jacket leaks and loses cooling efficiency	Reactor temperature increases and Chemical E is produced if temperature is increased above threshold	High severity	Weld failure	1/1 000 operating hours. Data from operating experience	Weld inspection every 12 months	Very critical	9	Increase weld inspection frequency to one in six months
	Valve V-12 leaks and causes loss in cooling efficiency	Reactor temperature increases and Chemical E is produced if temperature is increased above threshold	High severity	Seal failure	1/10 000 valve operations based on vendor information on valves	Preventative maintenance is performed every 24 months as per vendor instructions	Very critical	8	Perform preventative maintenance on valve seals every 12 months
	Heat exchanger leaks and causes loss in cooling efficiency	Reactor temperature increases and Chemical E is produced if temperature is increased above threshold	High severity	Weld failure	1/1 000 operating hours. Data from operating experience	Weld inspection every 12 months	Very critical	9	Increase weld inspection frequency to one in six months

Reactor temperature is not controlled	Reactor temperature increases and Chemical E is produced if temperature is increased above threshold	High severity	Flow controller flow D fails to allow proper amount of cooling water to cooling jacket	1/5 000 operating hours based on vendor data	Instrument technicians perform diagnostic tests every 4000 h	Very critical 9	Increase diagnostic inspection frequency to every 2 500h
	Reactor temperature falls below temperature required to maintain reaction	Moderate severity	Flow controller flow D allows too much cooling water to flow through cooling jacket	1/25 000 operating hours based on vendor data	Instrument technicians perform diagnostic tests every 4 000 h	Low criticality 5	Operating experience shows that this type of flow controller fails very seldom in this fashion

reliability model. So, to perform the analysis, there has to be a sequence of events. For the purposes of this example, the events of interest are:

- V-4 fails and begins to leak.
- Operator fails to detect change in Chemical A flow into the reactor.
- Pump A motor fails and overheats.
- Operator fails to detect Pump A motor failure.
- Fire erupts.

Table 29.8 shows the event tree for this sequence.

29.4.4 Sample Fault Trees

There are possibly up to a 100 individual fault trees that can be developed for various top events for this process plant. They, of course, can all be rolled into one large fault tree. Figures 29.4 and 29.5 present two sample fault trees. Figure 29.4 shows a fault tree for failing to pump Chemical A into the reactor. Figure 29.4 shows a fault tree for a bad batch of chemical.

29.4.5 Sample PRA

PRAs are performed on only the highest risk, most complex, and/or most valuable assets. Most likely a PRA would not be performed on something as simple as our

TABLE 29.8

Fire Event Tree for Process Analysis

Fire Event Sequence Initiation	Valve V-1 Leaks	Operator Fails to Detect Change in Chemical A Flow	Pump A Overheats	Operator Fails to Detect Pump A Over Heating	Fire Initiates	Outcome
	No					Normal Operations
		No				Chemical A Spill
	Yes		No			Chemical A Spill
		Yes		No		Chemical A Spill and Damaged Pump
			Yes		No	Chemical A Spill and Damaged Pump
				Yes		
					Yes	Catastrophic Fire

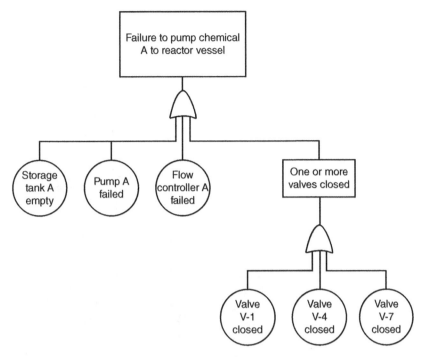

FIGURE 29.4   Fault tree for failing to pump Chemical A into the reactor.

example process plant. However, the following provides a flavor of how a PRA would be performed on a process plant.

A fire and subsequent release of chemicals into the atmosphere would most likely be the top event of interest for this plant. Chemicals C and E would be the really bad actors that any prudent chemical plant operator would not want in the environment. Therefore, this sample PRA will focus on a fire and subsequent release of Chemical C. Once again, this is only a sample of the type of analyses that would be performed and not a thorough and complete analysis.

The accident sequence to be monitored is:

- Chemical A or B begins to leak.
- The operator fails to detect the leak.
- An overheated pump or static spark ignites the spilled chemical.
- The operator fails to detect the fire.
- The fire damages either the storage tank, piping, pumps, or valves of the Chemical C process line.

Table 29.9 shows the event tree sequence for this event.

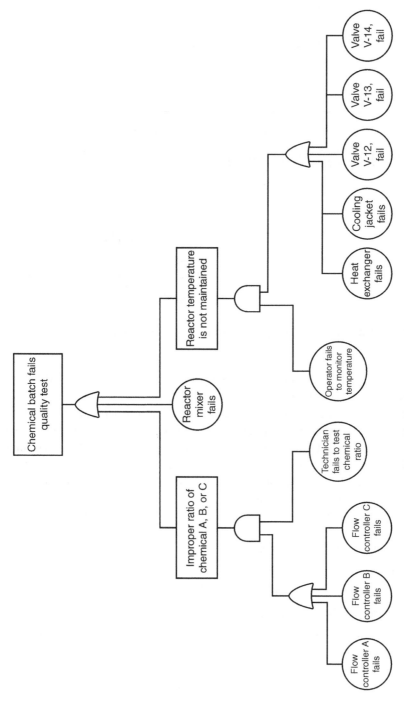

FIGURE 29.5 Fault tree for a bad chemical batch.

TABLE 29.9

Event Tree Sequence for Chemical Fore and Subsequent Release of Chemical C

Event Sequence Initiates	Chemical Spill	Operator Fails to Detect Leak	Ignition Source Present	Operator Fails to Detect Fire	Fire Damages Chemical C Process Equipment	Outcome
	No					Normal Operations
		No				Chemical A or C Spill
	Yes		No			Incipient Fire
		Yes				
				No		Major Chemical Fire
			Yes			
					No	Major Chemical Fire
				Yes		
					Yes	Chemical C Release

Fault trees are then developed for each of the major events in the sequence. Obviously, the operator actions are not complex, so human reliability analysis (HRA) event trees would not be developed. Also, if a major fire developed, it is almost certain that Chemical C process equipment will be affected. Therefore, fault trees would be developed for:

- Chemical spill
- Ignition source

Figures 29.6 and 29.7 are the fault trees for these three events.

Failure probabilities are assigned to the basic events in the fault trees. Table 29.10 shows these data, as well as the overall failure probability for the individual trees.

The next step is to combine the failure probabilities. Table 29.11 shows the failure rates for the events in the sequence and the overall failure rate. Rates are combined using Boolean AND logic since each of the events has to occur for the sequence to progress. By the time the failure rates are rolled up, the risk of the Chemical C process equipment being damaged by a fire becomes negligible.

Obviously, this is a simple analysis and error bounds have been omitted. If an analyst is going to perform a PRA on a system such as a nuclear power plant or a space vehicle, then, of course, bounding the error rates is absolutely important. For a simple analysis most operators want to know what are the prime drivers for failure. The failures with the highest priority can then be further analyzed, the equipment

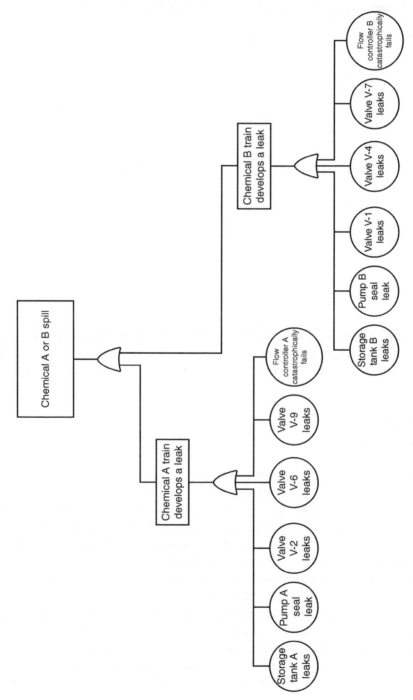

FIGURE 29.6 Sample fault tree for Chemical A or B spill.

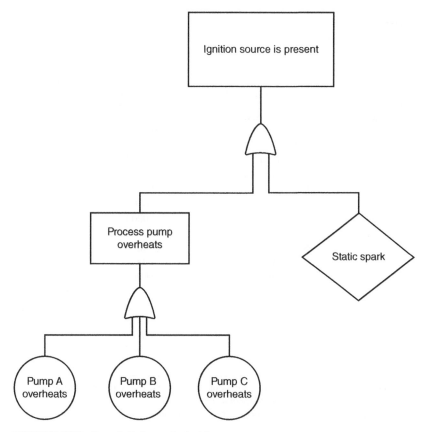

FIGURE 29.7   Sample fault tree for ignition source.

modified/replaced, or other mitigation factors implemented to reduce or eliminate the risk.

29.5 SUMMARY

This example shows the progress of a risk assessment from the very simple PHA to the more complex PRA. As shown, a wealth of information can be gained from each aspect of the analysis. In many cases an FMEA can provide the information necessary to make informed decisions concerning risk associated with a process. Analysis techniques such as PRA should be reserved for the most complex systems or those systems that have the highest risk.

TABLE 29.10
Failure Rate Data

Fault tree	Basic event	Failure rate	Failure rate for the fault tree
Chemical spill	Valve V-1 failure	0.000 1	
	Valve V-4 failure	0.000 1	
	Valve V-7 failure	0.000 1	
	Storage tank failure[a]	0.000 033	
	Pump A seal failure	0.002	
	Flow controller A failure	0.000 5	
	Valve V-2 failure	0.000 1	
	Valve V-5 failure	0.000 1	
	Valve V-8 failure	0.000 1	
	Storage tank failure	0.000 033	
	Pump B seal failure	0.002	
	Flow controller B failure	0.000 5	
			0.0057
Ignition source	Pump A overheats	0.001	
	Pump A overheats	0.001	
	Pump A overheats	0.001	
	Static spark[b]	0.000 1	
			0.0031

[a]Storage tank failure was listed as 1/30 000 operating hours. It was assumed that at any point in time, there is a 1/30 000 chance of a failure.
[b]A value that is reasonable can be used, if specific data is not available.

TABLE 29.11
Combined Failure Rates

Failure contribution	Sequence designator	Failure rate	Sequence	Sequence probability
Chemical spill	A	0.0057	A	0.005 7
Operator fails to detect leak	B	0.001	A × B	0.000 057
Ignition source	C	0.0031	A × B × C	0.000 000 018
Operator fails to detect fire	D	0.001	A × B × C × D	Negligible
Probability fire damages process equipment Value developed based on operating experience	E	0.01	A × B × C × D × E	Negligible

REFERENCES

1. Federal Emergency Management Agency, United States Fire Administration, National Fire Data Center (1989). Phillips Petroleum Chemical Plant Explosion and Fire, Pasadena, Texas.
2. EASHW. European Agency for Safety and Health (2011). Healthy Workplaces Campaigns. http://osha.europa.eu/en/campaigns/hw2010/maintenance/accidents/4-houston.pdf (accessed 30 September 2011).
3. Nova Chemicals (2011). Product risk profile, ethylene. http://www.novachem.com/Product%20Documents/Ethylene_RP_AMER_EN.pdf (accessed February 2019).
4. Airgas (2011). Isobutane, MSDS. https://www.airgas.com/msds/001030.pdf (accessed February 2019).
5. US Fire Administration (1989). Phillips Petroleum Chemical Plant Explosion and Fire, Pasadena, Texas. http://www.usfa.dhs.gov/downloads/pdf/publications/tr-035.pdf (accessed 25 August 2011).
6. Praxair (2011). Butadiene, MSDS. https://www.praxair.com/-/media/corporate/praxairus/documents/sds/1-3-butadiene-c4h6-safety-data-sheet-sds-p4571.pdf?la=en&rev=0b80ec176e5d4d8d9bbb6af3747ca606 (accessed February 2019).
7. OH (2000). Phillips provides update on K-Resin plant accident. *EHS Today* (17 April 2000). http://ehstoday.com/news/ehs_imp_33217 (accessed 26 August 2011).
8. Wogalter, M.S. (2006). *Handbook of Warnings*. CRC Press.
9. United States Chemical safety Board (USCSB). Reactive hazards. https://www.csb.gov/reactive-hazards (accessed February 2019).
10. BP (2005). Isomerization unit explosion. http://www.bp.com/liveassets/bp_internet/us/bp_us_english/STAGING/local_assets/downloads/t/final_report.pdf (accessed 30 August 2011).
11. United States Chemical Safety Board (USCSB) (2007). Refinery Explosion and Fire. Investigation Report. Report No. 2005-04-I-Tx. http://www.csb.gov/assets/document/CSBFinalReportBP.pdf (accessed 30 August 2011).
12. USDOL. US Department of Labor OSHA (2011). Timeline of events related to the BP Texas City refinery monitoring inspection. http://www.osha.gov/dep/bp/Timeline_BPTCR_Monitoring_Inspection.html (accessed 30 August 2011).
13. Greenhouse, S. (2009). BP Faces Record Fine for '05 Refinery Explosion. *New York Times* (30 October 2009). http://www.nytimes.com/2009/10/30/business/30labor.html (accessed 3 August 2011).

CHAPTER 30

Risk Assessment Framework for Detecting, Predicting, and Mitigating Aircraft Material Inspection

A Case Study

30.1 INTRODUCTION

The research to be described was initiated by the NASA Aviation Safety Program, Aircraft Aging and Durability (AAD) Project, which focused on the production of multidisciplinary analysis and capabilities that will enable development of system-level integrated methods for the detection, prediction, and management of aging-related hazards for current and future aircraft.

As a part of this project, NASA contracted to investigate and document a methodology to establish rigorous risk-based models for assessing damage detection capabilities in aircraft maintenance and repair operations (MROs). For purposes of this report, MROs provide major or minor maintenance, modification, refurbishment, or repair to aircraft.

The significance of the risk-based model is the incorporation of probability values for damage detection. Many factors can influence these probabilities in operational facilities that improve or hinder the ability to detect damage on an aircraft.

Risk Assessment: Tools, Techniques, and Their Applications, Second Edition. Lee T. Ostrom and Cheryl A. Wilhelmsen.
© 2019 John Wiley & Sons, Inc. Published 2019 by John Wiley & Sons, Inc.
Companion website: www.wiley.com/go/Ostrom/RiskAssessment_2e

We conducted a three-year study of aircraft MRO facilities to:

- Identify the operational factors (OFs) that are associated with the inspection process.
- Determine the extent to which relationships exist between OFs and inspection performance (i.e. probability of detection (POD).
- Develop probabilistic risk models based on the detection data.
- Develop generic methodology (framework) to assess the risk of OFs on inspection performance.

This report represents the final deliverable for the study and summarizes the data collection and analysis, and it presents the risk assessment framework for probabilistic risk models.

30.2 BACKGROUND

Real-world OFs, environmental conditions, and procedural variations influence the ultimate effectiveness of damage detection capabilities as implemented in commercial airline MROs. The AAD Project contracted the development of a risk-based methodology for assessing damage detection capability from the investigation team.

Since there are numerous possible applications (e.g. metallic airframe structures, engines, and wiring systems), this contract focused on a selected case study of the visual inspection of composite structure to develop and demonstrate a methodology that can be extended to other applications. Composite generally refers to the engineered materials made from several constituent materials with significantly different properties that remain separate and distinct within the finished structure. Structure refers loosely to large areas of primary or load-bearing structure, such as the fuselage of the B787, as opposed to small parts of the engine or cabin interior.

The goal of the first-year research was to identify OFs, including human factors associated with aircraft maintenance and inspection, which affect the visual detection of damage in the composite structure of aircraft. Examples of where human factors may increase or decrease the probability of damage detection include aspects of the workplace, task procedures, or tools that are used to detect aircraft damage. OFs that affect damage detection were identified through site visits and data collection at MRO facilities.

The goal of the second year was to conduct a visual inspection experiment that builds on the first year's results. OFs identified in the first year's research were utilized as independent variables that were systematically manipulated in an experimental design. This visual inspection experiment utilized aircraft maintenance personnel as subjects. The results of the visual inspection experiment were incorporated into POD curves, which are standard in the maintenance engineering community.

The goal of the third and final year was to develop and validate risk models associated with aircraft damage detection. Risk models were developed from the results of the visual inspection experiment and the resulting POD curves.

Additionally, a goal of the final year was to develop a generic methodology (i.e. framework) to develop risk models that can be extended to other areas of aircraft maintenance and repair. As a result of this research, an overall methodology was developed that can be used by researchers and MRO managers to assess the capability to detect damage in composite structure and other applications.

30.3 SUMMARY OF YEARS ONE AND TWO

30.3.1 Year One: Identify OFs

The goal of year one was to identify OFs, including human factors associated with aircraft maintenance and inspection that affect the visual detection of damage in the composite structure of aircraft. Examples of where human factors may increase or decrease the probability of damage detection include aspects of the workplace, task procedures, or tools that are used to detect aircraft damage. Research was conducted to identify OFs that affect damage detection on composite structure through site visits and data collection at maintenance, repair, and overhaul (MRO) facilities. Data were also collected, to a lesser extent, on OFs relating to repair of aircraft composite materials. It is difficult to completely separate the inspection and repair activities because the repair of the damaged structure depends on the results of the inspection.

Site Visits

Four MRO facilities were visited over the course of the year, in addition to a major commercial airplane manufacturer for the purpose of data collection regarding OFs. These site visits can be categorized as follows:

- A major overseas MRO facility.
- A major domestic airline MRO facility.
- A major domestic airline regional MRO facility.
- An intermediate military maintenance facility.

The purpose of the visits was to collect composite material inspection process data primarily to determine the OFs that are associated with the inspection process. Data collected from these site visits were utilized to determine the extent of relationships between OFs and inspection performance (i.e. POD).

Overseas Maintenance Site Visit   An overseas MRO facility was visited in November 2010. The facility employed 30 visual inspectors and 6 nondestructive testing (NDT) technicians. A data collection form was developed to guide the interviews. The first 30 questions pertained to the inspector and the last 5 questions pertained primarily to the supervisor.

The following seven inspection processes were observed during the site visit:

- Visual inspection.
- Tap test using an automated tap tester (Figure 30.1).
- Tap test using tap hammer.
- Vacuum test using manual vacuum device (Figure 30.2).
- Vacuum test using semiautomatic device.
- Thermograph testing (Figure 30.3).
- Pulse/echo ultrasound (Figure 30.4).

The project team observed a technician performing an NDT inspection with the "woodpecker" (an automated tap tester) and manual tap inspection of the underside elevator of a six-year-old aircraft (Figure 30.1). The area to be inspected was demarcated with tape into approximately 1- to 2-ft areas.

FIGURE 30.1   Tap testing with the "woodpecker" instrument.

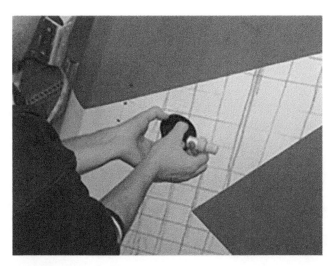

FIGURE 30.2   Vacuum testing.

FIGURE 30.3   Thermograph testing.

Operational Factors

The following is a list of the OFs that were observed:

1. Access to inspection areas
   - Frustration from difficult physical access to areas requiring inspection. NDT technicians indicated that obtaining physical access to areas being inspected was the most frustrating aspect of their work.

FIGURE 30.4 Pulse/echo testing.

2. Organizational factors
   ◦ Uncertainty in responsibility for dealing with defects or damage not identified in customer agreement.
   ◦ The aviation maintenance technicians (AMTs) indicated that a policy was implemented where damage is only fixed if it is stated on the job card. One worker indicated the situation could be "touchy" when defects are found that are not specified under the contract/agreement with the customer. (This might be a factor affecting risk related to discovery and repair of damage or defects.)

3. Environmental factors – noise
   ◦ Work other than inspection activities occurring simultaneously in the same physical area leads to distractions and noise.

4. Fatigue
   ◦ Fatigue was recognized as a factor affecting all work in the facility. (The policy allows workers to call their own breaks and rest when they deem necessary.)

Domestic Airline MRO Site Visit

The domestic airline MRO was visited in January 2011. This airline had recently merged with another large airline and is currently the largest MRO facility in North America. The airline employed in excess of 50 inspection personnel.

During the two-day site visit, data were collected from the following work groups utilizing the survey form, personnel interviews, and direct observations:

- Composite shop inspectors.

- Composite shop mechanics.
- Composite repair trainer.
- Line inspectors.
- Inspection managers.

Operational Factors

A significant amount of data was collected at this facility and indicated that the following OFs were the most significant:

1. Access to inspection areas
   - In many instances, the inspectors had to wait to inspect an area until AMTs had time to remove access panels. This created delays and it was also noted that sometimes areas were not inspected because the inspectors were not provided access.
2. Training
   - At this MRO, a composite inspection class was provided, but inspectors could perform composite inspections without the class.
3. Environmental factors – lighting
   - Lighting was noted by almost all the inspectors as a potential OF. However, all the inspectors carried high-powered flashlights and the inspectors said they used their flashlights no matter what the lighting levels to ensure they could see the area to be inspected.
4. Organizational factors
   - The airline had recently undergone a merger and there had been some issues resulting from differences in the manner that inspections had been performed in the two former airlines. Management was working with the inspectors to resolve these differences.
5. Procedures
   - Due to the two airlines merging, there were differences in the procedures being used to perform the inspections. This issue was also being worked.
6. Time
   - The inspectors were under time pressure to perform the inspections.

An important point was noted during the site visit regarding inspection and maintenance activities. Although they are separate operations, it is difficult for the inspectors and AMTs to completely separate the inspection and maintenance tasks because of the interdependency of the two operations. Many times, the activities cannot be separated.

Domestic Airline Regional MRO Site Visit

A large airline regional MRO facility was visited in April 2011. The airline had recently acquired a large airline and was in the midst of integrating the two operations. The airline employed in excess of 25 inspection personnel. The facility provided a large range of MRO capabilities, including line station-based services such as on-wing repair, engine change, A-checks, layover checks, inventory support, and NDT.

Over the two-day period, 13 individuals were observed performing their work duties. Personnel observed were as follows:

- Inspectors.
- Maintenance technicians.
- Maintenance leads.

The following processes were observed during the site visit:

- Visual inspection.
- Tap testing.
- Eddy-current application.
- Various maintenance activities, such as installation of equipment requiring fuselage penetration.

Activities where inspection was not the primary purpose were observed to understand how participants handle the discovery of damage and other problems. Additionally, the composite fabrication/repair shop was toured.

Operational Factors
In planning and preparing for an inspection, the inspectors relied heavily on institutional knowledge about a particular aircraft model's history and susceptibility to damage or failures. Information stemming from this institutional knowledge may be acquired through required readings and training, but much appeared to be acquired by experiential learning and word of mouth. Inspection techniques and tools varied considerably between the individuals performing the work. The OFs identified at this facility were as follows:

1. Access to inspection areas
   ◦ Difficulty using reflected light on underside or closely confined skin surfaces.
2. Training/experience
   ◦ There was a great diversity in composite inspection knowledge at this facility. One reason appeared to be the merger of the two airlines. One of the two airlines had very limited experience inspecting aircraft with more composite material.

3. Environmental factors – lighting
    - Inspector visually focused on the reflections of existing overhead light fixtures on aircraft skin surfaces to identify areas with unusual appearance (One inspector indicated that he has little or no vision in one eye).
    - Inspector also used flashlight to cast light across the surface of the skin for the same purpose.
4. Environmental factors – noise
    - High noise levels where tap testing is being performed.
5. Facilities
    - Platform extenders on scaffolding are difficult to operate and affect safety and convenience of inspection access.
6. Procedures
    - Some inspection techniques appear to be inconsistent with procedural requirements.
    - Inspector tapped during a general inspection although he told us the manual required only visual inspection.
    - Inspector said he writes up what he knows to be lightning damage as static discharge to avoid doing a lightning inspection.
    - Procedures for "required inspections" are consistently applied. However, less formal inspections, such as having a "second set of eyes" perform an inspection appears to be more of an individual decision. For example, this depends upon the inspector's reputation for quality of work. One inspector noted it is "embarrassing" to find a problem in someone else's work.

A tour of the composite shop and discussion with composite shop personnel revealed other OFs. The shop performed repairs on composite parts and fabricated entire replacement parts.

- The most difficult challenge in repairing damage is determining what composite is involved. The composition has evolved over time and may be different for different parts of the aircraft.
- This appeared to be a craft requiring a high degree of knowledge, experience, and skills. No younger personnel were noted in the shop. Personnel verified that younger individuals were not being trained in this type of work at the facility.

Military Composite Shop Site Visit

The US Navy performs maintenance at three different levels to ensure aircraft are fully operational 365 days a year. The maintenance levels are identified as organizational, intermediate, and depot. Each level performs a specific function with the utmost regard to quality and safety. Some repairs being either too time-consuming or requiring expertise outside of the level is one of the primary reasons why maintenance is broken down into this three-part system.

1. Organizational (O-level)
   Maintenance of this type is defined by the efforts required by a specific unit on a day-to-day basis to keep an airframe in an operational condition. In addition to maintaining assigned aircraft and aeronautical equipment in a full mission-capable status, O-level is also responsible for the continual improvement of the local maintenance process.

2. Intermediate (I-level)
   Maintenance of this type is required in support of operational activities. This includes the repair of specific parts unable to be refurbished at the organizational level and detailed maintenance of electronic components requiring specific equipment not germane to O-level. I-level is also responsible for maintaining a constant flow of necessary materials to operational assets.

3. Depot
   Depot level repairs include major alterations or refurbishing of an asset due to age or upgrade far beyond the capabilities of the organizational level requiring industrial facilities not available elsewhere. Special depot-level maintenance (SDLM) commonly performs overhaul operations on airframes in need of such attention.

The site visited was considered I-level. It was a small shop and performed repairs on a wide range of composite aircraft parts, including advanced fighters. The shop had adequate lighting but was exceptionally noisy because of its proximity to the airfield. The composite shop had both military and nonmilitary personnel. On the day of the visit, there was only one active duty military structural repair AMT, called an AM-7232 advanced composite structural repair intermediate maintenance activity (IMA) technician, and one contractor (entitled an "artesian") on site. Three other military personnel from the shop were at a composite repair class. The personnel who work in the shop do changeover frequently. The shop had several nonadjustable tables used to lay up the composite, a downdraft sanding booth, a walk-in oven/autoclave for curing the composite resin, and several smaller sawhorse type fixtures for holding the parts.

When a part enters the shop, it has a repair order that indicated the type of repair needed. Figure 30.5 shows a typical damaged composite part. The shop personnel conducted an inspection of the part to confirm where the damage was located. The shop used an acoustic tap hammer as their primary NDT apparatus (Figure 30.6). The IMA technician did have a good level of knowledge about the use of the acoustic hammer. The composite shop used the facility's NDT inspectors if a more complicated inspection was needed. Those inspectors were not dedicated composite inspectors. In fact, on any given day, they could be inspecting metal bomb hanger parts, landing gear, aircraft engine turbine blades, and 20 mm gun parts. The composite shop personnel interviewed were not aware of any composite reference samples being used to calibrate the instruments.

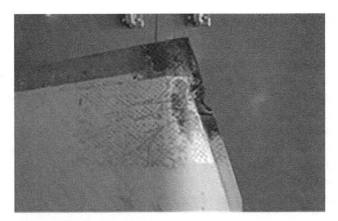

FIGURE 30.5   Military aircraft composite damage.

FIGURE 30.6   Acoustic hammer.

Operational Factors

These were the OFs identified during the visit to the military facility:

1. Training/experience
   - Personnel were not well trained on composite inspection or repair processes.
2. Environmental factors – noise
   - The shop had a very high, constant noise level due to its proximity to the airfield.
3. Personnel change
   - Personnel changed over frequently.
4. Facilities
   - The composite shop had less than adequate work benches.

Airplane Manufacturer Visit

A discussion was held with a commercial airplane manufacturer concerning aircraft composite inspection. The manufacturer stated that inspection of aircraft composite for the next-generation aircraft is that all initial inspections will be performed visually. NDT techniques will not be needed to characterize damage until the part has been designated as being damaged. For instance, tap testing will not be used to characterize damage and these types of techniques will not be in the original equipment manufacturer (OEM) procedures. If an operator uses such techniques, they will be performing the technique outside of recommended procedures. This was a very important point because all of the repair facilities that were visited used some sort of tap-testing technique in their initial inspection process.

30.3.2 Analysis and Discussion of Data

Figure 30.7 presents the high-level view of the inspection process observed at all facilities. From the airplane manufacturer's perspective, there is an important problem with this process. Once a visual inspection indicates an anomaly, the part is considered damaged and repair performed. Further NDT is not needed. All of the facilities performed extensive NDT on the part after it was found by visual inspection to be damaged. Therefore, in the airplane manufacturer's perspective, they are operating outside of procedures.

Table 30.1 lists the OFs that were identified at the four facilities. The table is arranged in order of OFs affecting the most facilities to OFs affecting the least facilities.

It was evident that no single OF was common to all of the sites visited. However, training, noise, and access to inspection areas were each identified as potentially affecting three facilities. Lighting, facilities, and procedures were identified as each potentially affecting two facilities.

The most common OFs identified at three MROs as potentially affecting inspection worker performance were as follows:

- Access to the area being inspected.
- Training.
- Noise.

The OFs identified at two MROs as potentially affecting inspection worker performance were as follows:

- Lighting.
- Facilities.
- Organizational factors.
- Procedures.

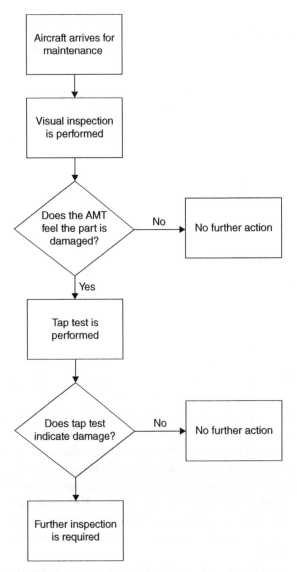

FIGURE 30.7   High-level view of the inspection process.

Finally, those OFs affecting only a single facility were as follows:

- Personnel turnover.
- Time.
- Fatigue.
- Distractions.

TABLE 30.1
Operational Factors (OFs).

Operational factor		Foreign maintenance, repair, and overhaul (MRO)	Domestic MRO	Domestic regional MRO	Military facility
Access to inspection areas		X	X	X	
Training/experience			X	X	X
Environmental factors	Noise	X		X	X
	Lighting		X	X	
Facilities				X	X
Organizational factors		X	X		
Procedures			X	X	
Personnel turnover					X
Time			X		
Fatigue		X			
Distractions		X			

Additionally, it was observed that all of the facilities and most of the inspectors performed inspections of composite materials differently. This raises the question: *Does the high variance in inspection techniques produce reliable (consistent) results?* Furthermore, there is a lack of data regarding the relationship between indicated surface anomalies and the inspector's ability to flag actual composite damage (1).

Surface indications of composite damage included dents, holes, tears, abrasions, ripples, depressions, raised areas, and other irregularities. However, there may or may not be a surface indicator for underlying composite damage. Overall, there is not an adequate body of information regarding the visual POD of composite material damage, nor the correlation of surface indicators and actual underlying composite damage. Therefore, a solid baseline on the reliability of the visual inspection of composite materials does not exist.

The information collected indicated that the inspectors performed their inspection tasks regardless of the operational environment. Regardless of the lighting conditions (low ambient light, moderate ambient light, or bright ambient light), the inspectors used their flashlight to aid in the inspection process. Differences existed with respect to how the inspectors might use their flashlights, but they routinely utilized a flashlight to aid in the inspection process. Therefore, the effect of lighting does not appear to affect the inspection task. Noise level was also noted as an issue by the inspectors when conducting a test that produced an audible sound, such as tap testing. However, inspectors often waited for high ambient noise levels to abate before performing these tasks or they used automated devices that were not affected by high ambient noise.

30.4 YEAR TWO: CONDUCT A VISUAL INSPECTION EXPERIMENT

The goal of the second year of the study was to conduct a visual inspection experiment. During the first year of the project, OFs were identified that they could be used as independent variables. Since the list of OFs was extensive, it was decided to narrow the experiment to using general visual inspection (GVI) protocol and detailed visual inspection (DVI) protocol as the independent variables. GVI was purely a visual task, whereas DVI had also tactile attributes associated with this protocol. The project team felt the study participants would better understand the distinctions between the two variables better than trying to use OFs that could be less well controlled. From the experiment, composite inspection probability data were developed that was incorporated into the risk assessment framework.

Participants performed inspections on the same test article under similar situational conditions. There were two experimental questions:

1. Will the participants consistently identify surface anomalies as indicative of damage?
2. Will the participants identify surface anomalies that correlate well with actual damage?

The test article was an inboard section of a composite horizontal stabilizer. The stabilizer had existing damage that was incurred during its service on a commercial Boeing 737 aircraft. Additional damage was inflicted on each side of the stabilizer in preparation for experiment. Figures 30.8–30.11 show the dimensions of the test article.

This experiment focused on two OFs, GVI, and DVI as independent variables, using the POD as the dependent variable.

For purposes of this study, GVI protocol did not allow participants to touch the test article. However, auxiliary lighting, such as a flashlight, could be utilized. Visual data are the only information collected.

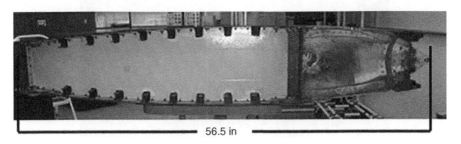

56.5 in

FIGURE 30.8   Inboard side of the test article.

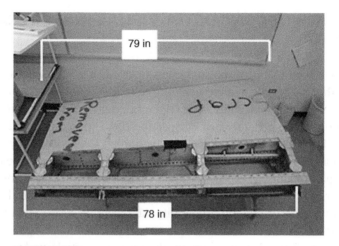

FIGURE 30.9   Bottom side of test article.

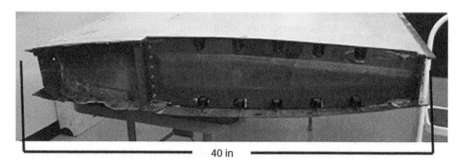

FIGURE 30.10   Topside of the test article.

FIGURE 30.11   Bottom side of the test article.

The DVI protocol allowed participants to touch the test article and use tools that contacted the test article to facilitate the process, allowing for visual, tactile, and auditory information to be collected. Visual techniques were as described for the GVI process. Tactile techniques involved contact with the test article, such as feeling the test article with the hand. Auditory techniques introduced mechanical energy into the test article (tapping) and received the sound energy that reflected from the piece.

The experiment was designed as follows:

1. Half of the participants were randomly assigned to inspect the topside of the test article under conditions intended to replicate GVI protocol. The other half of the participants were assigned to inspect the topside of the test article under conditions intended to replicate DVI protocol.
2. Similar random assignment and conditions intended to replicate GVI and DVI procedures were assigned for the bottom side of the test article.
3. The participants were asked to place a sticker on the areas where they perceived the test article was damaged. A green sticker was used for GVI protocol and a red sticker for DVI protocol.

Situational conditions were controlled with the intent of maximizing the opportunity for reliability and validity of the experimental inspections. The inspection occurred with the following conditions:

- Open access to the test article. These conditions varied depending on whether the GVI or the DVI process was applied to the individual test.
- Minimal distractions in the work area.
- Adequate ambient light.
- Auxiliary lighting (flashlights) available for use.
- Other tools available that are typically used for inspections that do not constitute NDT equipment (e.g. ultrasound equipment).
- No time limits on an inspection event.

30.4.1 Pilot Study

To test the experimental protocol, a pilot study was conducted at Idaho State University (ISU) Aircraft Maintenance Technician (AMT) School in Pocatello, Idaho. A total of 16 students participated in the pilot study. All of the students had been trained in visual composite inspection techniques, though not extensively experienced.

After the students had completed their inspections, all of the visual inspection experiment data were collected. The test article was marked with a 7-in-by-7-in grid as shown in Figures 30.12 and 30.13. The researchers assigned each of the participant's perceived indications of damage to a grid location. This was done with extreme care and all three researchers came to consensus as to the grid location of the perceived

FIGURE 30.12 Topside of the test article with grid lines.

FIGURE 30.13 Bottom side of the test article with grid lines.

damage. The 7-in-by-7-in grid dimension was selected because it matched the size of the aperture of the thermographic camera which NASA Langley personnel intended to use to collect data on the test article.

A factor for each quadrant was developed for each of the participant groups called the "positive perceived indication of damage" (PPID). This is simply the number of positive indications of damage divided by the total number of participants who inspected that quadrant for damage multiplied by 100. If, for instance, in a certain quadrant, three out of six participants indicated there was damage, then the PPID would be calculated as shown in the equation:

$$PPID = \frac{\text{Indication of damage}}{\text{Total number of participants}} \times 100 = \frac{3}{6} \times 100 = 50\%.$$

In this manner, statistical tests could be performed without the POD calculations. Qualitative data were also collected to the extent possible based on the availability of the participant to answer questions.

The pilot study went well and no significant changes were made to the data collection methods. The significant findings from the statistical analyses were as follows:

- There was no statistically significant difference between the results for GVI and DVI protocols when comparing the results for all of the quadrants, those with anomalies, and those without. This might indicate that, for inexperienced participants, the potential to find surface anomalies is independent of inspection protocol. The actual difference between GVI and DVI is 12%, with the GVI protocol having the larger number of positive quadrants.
- There were statistically significant differences for the comparisons of quadrants with anomalies to quadrants without anomalies for both GVI and DVI protocols.

There was no statistically significant difference between the results for GVI and DVI protocols when comparing the results for only those quadrants with anomalies. This might also indicate that, for inexperienced participants, the potential to find surface anomalies is independent of inspection protocol. The actual difference between the average GVI and DVI results was much smaller than for the test using all of the quadrants. In this case, the actual difference was only 6%, with GVI producing the higher value.

30.4.2 Full Study

Because the pilot study validated the experiment protocol as designed, no changes were made for the full study. The full composite visual study was conducted at the following:

- Airline Regional Facility MRO in Salt Lake City, Utah.
- General aviation (GA) – two fixed based operators in Idaho Falls, Idaho.

To increase the sample diversity and size, the data collected from ISU AMT School were included in the full study.

Airline Regional Facility MRO

At the airline regional facility MRO, data were collected from 10 airline inspectors with 20 samples collected.

The test article was transferred to the facility and placed in the airline facility hangar available for the midnight shift and the day-shift participants. Each participant verbally received the instructions regarding the GVI and DVI processes and auxiliary lighting and tools they could use.

Each participant inspected both sides of the test article, one side using GVI techniques and the other side using DVI techniques. These presentations were randomized to the participants. The participants were asked to place a sticker on the areas where they perceived the test article was damaged. A green sticker was used for GVI protocol and a red sticker for DVI protocol. Photos were taken after each participant completed inspecting each side of the test article. Figure 30.14 shows the use of a flashlight for additional lighting and Figure 30.15 shows the participant tapping with a coin.

General Aviation (GA)

The full study included two GA fixed-base operators (FBO). Data were collected from three instructor helicopter pilots, a mechanic, and a ramp agent at an Idaho Falls FBO with 10 samples. In addition, data were collected from another Idaho Falls FBO that included an AMT and two ramp agents with six samples.

Full Study Conclusions

The two OFs identified as important during the full study were access to inspection areas and environmental factors. All figures and tables containing the full results of the experiments are contained in the year-two report. The experiment showed that all groups did systematically examine the test article and found higher percentages of perceived damage in the quadrants with surface anomalies than in those quadrants without surface anomalies. This was statistically significant for all the study groups. This indicated the study did not produce random data, but useable results.

FIGURE 30.14   Participant utilizing the aid of a flashlight.

FIGURE 30.15   Participant utilizing a coin as he taps across the test article.

Several significant findings were made as follows:

1. Inspectors using the GVI protocol produced more positive indications of perceived damage as compared with the DVI protocol when inspecting the quadrant surface anomalies. This possibly indicates that GVI produces more false positives. There is a 60% reduction in the PPID using the DVI protocol as compared with the GVI protocol.
2. Students, in general, had higher percentages of PPID factors than did the airline group, but less than the GA study group.
3. A higher percentage of study participants perceived holes and cracks as damage. In fact, the surface anomalies that were holes and cracks were definitely damage. Paint chips were the second highest type of anomaly perceived as damage, followed by dents, and then other types of irregularities.

Following completion of all experiments, the test article was thoroughly examined by NASA Langley personnel using NDT/nondestructive examination (NDT/NDE) techniques to determine where damage actually existed. The experiments provided valuable insights into the visual inspection process.

30.5 AIRCRAFT COMPOSITE VISUAL INSPECTION RISK FRAMEWORK DEVELOPMENT

There are many ways to define a risk framework depending on the industry. For instance, the financial industry's definition of *risk framework* is different from that of a utility that operates nuclear power plants' definition. A risk framework is not a risk assessment per se; however, many use the terms interchangeably.

For the purpose of the visual inspection of composite materials, *risk framework* is defined as follows:

- A bounded set of activities associated with the visual inspection of composite materials that can generate risk.
- The ways in which these activities can fail.
- The failure/error rates for these activities.
- The major influences (for instance, performance shaping factors [PSF]) on these activities that can increase or reduce the risk.
- The potential ways in which the risk can be mitigated or eliminated.

The following discusses the development of the aircraft composite visual inspection risk framework.

30.5.1 Description of the Aircraft Composite Visual Inspection Process

Figure 30.16 shows the basic model developed for visual inspection of aircraft composites based on the data collected from the site visits during the first year of this study. In the second year of the study, quantitative data based on the visual inspection experiment were collected.

The differences between GVI and DVI protocols are explained in sections that follow. Figure 30.16 shows the basic flow of the inspection process. Whether an inspector

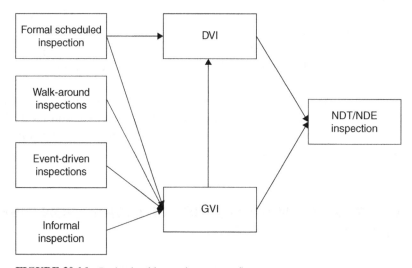

FIGURE 30.16   Basic visual inspection process flow.

is required to use either the GVI or DVI protocol for conducting an inspection depends on whether the inspector perceives a part to be damaged. If so, it will most likely need to be examined using an NDT/NDE technique before a repair can be performed. Only if the part is clearly damaged beyond repair will it not need to have an NDT/NDE.

The following four basic reasons a visual inspection might be performed were identified:

- *Formal scheduled inspection.* Inspections mandated by the maintenance manuals for the aircraft. These can be GVI or DVI.
- *Walk-around inspections.* Inspections that occur when a pilot or other airline personnel notice an anomaly on an aircraft. Maintenance is notified, who then send a mechanic or inspector to evaluate the anomaly. These inspections most likely start as a GVI and can progress to a DVI.
- *Event-driven inspections.* Inspections that result from incidents such as a bird strike, lightning strike, or hail damage. These inspections most likely start as a GVI and can progress to a DVI.
- *Informal inspections.* Similar to walk-around inspections, these inspections occur when a ramp agent or other airline personnel notice an anomaly and notifies maintenance, who then sends a mechanic or inspector to evaluate the anomaly. These inspections most likely start as a GVI and can progress to a DVI. For the purpose of this study, it is assumed that once an inspection is initiated, it will progress to either a GVI or DVI. The part of the overall inspection process designated as the "inspection space" (Figure 30.17) pertains only

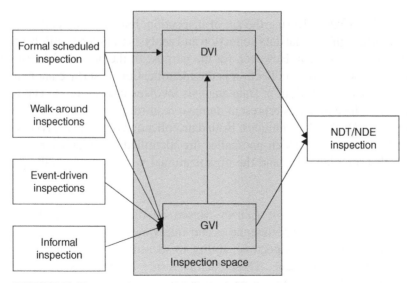

FIGURE 30.17   Inspection space. NDT, nondestructive testing; NDE, nondestructive examination.

to the GVI or DVI process. This inspection space represents the tasks that were analyzed.

30.5.2 How Data Collected During Year One and Two of the Project Were Used in Support of the Risk Framework

The first three chapters of this report discussed, in general, the data collection process and results of the data collection efforts. Many interesting findings and observations were made. This section discusses how the data were used to develop the risk framework.

Qualitative Analysis and the Development of Human Error Types

A fitting analogy for the inspection process is to consider the human as the instrument for collecting and analyzing data (2). A prevailing view is that human error must be considered in the context of critical process parameters and the organizational system in which errors occur (3). The visual inspection experiment found examples of PSFs residing in the inspection event itself, the process in which the inspection event occurs, and in the organizational system. The identified risk framework model (Figure 30.16) encompasses all three. The inspection event and process parameters reside within the model's "inspection space," while the system includes the following:

- Organizational culture.
- Organizational knowledge and tacit knowledge.
- Rules, procedures, and guidelines.

The Inspection Event During the act of inspection, humans both collect and analyze data. It is recognized that data collection and analysis cannot be entirely separated during the inspection event. However, for the purpose of the present risk framework, the primary purpose of data collection is the identification of surface anomalies (i.e. "Something doesn't look right"). Data analysis involves inspectors' perception as to whether or not the anomalies represent damage requiring repair or requiring further analysis using NDE/NDA techniques. Both data collection and data analysis are influenced by the process by which anomalies are identified, the characteristics of the anomalies that are identified, and the organizational system in which the processes reside.

Factors Affecting Inspection Errors Based on the observation of inspection events – both actual and experimental – and interviews with selected participants, several factors can affect inspectors' ability to identify surface anomalies and to perceive anomalies as damage. These factors reside in the physical condition of the inspector, conditions of the inspection event, nature of the anomalies, process parameters, and organizational system. Following is a discussion of these factors as they might affect data collection (detection of anomalies) and data analysis (perception of damage).

Event Conditions
The condition of the inspector (fatigue and sensory impairment) may contribute to errors in detection of anomalies (i.e. data collection). While fatigue is not necessarily an observable condition and none of the participants claimed to be fatigued, one inspector who was interviewed claimed to have little or no vision in one eye.

Lighting and access to inspection areas are conditions in the operating environment that may contribute to errors in identifying anomalies. Upon reviewing the results obtained in the first year of this study, it became apparent that the inspectors performed their inspection tasks regardless of the operational environment. The best example of this is that no matter the lighting conditions (low ambient light, moderate ambient light, or bright ambient light), the inspectors used a flashlight to aid in the inspection process. There were differences in how the inspectors used flashlights, but all routinely used a flashlight to aid in the inspection process. Therefore, the level of lighting does not appear to affect the risk of error in the task.

Tap testing appears to be a nearly universally applied technique for determining whether or not a surface anomaly is, or may be, damage (i.e. data analysis). Therefore, noise level in the operating environment may affect perceptions of damage. Noise level was noted as an issue by the inspectors interviewed during the first year of this research, but inspectors often waited until high ambient noise levels abated before they performed tasks such as tap testing, or they used automated devices that provided visual rather than auditory stimuli, which were not affected by high ambient noise. It should be noted, however, that even brief exposures to noise can shift hearing thresholds significantly, requiring hours or days to recover (4). One inspector was observed wearing hearing aids. So, simply inspecting in a low ambient noise environment may not adequately mitigate the risk of error if the inspector has recently been exposed to high noise levels.

Some inspectors that were interviewed in the first year indicated that difficult access to inspection areas caused frustration, increased the complexity of the inspection event, and sometimes affected their ability to perform the work. While formal inspections are generally scheduled to minimize interfering activities in the same workspace, it was concluded that access to the inspection area is a contributor to risk of inspection errors.

Characteristics of Surface Anomalies
The experimental data from year two demonstrate that failure to perceive anomalies as damage is lowest for holes/cracks in the test article. The failure rate increases progressively through paint chips, dents, and irregularities. Section 30.5.2.2 describes these results and the method for calculating failure rates among the study groups in detail.

Process Parameters

During the study's second-year experiment, participants were requested to identify damage to the test article. Therefore, participants both collected data (i.e. "Something doesn't look right") and analyzed data (i.e. "This surface anomaly is, or might be, damage"). Also, participants used GVI techniques and DVI techniques. Most participants performed both techniques, one on each side of the test article. Therefore, the inspection events were embedded in either the GVI or DVI process parameters. These parameters influenced the inspection techniques used by participants as well as their perceptions of damage.

When conducting GVI, participants were not allowed to touch the test article. However, they were permitted to use auxiliary lighting, such as a flashlight. For DVI, participants could touch the test article and use tools that contacted the piece to facilitate the process. In neither case was a time constraint imposed on the inspection events.

General Visual Inspection (GVI)

For GVI, the only input was visual. Participants approached the inspection in various ways – some systematically moved horizontally across the piece from top to bottom, while others appeared to randomly scan the whole piece and focus attention on an apparent anomaly. Regardless of approach, once attention was focused, the participants often changed their angle of view to include different perspectives around the apparent anomaly. Some used the light of a flashlight directly on the piece, others took advantage of light reflected on the piece from some other source in the area, and others used no auxiliary lighting.

Detailed Visual Inspection (DVI)

DVI involved physical contact with the test article. The participants conducted inspections in a variety of ways using three sensory inputs. Visually, the approach was similar to the GVI process. In many cases, the visual input appeared to be the first input relied upon, functioning as a "screening" activity. Tactile input, or feeling the test piece with the hand, was a frequently used input. Participants were observed rubbing their fingertips over the surface, appearing to confirm results from a different sensory input, and a few appeared to rely on tactile input first. In two cases, participants rubbed the test piece with the back of their hand rather than their fingertips.

Auditory input was produced by introducing mechanical energy into the test piece (tapping) and receiving the sound energy reflected from the piece.

Participants who used tapping as part of the event used a variety of tapping tools such as coins or coin-shaped disks, other small metal objects, flashlight (the instrument itself), and, in a few cases, fingertips or knuckles. Generally, tapping appeared to be used most often to confirm an anomaly discovered by visual or tactile means. One participant attempted to use systematic tapping as the only means of data collection, but after a few minutes introduced tactile and visual methods into the event.

During DVI, participants rarely relied on only one type of sensory input. At any given time during the DVI process, participants were collecting data via two sensory inputs in parallel – most often visual and tactile, but sometimes using any combination of the three (visual, tactile, and auditory).

Based on the observations during the MRO site visits and the visual inspection experiment, it is concluded that the inspection event is highly individualized by those performing the work. While inspectors collected damage data using visual, tactile, and auditory sensory input, the application of these techniques varies by each individual. It is concluded that GVI is primarily a data collection process, while DVI is primarily one of data analysis. Significant quantitative data differences exist between GVI and DVI, which are discussed in detail in Section 30.5.2.2.

Organizational System
Within this category of PSF, the model incorporates rules, procedures, guidelines, organizational culture, and organizational knowledge.

Individual inspection events may be driven by rules, procedures, and guidelines issued by airlines, regulators, and aircraft manufacturers. Inspectors that were observed and interviewed displayed understanding of the requirements and the sources of requirements associated with the conduct of their work. An administrative situation at one airline MRO facility was observed that was potentially inducing errors in the inspection process. A merger between two airlines was still being accomplished administratively. As a result, inspectors and their supervisors were utilizing corporate rules, procedures, and guidelines that were artifacts of the two airlines prior to the merger, as well as some new documents that were being created from the artifacts. Interviewees described some confusion about applicability of three different partial sets of documents, which could result in misapplication of inspection requirements, potentially unidentified anomalies, or delays in identifying anomalies. No cases were identified where this actually occurred.

Generally, a high degree of morale and pride in workmanship existed among the aircraft industry employees that were interviewed and observed. Individuals assigned the role of "inspector" tended to have a high average number of years of experience. Their training to the various levels of qualification provided the explicit knowledge necessary to conduct their work, and it is assumed that inspectors claiming qualification have received the requisite training and have demonstrated their knowledge and skills satisfactorily.

However, there exists among inspectors a high degree of "institutional knowledge" – also known as *tacit knowledge* (5). This is expertise gained primarily from experience. Explicit knowledge is delivered primarily through training and other formal learning. Unlike explicit knowledge, which can be codified outside the knower, tacit knowledge represents knowledge that is unique to the knower and is highly internalized by the knower.

One inspector explained his practice of focusing on the underbelly of the aircraft, especially near landing gears. He assumes a higher probability of damage in these areas since an aircraft picks up debris from the runway during takeoffs and landings. This assumption could increase the probability that he would perceive an anomaly in this area as damage.

Tacit knowledge is probably shared informally among inspectors and other occupations involved with aircraft maintenance and operations. The assumption about focusing on areas of high damage probability might increase inspection efficiency, but could increase false-positive results and cause anomalies to be missed on other parts of the aircraft. Erroneous facts, assumptions, and opinions in this flow of tacit knowledge may increase the probability of errors in detection of anomalies, as well as perceptions of damage.

A noteworthy point is that a large proportion of tacit knowledge is invisible to the system in which it flows because the system can only recognize codified knowledge. This contributes both positively and negatively to the risk of inspection errors.

Qualitative Data Summary
Qualitative data was collected through interviews with inspectors and observing inspectors doing their work. Both qualitative and quantitative data were collected through the experiment conducted during the second year of the project. The inspectors indicated that noise in the inspection environment and difficult access can affect performance. These risk factors are partially mitigated by refraining from scheduling inspections around times of high ambient noise and other work activities in the inspection area. Although it is reasonable to expect that the availability of light in the inspection area might be a significant PSF, this risk factor is mitigated for the most part by inspectors' nearly universal use of auxiliary light sources.

During the second year of the experiment, qualitative data were collected by observing the strategies used by participants to identify surface anomalies and suspected damage on a test article using both GVI and DVI techniques. During GVI, inspectors were allowed to use only visual inputs with auxiliary lighting. During DVI, participants used visual, auditory, and tactile inputs, as well as auxiliary lighting to identify anomalies and suspected damage. Qualitatively, participants were observed using various strategies and approaches in conducting GVI and DVI. For the GVI and DVI techniques, the qualitative data were compared with the quantitative data collected during the experiment. The false-positive damage detections were significantly higher with GVI (where only visual input was used) than in DVI (where multiple sensory inputs were used).

From a qualitative research perspective, the inspection event displays features that are unique to each inspector. They utilize tools, techniques, and strategies that have evolved from their own experience, all of which increase the difficulty of identifying generally applicable factors that shape performance. The inspectors' apparent high degree of dependence on tacit (uncodified) knowledge contributes to this difficulty. While the application of tacit knowledge may increase the efficiency of inspections, it may also increase the probability of errors in an individual inspection event and cause inconsistency in damage detection among inspectors.

The following section describes in detail the quantitative data related to surface anomalies on the experimental test article and the process parameters used to detect anomalies and determine damage.

Quantitative Data

Quantitative data collected during visual inspection experiment were used to develop failure rate data. It is important to note that the experiment conducted during year two of the study was not a POD study, per se, because the participants were not asked to find all the dents. Participants were asked to determine whether surface anomalies were actual damage. As was pointed out by Boeing, a dent around a rivet might just indicate that when the rivet was tightened up it drew in the composite. Figure 30.18 shows these types of dents. Though thermographic examination (Figure 30.19) shows what might be corrosion around the rivets, there was no composite damage. None of the experienced airline inspectors indicated that these types of dents were damage. Only 20% of the students and 33% of the GA participants indicated these dents were damage. Therefore, the experienced airline participants correctly categorized this surface anomaly, whereas a percentage of the other two participant groups did not. Certain types of surface anomalies are most certainly damage. Figure 30.20 is a crack in Quadrant R3. Thermographic examination confirms it is damage (Figure 30.21).

The most common types of subsurface damage found include the following:

- Holes/cracks. Extend from the surface into the interior of the composite material.
- Porosity. Defined by the NDT expert as "light" to "heavy." The manufacturer indicated that this porosity might even be an artifact from the manufacturing methods used at the time of fabrication.
- Corrosion. The corrosion of metal fasteners.
- Cracking. Small cracks around fasteners that were not visible on the surface.

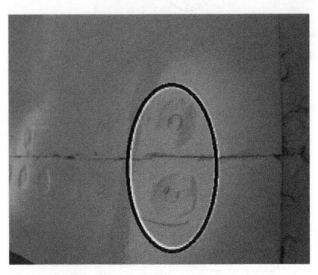

FIGURE 30.18   Dents around rivets.

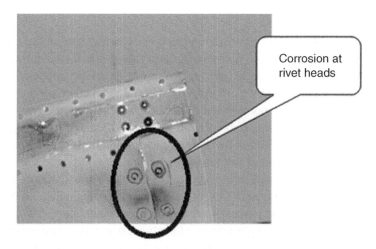

FIGURE 30.19 Thermograph of damage around rivets.

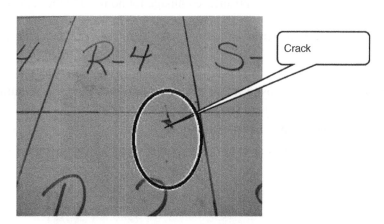

FIGURE 30.20 Composite crack in Quadrant R3.

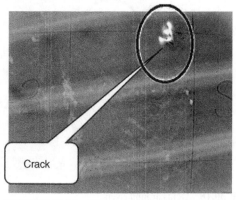

FIGURE 30.21 Thermographic image of a composite crack in Quadrant R3.

- Uneven gluing. Manufacturing defect.
- Unidentified damage. Damage that could not be categorized.

There is a question as to what underlying composite damage inspectors can detect using GVI/DVI techniques. This will be explored further in Section 30.5.6.

Table 30.2 shows the generalized PPID results from the experiment conducted in year two. This is an indicator of successfully finding composite damage. Usually the failure rate is calculated for a risk assessment. To determine a rough estimate of the failure rate, the PPID was subtracted from 1.0.

$$1 - \text{PPID} = \text{Failure rate.}$$

Table 30.2 is a combination table and shows the failures associated with each of these anomaly type categories using the data from the second part of Table 30.2.

However, the results in Table 30.2 were developed using the data from the experimental studies prior to determining whether all the surface anomalies were actually damage.

An attempt was made to determine how well inspectors correctly differentiated between damaged and non-damaged quadrants using GVI/DVI techniques.

TABLE 30.2

Generalized Year Two Positive Percent Indication of Damage Results.

Anomaly type	Inspection protocol	Study group		
		Students (%)	Airline (%)	General aviation (GA) (%)
Holes/cracks	General visual inspection (GVI)	57	74	67
	Detailed visual inspection (DVI)	61	87	67
Paint chips	GVI	28	33	42
	DVI	27	53	42
Dents	GVI	15	5	33
	DVI	6	17	33
Irregularities	GVI	8	14	36
	DVI	16	27	26

Failure rates developed from generalized year two positive percent indication of damage results

Holes/cracks	GVI	43	26	33
	DVI	39	13	33
Paint chips	GVI	72	67	58
	DVI	73	47	58
Dents	GVI	85	95	67
	DVI	94	83	67
Irregularities	GVI	92	86	64
	DVI	84	73	74

TABLE 30.3
PPID by Quadrant.

Result category	Color representation in table
Lightest grey – correctly identified quadrant without damage	
Light grey – false positive	
Medium grey – correctly identified quadrant with damage	
Darker grey – incorrectly identified quadrant with damage	
Dark grey – results are questionable	

Quadrant	Surface anomaly	Underlying damage	PPID					
			Students		Airline		GA	
			GVI	DVI	GVI	DVI	GVI	DVI
A1	No	Repair, disbonding around right bolt hole, and light porosity	11	20	0	0	0	0
A2	No	No damage	0	0	0	0	0	0
A3	No	Light porosity	0	0	0	0	0	25
B1	Small crack	Crack, medium porosity	11	20	20	60	25	0
B2	Large hole	Hole, medium porosity	78	80	100	100	75	100
B3	Dent	Impact damage, medium porosity	67	0	0	80	75	0
B4	No	Medium porosity	22	20	0	20	0	0
B5	No	Light porosity	0	0	0	0	0	0

Table 30.3 shows examples of the quadrants and whether damage was detected by NDT in that area and whether the participants perceived the quadrant to be damaged.

This table clearly shows that participants in this study were not consistent in perceiving surface anomalies as damage, nor were they consistent in determining whether a quadrant without a surface anomaly had underlying damage using GVI/DVI techniques. The table also shows that DVI is less prone to false positives than GVI. Tables 30.3 and 30.4 show examples of the data collected.

30.5.3 Probability of Detection

POD curves were developed using the generalized data. Figure 30.22 shows an example. This is the overall dent detection POD curve. Once again, since the study

TABLE 30.4
Positive Percent Indication of Damage by Quadrants with Surface Anomalies.

Quadrant	Surface anomaly	Underlying damage	Positive perceived indication of damage (PPID)					
			Students		Airline		General aviation (GA)	
			GVI	DVI	GVI	DVI	GVI	DVI
B1	Small crack	Crack, medium porosity	11	20	20	60	25	0
B2	Large hole	Hole, medium porosity	78	80	100	100	75	100
B3	Dent	Impact damage, medium porosity	67	0	0	80	75	0
D5	Paint chip	Paint chip, medium porosity	33	0	20	40	25	75
E2	Dent	Low porosity, cracking around bolt	0	0	0	0	0	50
E3	Small dent	Sub-surface damage	0	0	0	0	0	75

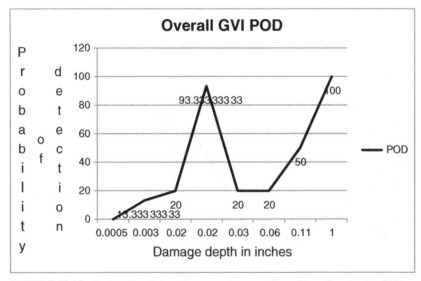

FIGURE 30.22 Probability of detection curve for anomalies with surface depths. GVI, general visual inspection; POD, probability of detection.

participants were asked to look for damage rather than to find all the dents, this curve reflects that study objective. When only those dents that were actually detectable damage were included in the analysis, much more traditional POD curves could be developed. These are shown in Figures 30.23 and 30.24.

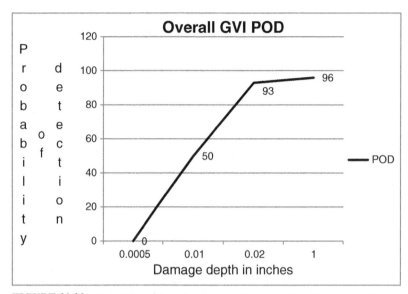

FIGURE 30.23 Probability of detection for overall general visual inspection of surface anomalies that are actually damage.

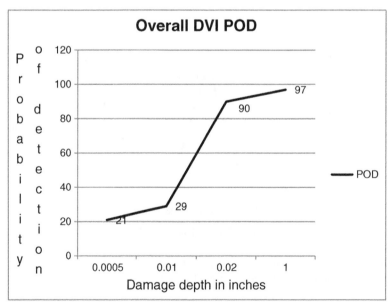

FIGURE 30.24 Probability of detection for overall detailed visual inspection of surface anomalies that are actually damage.

30.5.4 Development of Failure Rates

Using the results from the NDE/NDT and results from the experiment performed in year two, a set of failure rates for detecting aircraft composite damage was developed. From these results, the human error types were developed and are contained in Table 30.5.

Modeling Aircraft Composite Visual Inspection Risk

To model the risk associated with the visual inspection of composite materials, the following were used:

- Event trees.
- Fault trees.
- Human reliability assessment/analysis (HRA) event trees.

TABLE 30.5
Failure Rate Table.

Human error type	Average failure rate (%)	Organization		
		Student (%)	Airline (%)	GA (%)
Failure to perceive a large hole as damage using GVI	16	22	0	25
Failure to perceive a large hole as damage using DVI	7	20	0	0
Failure to perceive a crack as damage using GVI	49	60	40	62
Failure to perceive a crack as damage using DVI	35	37	20	50
Failure to perceive any paint chip as damage using GVI	34	28	30	42
Failure to perceive any paint chip as damage using DVI	40	27	30	42
Failure to perceive a large dent as damage using GVI	17	23	4	25
Failure to perceive a large dent as damage using DVI	13	20	3	15
Failure to perceive other irregularities as damage using GVI	70	83	60	70
Failure to perceive other irregularities as damage using DVI	36	40	20	80
False positive rate for GVI	7	7	7	9
False Positive rate for DVI	6	6	6	7

Although a risk model is not a risk framework, risk models do help elucidate where the largest contributors to risk lie.

In general, there are four possible outcomes to a GVI or a DVI for a surface anomaly. These include the following:

1. A surface anomaly is not actual damage, but the inspector perceives it as damage. This is considered to be a *false positive*.
2. A surface anomaly is not actual damage, and the inspector does not perceive it as damage. This is considered to be a *safe condition*.
3. A surface anomaly is actual damage that the inspector does not perceive as damage. In this condition, the damage goes *undetected*.
4. A surface anomaly is actual damage, and the inspector perceives it as damage. The damage is detected, and the process then progresses to another level of inspection. In the case of GVI, this could be a DVI; or, in the case of DVI, it could be NDE/NDT.

Figure 30.25 shows a very simple event tree for a GVI. Figure 30.26 shows a GVI event tree followed by a DVI event tree.

A simple fault tree can also be developed for the process. However, a fault tree only shows the paths of failure for the process and no other potential paths. Figure 30.27 shows a simple fault tree for the process.

HRA event trees can be developed for GVI and DVI. Figures 30.28 and 30.29 show two HRA event trees that can be developed for the processes. Failure rates for the paths for the HRA event tree in Figure 30.29 can be calculated. Table 30.6 contain these calculations.

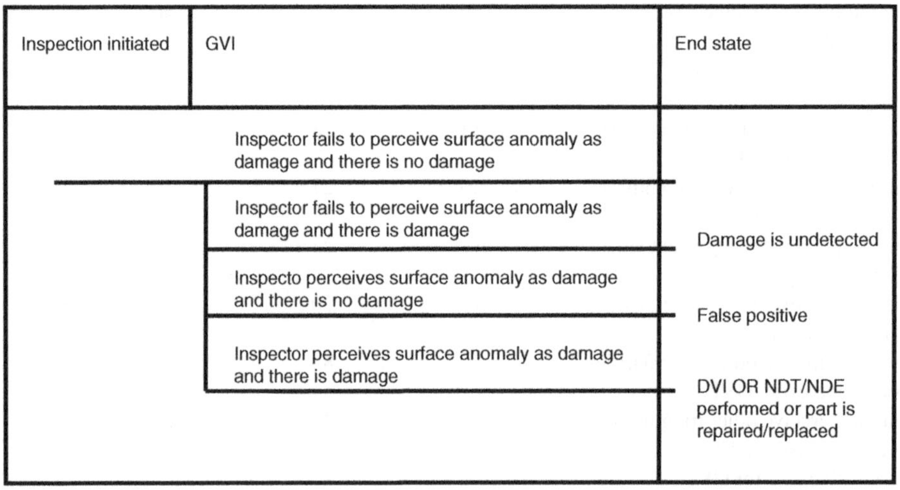

Inspection initiated	GVI	End state
	Inspector fails to perceive surface anomaly as damage and there is no damage	
	Inspector fails to perceive surface anomaly as damage and there is damage	Damage is undetected
	Inspecto perceives surface anomaly as damage and there is no damage	
	Inspector perceives surface anomaly as damage and there is damage	False positive
		DVI OR NDT/NDE performed or part is repaired/replaced

FIGURE 30.25 Event tree for GVI.

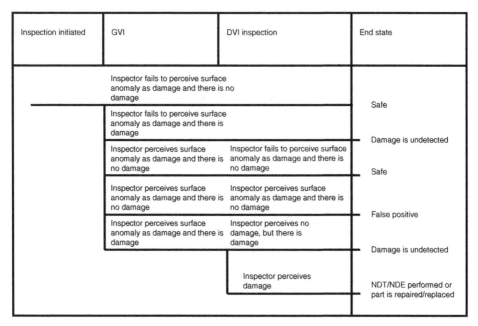

Inspection initiated	GVI	DVI inspection	End state
	Inspector fails to perceive surface anomaly as damage and there is no damage		Safe
	Inspector fails to perceive surface anomaly as damage and there is damage		Damage is undetected
	Inspector perceives surface anomaly as damage and there is no damage	Inspector fails to perceive surface anomaly as damage and there is no damage	Safe
	Inspector perceives surface anomaly as damage and there is no damage	Inspector perceives surface anomaly as damage and there is no damage	False positive
	Inspector perceives surface anomaly as damage and there is damage	Inspector perceives no damage, but there is damage	Damage is undetected
		Inspector perceives damage	NDT/NDE performed or part is repaired/replaced

FIGURE 30.26   Event tree for GVI followed by DVI.

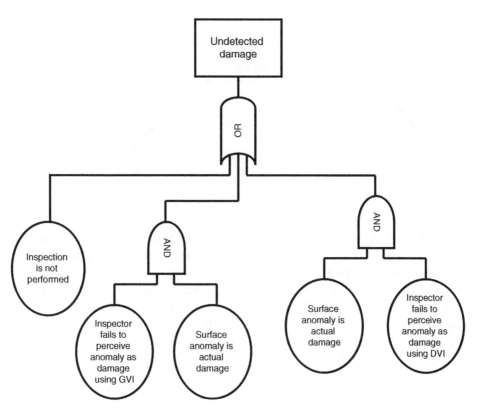

FIGURE 30.27   Fault tree for GVI and DVI.

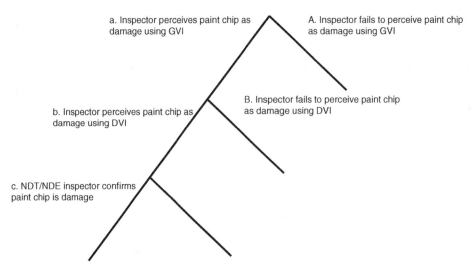

FIGURE 30.28 Human reliability assessment/analysis (HRA) event tree for condition in which a part is damaged.

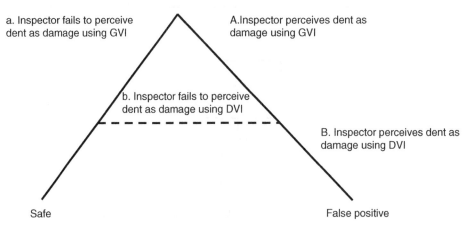

FIGURE 30.29 HRA event tree for condition in which a part is not damaged.

The analyses illustrated earlier are based on finding one anomaly on a part. However, in real inspection situations, an inspection zone or part on a composite aircraft might have multiple surface anomalies. In addition, if an inspector finds a certain type of anomaly (i.e. a large hole), then the inspection might end because a hole could be enough to fail the zone/part. The surface anomalies (except for two sites) on the test article used in the experiment were there when the project obtained it. On the topside of the test article, there were 24 surface anomalies. On the bottom side of the test article, there were six surface anomalies.

Using the bottom side of the test article as a test case, a risk model was developed for the six surface anomalies using HRA event trees. A fault tree will be

TABLE 30.6
Failure Rate Calculations for Human Reliability Event Tree in Figure 30.29.

Item	Overall	Student	Airline	GA
Failure to perceive surface anomaly as not being damaged using GVI	0.07	0.07	0.07	0.09
Successfully determining part is not damaged using GVI	0.93	0.93	0.93	0.91
Failure to perceive surface anomaly as not being damaged using DVI	0.06	0.06	0.06	0.07
Successfully determining part is not damaged using DVI	0.94	0.94	0.94	0.93
Tree path totals				
A – success path	0.93	0.93	0.93	0.91
AB – failure path	0.07×0.06 $= 0.0042$	0.07×0.06 $= 0.0042$	0.07×0.06 $= 0.0042$	0.09×0.07 $= 0.0063$
AB – success path	0.07×0.94 $= 0.066$	0.07×0.94 $= 0.066$	0.07×0.94 $= 0.066$	0.09×0.93 $= 0.084$

used to link the HRA event trees. The following assumptions in this model are as follows:

- Every surface anomaly needs to be investigated.
- Every surface anomaly is independent from the others.
- GVI will be used first, followed by DVI.
- False positives will be considered for each quadrant.
- Airline inspectors will be performing the inspections. One inspector will perform the GVI and a different inspector will perform the DVI.
- There are 6 quadrants with damage and 42 quadrants without surface anomalies. K2 is excluded because there are questions as to its disposition.

The surface anomaly types on the bottom side of the test article are listed in Table 30.7.

The related failure rates for these failure types are shown in Table 30.8. Note that in the case of this model, none of the airline participants noted that the dent on E2 was damage. This failure rate was added.

The first analysis is considering GVI only. HRA event trees do not need to be developed for this example of GVI analysis because they are single point failures. Table 30.9 shows the results of this analysis.

TABLE 30.7
Surface Anomaly Types of Bottom Side of Test Article.

Quadrant	Surface anomaly	Underlying damage
B1	Small crack	Crack, medium porosity
B2	Large hole	Hole, medium porosity
B3	Dent	Impact damage, medium porosity
D5	Paint chip	Paint chip, medium porosity
E2	Dent	Low porosity, cracking around bolt
E3	Small dent	Subsurface damage

TABLE 30.8
Failure Rates Used in Risk Analysis of Bottom Side of Test Article.

Human error type	Airline Failure rate (%)
Failure to perceive a large hole as damage using GVI	0
Failure to perceive a large hole as damage using DVI	0
Failure to perceive a crack as damage using GVI	40
Failure to perceive a crack as damage using DVI	20
Failure to perceive any paint chip as damage using GVI	30
Failure to perceive any paint chip as damage using DVI	30
Failure to perceive a large dent as damage using GVI	4
Failure to perceive a large dent as damage using DVI	3
False-positive rate for GVI	7
False-positive rate for DVI	6
Failure to determine small dent on E2 as damage using GVI or DVI	0

TABLE 30.9
Failure Rates for GVI of the Bottom Side of the Test Article.

Quadrant	Source of data	Failure rate calculations	Total failure rate
Quadrant B1	Failure to perceive a crack as damage using GVI	0.40	0.40
Quadrant B2	Failure to perceive a large hole as damage using GVI	0	0
Quadrant B3	Failure to perceive a large dent as damage using GVI	0.04	0.04
Quadrant D5	Failure to perceive any paint chip as damage using GVI	0.30	0.30
Quadrant E2	No damage	0	0
Quadrant E3	Failure to perceive a large dent as damage using GVI	0.04	0.04
False positive	False-positive rate for GVI	0.07×42 quadrants $= 2.94$	2.94
False positive	False-positive rate for GVI	0.07	0.07

TABLE 30.10
Failure Rates for DVI of the Bottom Side of the Test Article.

Quadrant	Source of data	Failure rate calculations	Total failure rate
Quadrant B1	Failure to perceive a crack as damage using GVI	0.20	0.20
Quadrant B2	Failure to perceive a large hole as damage using GVI	0	0
Quadrant B3	Failure to perceive a large dent as damage using GVI	0.03	0.03
Quadrant D5	Failure to perceive any paint chip as damage using GVI	0.30	0.30
Quadrant E2	No damage	0	0
Quadrant E3	Failure to perceive a large dent as damage using GVI	0.03	0.03
False positive	False-positive rate for GVI	0.06	0.06

This analysis shows that the largest contributor to risk is the small crack in Quadrant B1 (40%) followed by the paint chip in D5 (30%). False positives can be calculated in different ways. If, in each quadrant without damage, there is a potential for a false positive, then there are 42 potential false positives. This creates a combined failure rate much greater than 1.0 because the rate calculated was on a per quadrant basis. In reality, this value would probably be much less. If only one false positive is considered, the result would be different. It then would become 0.07 or approximately 7%. Each of the contributors was combined to those with a Boolean OR and the risk of failure becomes 85% from either missing damage that is associated with a surface anomaly or a false positive.

The next analysis considers DVI only, with the results shown in Table 30.10. The individual failure rates are combined using a Boolean OR. The analysis shows the overall failure rate from a DVI only is about 62%, which is about 28% less than GVI only.

An analysis of a GVI followed by DVI was performed as a comparison to GVI and DVI only. The HRA event trees developed for the six surface anomaly areas are shown in Figures 30.30 through 30.35. Figure 30.36 is the generic HRA event tree for

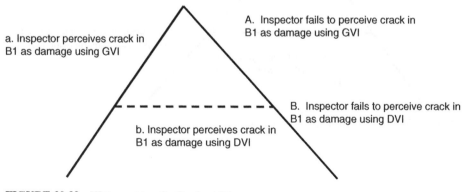

FIGURE 30.30   HRA event tree for Quadrant B1.

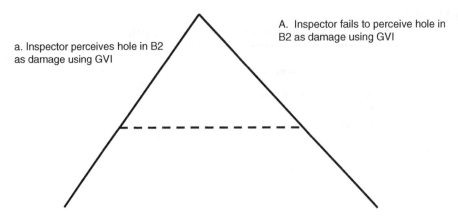

FIGURE 30.31   HRA event tree for Quadrant B2.

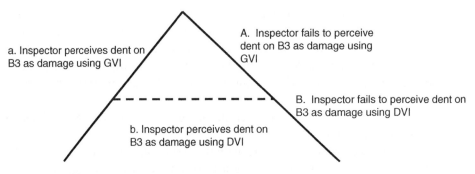

FIGURE 30.32   HRA event tree for Quadrant B3.

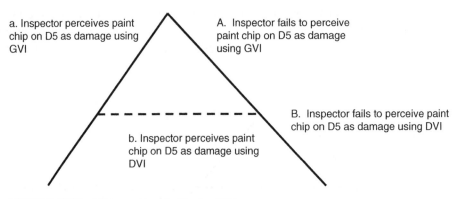

FIGURE 30.33   HRA event tree for Quadrant D5.

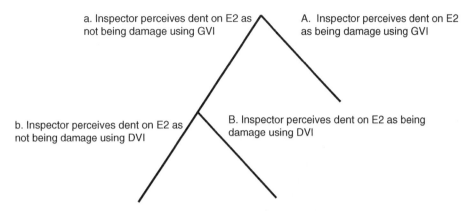

FIGURE 30.34   HRA event tree for Quadrant E2.

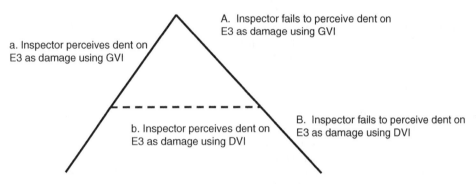

FIGURE 30.35   HRA event tree for Quadrant E3.

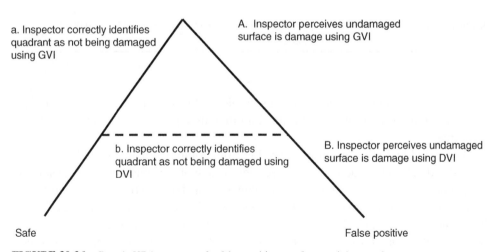

FIGURE 30.36   Generic HRA event tree for false-positive rate for remaining quadrants.

TABLE 30.11
Failure Path Calculations.

HRA event tree	Failure paths	Source of data	Calculations	Total failure rate for HRA event tree
HRA event tree for Quadrant B1	AB	Failure to perceive a crack as damage using GVI and DVI	$0.40 \times 0.20 = 0.08$	0.08
HRA event tree for Quadrant B2	AB	Failure to perceive a large hole as damage using GVI and DVI	0	0
HRA event tree for Quadrant B3	AB	Failure to perceive a large dent as damage using GVI and DVI	$0.04 \times 0.03 = 0.0012$	0.0012
HRA event tree for Quadrant D5	AB	Failure to perceive any paint chip as damage using GVI and DVI	$0.30 \times 0.30 = 0.09$	0.09
HRA event tree for Quadrant E2	AB	No damage	0	0
HRA event tree for Quadrant E3	AB	Failure to perceive a large dent as damage using GVI and DVI	$0.04 \times 0.03 = 0.0012$	0.0012
HRA event tree for false positives	AB	False-positive rate for GVI and DVI	$0.07 \times 0.06 = 0.0042$	0.0042 or $0.004 \times 42 = 0.168$ or 0.17
One false positive	AB	False-positive rate for GVI and DVI	$0.07 \times 0.06 = 0.0042$	0.0042

the quadrants without surface anomalies. Table 30.11 contains the failure path data for each of the trees. Figure 30.37 is the fault tree linking each of the individual HRA event trees.

The individual HRA event trees are logically combined using a Boolean logical OR and added because each of the inspections of each of the quadrants was considered independent. This analysis shows that the largest contributor to risk is the potential for false positives when considering all 42 quadrants. Using the failure rate developed, there is an approximately 17% chance of failing the part due to a false positive. If only the potential for one false positive is considered, then the false-positive rate becomes 0.4%. The next highest risk is missing a paint chip at 9% and the potential for a small crack at 8%. The risk of a failure of a GVI followed by a DVI is approximately 18%. Using this analysis, it is clear that GVI by itself is significantly riskier than a GVI

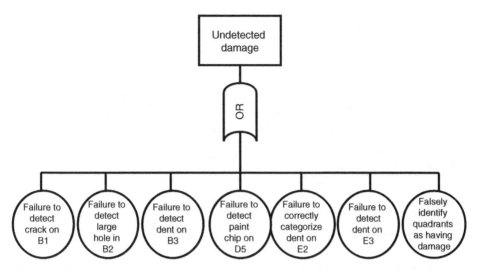

FIGURE 30.37   Fault tree linking HRA fault trees together.

TABLE 30.12
Failure Rate Comparison.

Inspection protocol	Overall failure rate (%)
GVI only	85
DVI only	62
A GVI followed by a DVI	18

followed by a DVI. In fact, it is approximately 25% of the risk of a GVI only and approximately 33% of a DVI by itself. The DVI is the recovery for a GVI. This is shown in Table 30.12.

These risk values provide a relative risk for the contributions to risk by the various types of surface anomalies.

30.5.5 Aircraft Composite Visual Inspection Risk Framework

A risk framework is basically the context in which safety/risk is considered for a system and/or organization. Several different alternatives for depicting the risk framework for visual inspection were considered. Figure 30.38 shows one depiction. The overarching factor in this depiction is the organizational culture.

In this framework, there are four generalized entry points into the visual inspection space. These include the following:

- Formal scheduled inspections.
- Walk-arounds.

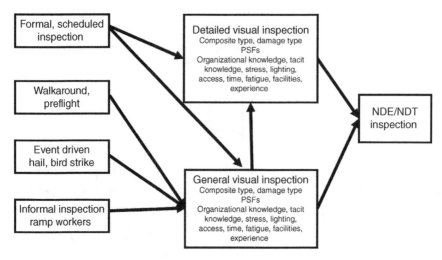

FIGURE 30.38 Risk framework.

- Event-driven inspections.
- Informal inspections.

The inspection process flows to either a DVI or a GVI based on several factors as follows:

- Procedures.
- Rules.
- Guidelines.
- Visual cues.

A GVI or DVI is performed based on the procedures outlined by the organization. Numerous OFs, in other terms PSFs, affect the execution of the inspections. For GVI, these PSF include the following:

- Organizational knowledge/tacit knowledge.
- Lighting.
- Stress.
- Access.
- Time.
- Fatigue.
- Facilities.
- Distractions.
- Experience.

For DVI, these PSF include the following:

- Organizational knowledge/tacit knowledge.
- Lighting.
- Stress.
- Access.
- Noise.
- Tactile response.
- Time.
- Fatigue.
- Distractions.
- Facilities.
- Experience.

In addition, the inspectors perceive certain types of surface anomalies as being damage more than other types. The PSF and anomaly type affect the outcome of the inspection.

If the inspector perceives that a surface anomaly is actual damage using GVI techniques, then in some cases a DVI is performed and in other cases an NDE or NDT is used to help ensure the part is actually damaged. If the inspector perceives that a surface anomaly is actual damage using DVI techniques, then the part is examined using NDE/NDT techniques. The results of the NDE/NDT usually determine whether the part is to be repaired or replaced.

There are several probabilistic aspects to this risk framework. These include the following:

- Whether or not the inspector perceives an anomaly as damage using either DVI or GVI.
- The false-positive rate for GVI versus DVI – GVI appears to lead to more false positives.
- The POD for a type of anomaly.

In this regard, there are also certain error types associated with GVI and DVI.

The inspection can be successful if an inspector finds an anomaly and determines it to be either truly damage or not damage. An inspection fails if an inspector does not find anomalies and/or does not properly determine whether they are truly damage. There is a risk in both events occurring.

The depiction of the preferred visual risk framework is shown in Figure 30.39. All the same elements appear in this framework as are depicted in Figure 30.38, but this is a cleaner way of depicting them. This framework is based on the onion risk model concept (6).

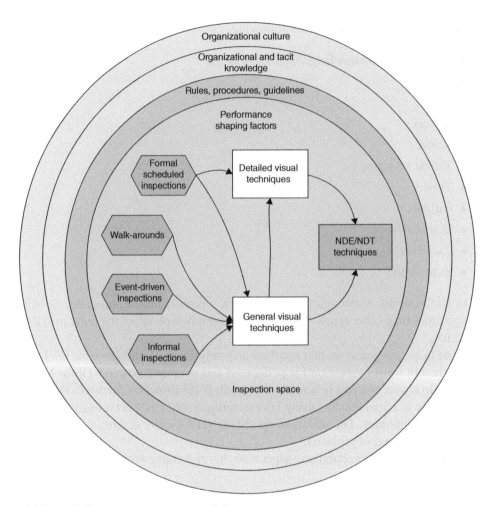

FIGURE 30.39 Onion model risk framework.

This framework illustrates that the organizational culture is the very basic driver for safety/risk within a system. The next level is the organizational and tacit knowledge. For this system, there was a distinction between organizational knowledge and tacit knowledge.

Organizational knowledge comes from formal professional training and training conducted within the airline or MRO facility. Tacit knowledge is gained from many years of experience on the job. The next level is includes rules, procedures, and guidelines that pertain to how work is performed within the organization. The next level is the inspection space and the safety/risk governed by the activities performed, along with the individual PSF. The entry points to the visual inspection, the basic flow into the inspection technique, the PSF, and error types are the same for these two risk frameworks. It is the depiction that is different.

30.5.6 Summary of Experimental Findings

Following are the three primary findings from the research that was conducted:

1. GVI is error prone. Because GVI only requires visual observation of the composite surfaces, there is a much higher potential for missing true damage and for indicating false positives. In real inspections, it appears that GVI is never truly used by itself. When observing inspectors in the field, it was always indicated that if they saw a surface anomaly, they then used some other technique, such as touching or tapping the surface to get a better understanding as to whether it was true damage or not. GVI might be useful for metal-skinned aircraft, but composites can be damaged and show very little on the surface. From a risk assessment perspective, a GVI by itself has a single point of failure; there is no second check to reduce the risk.

   A DVI, on the other hand, has a second check built into it to an extent. Inspectors usually look at a surface and then touch or tap it with a tap hammer or a coin/token to get a better understanding as to whether it is damaged. There is the least amount of risk when a GVI is performed and then a DVI. However, once again, the inspectors do not solely rely on their vision to determine if a composite component is damaged in most real-life inspections.

2. The more sensory input the inspectors can bring to bear on the inspection task, the higher likelihood of a positive outcome. That is, the inspectors identified fewer false-positive detections of damage, and the additional sensory input produced results more consistent with actual damage.

3. The qualitative and quantitative data collected complement each other. There is a strong relationship between the protocol used by inspectors and their perceptions of the amount of damage. Qualitatively, the primary difference between the GVI and DVI protocols is the variety of sensory input utilized, with GVI relying on visual data only and DVI using visual, auditory, and tactile inputs. Evaluating the qualitative and quantitative data together yields results that run counter to what might be expected – that the detailed inspection (DVI) would produce more damage detections than the general inspection (GVI). However, the opposite appears to be the case. The use of multiple sensory inputs allows inspectors to perceive fewer surface anomalies as actual damage, which appears to be a more accurate assessment of the situation.

30.6 RISK FRAMEWORK AND MODEL DEVELOPMENT PROCESS

This chapter presents a high-level, generic approach to developing a risk framework and risk models. This approach was used to develop the risk framework for this project. The majority of the individual techniques discussed in the chapter are found in Ostrom and Wilhelmsen (7). These are the basic steps in the process:

1. Identify the system and the boundaries of the system.
2. Define the system.
3. Collect data on the system.
4. Model the system.
5. Develop failure/error types.
6. Develop failure rates.
7. Develop failure scenarios.
8. Develop integrated risk models.
9. Develop risk mitigation strategies.
10. Develop a risk framework.

The following presents a high-level overview of these process steps.

30.6.1 Identify the System and the Boundaries of the System

Though it might seem like the easiest of tasks to identify the system of interest, in reality, it can be the step that complicates the assessment the most. The inspection of commercial aircraft is a good model to help illustrate this. If all of the inspection processes are to be examined, then the analysis task will be very involved and require many years of effort. However, if a select set of inspection processes are to be assessed, then the task is manageable.

The tasks to be analyzed should be related. For instance, inspection tasks for structure can be grouped together and inspection tasks for engine components can be grouped together. Combining the two into one analysis would not make logical sense because they require different sets of tools and, in general, the inspectors are different for the two different sets of tasks.

The system to be analyzed should be bounded by distinct start and end points for the analysis. If this is not done, then the analysis can expand rapidly and can become unmanageable. Figure 30.16 displays a good method of illustrating the boundaries of an analysis. It is an example of how a block flow diagram can help analysts show the boundaries of a risk assessment.

30.6.2 Define the System

The definition of a system entails a written and illustrated description of the system. This should include the inputs and outputs of the system and how they interconnect. The definition should explain to the customer and other stakeholders the pertinent information about the system of interest. It should only include the information that is germane to the system and not the information about other systems, unless those systems provide input to or receive output from the system of interest.

30.6.3 Collect Data on the System

There are two major categories of data: qualitative and quantitative data. Both of these types of data are essential for performing risk assessments. Qualitative data provides the context of the process being analyzed, while quantitative data provides the basis for the probabilistic aspects of the potential for success and/or failure for the process.

Data collection is performed in a number of ways. These can include the following:

- Direct observation of tasks.
- Interviews (open-ended, guided, or highly structured).
- Surveys.
- Experimentation.

Qualitative and quantitative data can be generated by any of these techniques if the data collection methods are designed appropriately. Ostrom and Wilhelmsen (7) present a complete discussion on collecting data for risk analyses. The overall process for collecting data for HRAs is task analysis. A *task analysis* is any process for assessing what a user does (task), how the task is organized, and why it is done in a particular way and then using this information to design a new system or analyze an existing system. It is an investigative process of the interaction of operators and the equipment and/or machines they utilize. It is the process of assessing and evaluating all observable tasks and then breaking those tasks into functional units. These units allow for the evaluators to develop design elements and appropriate training procedures and identify potential hazards and risks. *Task analysis* has been defined as "the study of what an operator (or team of operators) is required to do, in terms of actions and/or cognitive processes, to achieve a system goal" (7).

The following is an example of how the data collection process can be accomplished and how it fits into a risk assessment. For this example, a risk assessment team is tasked with determining the steps in an inspection process. The team sets up data collection in the following manner.

The team first observes the inspection process by shadowing inspectors as they examine aircraft components. The team performs the shadowing as many times as needed until they feel they have data saturation. This means that the team will not gain any more data by participating in additional job shadowing. Next, the team uses the data collected in the job shadowing activity to design interviews for a fresh set of inspectors. The questions are usually open ended so that additional information can be collected and not limited to what can be gained by close-ended questions. The team then decides to use the results from the interview data to develop a survey that they will send out to inspectors in a number of airlines to determine if the practices seen at this location are common to other airlines. The survey results can be meaningful by themselves or they can be used to help develop one or more experiments. The survey data and experimental data will aid in further defining the process and will determine the potential failure points of the inspection system. In addition, the data can also be used to help develop failure rate data for a task.

These are other types of data that can be used in risk analyses:

- Accident data.
- Operating data.
- Industry data.
- Data reported in research journals.
- Repair and maintenance data.
- Manufacturer's data.
- The sources of data can be quite extensive.

30.6.4 Model the System

In this step, the system is thoroughly modeled using the data collected. These models can be a thorough written description of the process or they can be highly graphical. Graphical models can be easier to follow and can be accomplished by using something like process mapping or a detailed process block flow diagram, similar to what is shown earlier in the report. This system model is used to help identify the points in the system with vulnerabilities.

30.6.5 Develop Failure/Error Types

The failure or error types are developed based on the data collected in previous steps of the process. Failure types are considered hardware or software in nature and error types are considered human in nature. These failure or error types may be found during the interviews of the task performers or from the job shadowing. They might also be determined from accident data, operating data, and so on. Brainstorming and techniques like the Delphi technique and critical incident technique (7) can also be used to develop failure or error types. These failure or error types might take the form of a list of the types of failures or errors that can occur as shown in Table 30.13. It shows the process step, the potential failure mode, and the effect if the failure occurs.

30.6.6 Develop Failure Rates

Failure rates are developed from the data collected in previous steps of the process or from new data collection efforts undertaken because data were not previously found that could be used to develop the failure rates. Ostrom and Wilhelmsen (7) present a detailed examination of this topic. However, a brief overview is presented here.

TABLE 30.13
Example Failure and Error Types.

Inspecting chip detector		
Process steps	Potential failure modes	Potential failure effects
Remove chip detector	Improper removal can remove debris from chip detector and cause false reading. Chip detector can be damaged if improperly removed	Engine could fail if chips are not properly detected Added cost to replace damaged chip detector
Examine chip detector	Aircraft maintenance technician (AMT) fails to notice debris on chip detector	Engine could fail if chips are not properly detected
Clean chip detector	AMT fails to properly clean chip detector	Debris could be placed back into engine
Replace chip detector	AMT fails to properly install chip detector	Oil could leak past chip detector Threads of chip detector could be damaged
Lock wire chip detector	AMT fails to properly lock wire chip detector	Chip detector could become loose and fall out, leading to loss of engine oil
Replace oil drain plug	AMT fails to properly install oil drain plug	Engine oil could leak out Oil drain plug could become damaged
Lock wire oil drain plug	AMT fails to properly lock oil drain plug	Oil drain plug could become loose and fall out Oil drain plug could become damaged
Replace oil	AMT fails to properly replace oil	Engine could fail

Sources of failure rate or error rate data can include the following:

1. Hardware failure rate data

   In the truest sense, hardware failure data is much easier to obtain than human error probabilities. In many cases, failure rate data is available at the system, as well as at the subsystem and component levels. Hardware failure rate data can be obtained from the manufacturer, historical data, government and military handbooks, commercial data, or generated from testing by the user.

   (a) Manufacturer data

   Failure rate data can be obtained from the manufacturer for certain pieces of industrial equipment such as pumps, valves, motors, electrical panels, controllers, and even for components such as chips, diodes, and resistors. These data are usually supplied on a product data sheet or can be requested from the manufacturer. In addition, the product data sheets will sometimes supply failure modes for the equipment.

(b) Historical data

Many organizations maintain internal databases of failure information on the devices or systems that they produce that can be used to calculate failure rates for those devices or systems. For new devices or systems that are similar in design and manufacture, the historical data for similar devices or systems can serve as a useful estimate.

(c) Government and military handbooks

(d) Commercial data sources

There are many commercially available failure rate data sources. Loss prevention handbooks, insurance companies, data mining organizations, and trade organizations can be sources of data for use in inclusion in risk assessments.

(e) Operational data and testing

Within an organization, failure rate data can be calculated from failures of components within a facility or multiple facilities. Accurate records as to the failure need to be kept for this data to be useful. These are the types of data that aid in a risk assessment:

- How many hours, demands, or miles of travel were on the device when it failed?
- What other factors were involved?
  - Was the environment hot, cold, wet, or dry?
  - Was periodic maintenance performed or not?
  - How was the system operated?

2. Accident data

Failures that lead to or are caused by accidents can aid in determining failure rates as well. The most accurate source of data is to test samples of the actual devices or systems in order to generate failure data. This is often prohibitively expensive or impractical, so that the previous data sources are often used instead.

There are numerous organizations that maintain accident data:

- Bureau of Labor Statistics.
- Chemical Manufacturers Association.
- Institute of Nuclear Power Operations (INPO).
- National Transportation Safety Board (NTSB).

These data can be used directly to develop failure or error rates or can be combined with other data to generate failure rates. Other failure rate and error rate development techniques include the following:

- Monte Carlo simulation.

- Human reliability data sources – techniques for human error rate prediction (THERP).
- Delphi technique.

The failure rate or error rate data can take many forms, but the two most common are failures over time or failures per demand. *Failures over time* include failures associated with light bulbs, engine operating hours, or crack growth in aircraft skin. *Failures per demand* include failures associated with turning on a pump or starting a diesel generator.

30.6.7 Develop Failure Scenarios

Failure scenarios are developed from the process model and the associated failure and error types. A *failure scenario* is a postulated accident sequence. In essence, it is the series of events that occur that lead to unwanted consequences. In risk assessments, these are developed in anticipation of something happening. In an accident investigation, these are what happened that led to the accident. These scenarios range from being simple and involving only an isolated portion of a process model to being complex and involving the entire process model. Event trees are a very easy way of illustrating a risk scenario. Figures 30.40 and 30.41 show examples of how risk scenarios can be presented using event trees.

30.6.8 Develop Integrated Risk Models

Integrated risk models tie everything together. The method most commonly used to accomplish this is a fault tree analysis (FTA). The FTA technique is proven to be an effective tool for analyzing and identifying areas for hazard mitigation and prevention

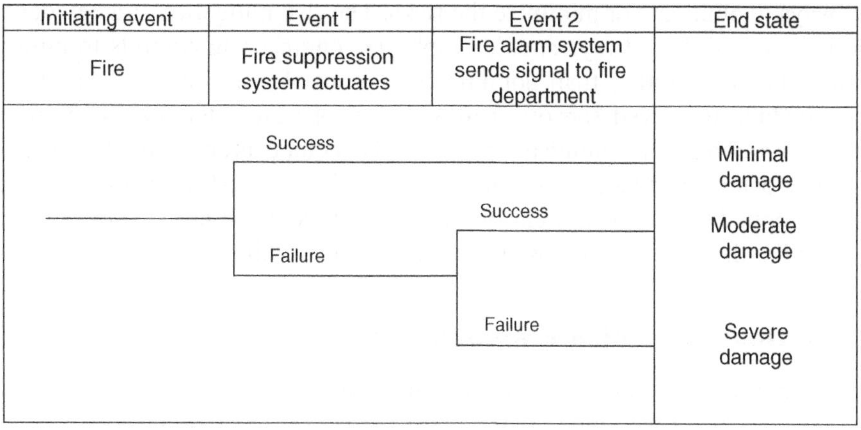

Initiating event
Event 1
Event 2
End state
Fire
Fire suppression system actuates
Fire alarm system sends signal to fire department
Success
Failure
Success
Failure
Minimal damage
Moderate damage
Severe damage

FIGURE 30.40 Example risk scenario 1.

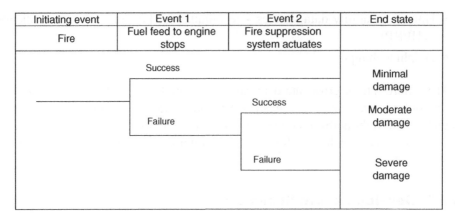

Initiating event	Event 1	Event 2	End state
Fire	Fuel feed to engine stops	Fire suppression system actuates	

FIGURE 30.41   Example risk scenario 2.

while in the planning phase or anytime a systematic approach to risk assessment is needed. FTA is used as an integral part of a probabilistic risk assessment (PRA), though they are used quite frequently without developing a full-blown PRA. The NASA *Fault Tree Handbook with Aerospace Applications* (8) is a complete guide to FTA. Table 30.14 shows the major symbols used in an FTA and Figures 30.42 and 30.43 illustrate how a simple FTA is typically shown.

Event trees in many cases can be all that is needed to thoroughly illustrate the accident pathways in simple accident scenarios. However, for more complex scenarios, fault trees are more effective. Event trees can also be used to link several fault trees together into a logical sequence.

30.6.9 Develop Risk Mitigation Strategies

The ultimate goal of performing a risk assessment is to find where risk exists and to find ways to eliminate or minimize the risks. The risk mitigation strategy depends on the risks found, but always should emphasize engineering controls to eliminate or reduce the risks, rather than administrative controls. Examples of engineering controls include improved sensors, hardware or software interlocks, and stronger structures. Examples of administrative controls include procedures, training, and supervisory oversight. Once a risk analysis is performed, risk mitigation strategies should be applied first to those areas of the process model that pose the most risk and then second to those areas that pose less amounts of overall risk.

30.6.10 Develop a Risk Framework

Section 30.5.5 of this chapter provided details regarding the process used to develop the risk framework for this study. As stated in this section, a risk framework is the context in which safety/risk is considered for a system and/or organization. This context

TABLE 30.14
Typical Fault Tree Symbols.

Symbol name	Symbol	Description
Basic event		A basic initiating fault (or failure event)
Undeveloped event		An event that is no further developed. It is a basic event that does not need further resolution
Output event		An event that is dependent on the logic of the input events
External event (house event)		An event that is normally expected to occur. In general, these events can be set to occur or not occur (i.e. they have a fixed probability of 0 or 1)
Conditioning event		A specific condition or restriction that can apply to any gate
Transfer		Indicates a transfer to a sub-tree or continuation to another location

should be developed so that it is logical for the system being analyzed. Another way of describing a risk framework is a specific set of functional activities or tasks and associated PSFs that define the risk management system in an organization and the relationship to the risk management organizational system. A risk framework defines the processes and the influences that affect risks and should be logical, transparent, and understandable to all the stakeholders.

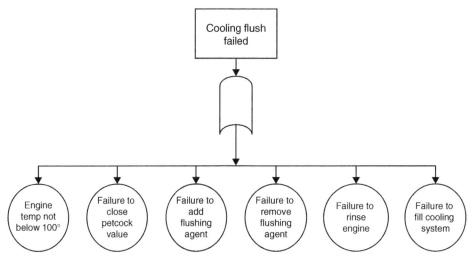

FIGURE 30.42   Example of a typical fault tree for a cooling flush task.

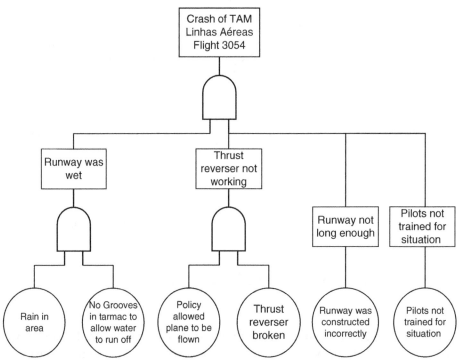

FIGURE 30.43   Example of a fault tree to illustrate the crash of Transporte Aéreo Militar (TAM) Linhas Aereas Flight 3054.

In the risk framework developed for this study and shown in Figure 30.39, the overarching influence on risk was determined to be the organizational culture. This was followed by the organizational and tacit knowledge, and then the rules, procedures, and guidelines used in the organization. The last of the major influences was the PSFs. These influences vary from organization to organization. It is the task of the analysts to determine what the major influences on the system are. These influences will be based on the data collected and the risk scenarios/models developed. No two systems will be alike. The goal is to determine the major influences so that appropriate risk mitigation strategies can be developed to eliminate or minimize the risks.

REFERENCES

1. Erhart, D., Ostrom, L., and Wilhelmsen, C. (2004). Visual detectability of dents on a composite aircraft control surface: an initial study. *International Journal of Applied Aviation Studies* 4 (2): 111–122.
2. Lincoln, Y. and Guba, E. (1985). *Naturalistic Inquiry*. Newbury Park, CA: Sage Publications.
3. Dekker, S. (2006). *The Field Guide to Understanding Human Error*. Burlington, VT: Ashgate Publishing.
4. National Institute on Deafness and Other Communication Disorders (2013). Noise-induced hearing loss. http://www.nidcd.nih.gov/health/hearing/noise.asp (accessed November 2013).
5. Minu, I. (2003). Knowledge sharing in organizations: a conceptual framework. *Human Resources Development Review* 2: 337–359.
6. Guldenmund, F.W. (2010). Understanding and exploring safety culture. PhD Thesis. Delft University. http://repository.tudelft.nl/view/ir/uuid%3A30fb9f1c-7daf-41dd-8a5c-b6e3acfe0023 (accessed November 2013).
7. Ostrom, L.T. and Wilhelmsen, C. (2012). *Risk Assessment Tools and Techniques and Their Application*. Hoboken, NJ: Wiley.
8. Vesely, W., Dugan, J., Fragola, J. et al. (2002). *Fault Tree Handbook with Aerospace Applications*. Washington, DC: NASA.

CHAPTER 31

Traffic Risks

31.1 INTRODUCTION

As with all things in life, being in traffic as a driver, a cyclist, or a pedestrian has risks inherent to the respective activity. In this chapter, we will explore several of these risks and explore recent studies on the subject. The chapter will conclude with risk mitigation factors.

31.2 UNDERSTANDING TRAFFIC RISKS

A recent study comparing traffic types across differing transit corridors revealed several issues to focus on when in traffic as well as risk reduction and mitigation strategies to improve safety. The Centers for Disease Control rank mortality from road traffic accidents low compared to other causes of death, at 2.1%, see Table 31.1 for perspective.

A few facts worth noting concerning the risk of traffic use are as follows:

- In 2015, Americans traveled just more than 3.0 T miles over 4.1 M miles of road.
- There are 225 821 241 Registered Vehicles in the United States.
- There were 35 092 fatalities and 2.4 M injuries in the 6.3 M crashes that same year.

While the percentages of accidents per miles driven or per vehicles on the road is small (See Figure 31.1), the consequence of injuries and fatalities are immense.

Traffic is typically considered as driving and congestion, however, traffic is any use of the public and private roadways by vehicles, pedestrians, and other assorted conveyances. This traffic is typically divided between rural and urban systems. Rural

Risk Assessment: Tools, Techniques, and Their Applications, Second Edition. Lee T. Ostrom and Cheryl A. Wilhelmsen.
© 2019 John Wiley & Sons, Inc. Published 2019 by John Wiley & Sons, Inc.
Companion website: www.wiley.com/go/Ostrom/RiskAssessment_2e

TABLE 31.1
US Mortality Rates (2002).

		Percent of deaths (%)		Deaths per 100 000		
Group	Cause	Group	Subgroup	All	Male	Female
–	All causes	100.0	100.0	916.1	954.7	877.1
A	Cardiovascular diseases	29.3	–	268.8	259.3	278.4
B	Infectious and parasitic diseases	23.0	–	211.3	221.7	200.4
A.1	Coronary artery disease	–	12.6	115.8	121.4	110.1
C	Malignant neoplasms (cancers)	12.5	–	114.4	126.9	101.7
A.2	Cerebrovascular disease (stroke)	–	9.7	88.5	85.4	95.6
B.1	Respiratory infections	–	7.0	63.7	63.5	63.8
B.1.1	Lower respiratory tract infections	–	6.8	62.4	62.2	62.6
D	Respiratory diseases	6.5	–	59.5	61.1	57.9
E	Unintentional injuries	6.2	–	57	73.7	40.2
B.2	HIV/AIDS	–	4.9	44.6	46.2	43
D.1	Chronic obstructive pulmonary disease	–	4.8	44.1	45.1	43.1
–	Perinatal conditions	4.3	4.3	39.6	43.7	35.4
F	Digestive diseases	3.5	–	31.6	34.9	28.2
B.3	Diarrhea diseases	–	3.2	28.9	30	27.8
G	Intentional injuries (suicide, violence, war, etc.)	2.8	–	26	37	14.9
B.4	Tuberculosis	–	2.8	25.2	32.9	17.3
B.5	Malaria	–	2.2	20.4	19.4	21.5
C.1	Lung cancer	–	2.2	20	28.4	11.4
E.1	Road traffic accidents	–	2.1	19.1	40.8	10.4
B.6	Childhood diseases	–	2.0	18.1	18	18.2
H	Neuropsychiatric disorders	2.0	–	17.9	18.4	17.3
A.3	Hypertensive heart disease	–	1.6	14.6	13.4	15.9
G.1	Suicide	–	1.5	14	17.4	10.6

Source: Adapted from Centers for Disease Control; Heron (4).

systems are complex, but are typically highways and freeways (Interstates). Urban roadways are the connecting streets that are used to travel from typical urban settings to other urban systems. To further complicate traffic is that there are many differing vehicle types and many road and weather conditions. For instance, the conditions encountered on a mountainous rural road will be vastly different than an inner city street. The common denominators for risk mitigation are safe traffic behaviors and properly operating conveyances. This chapter on traffic risks is not designed to be all encompassing, but more to develop the framework for a typical user or policy maker to reduce the probability of accident, injury, and fatality due to traffic engagement (Figure 31.2).

Traffic Hazards come in many types; these include but are not limited to the following:

1. The driver or traffic user.
2. Animals.

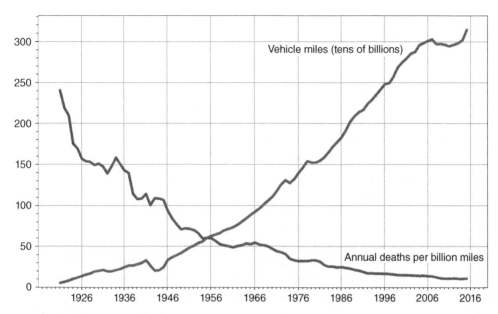

FIGURE 31.1   US vehicle miles traveled and proportionate fatality rate. *Source*: Adapted from (1).

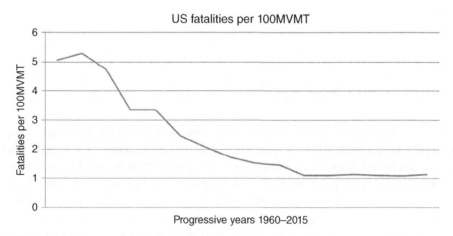

FIGURE 31.2   US Fatalities per 100MVMT. *Source*: Adapted from (2).

3. Weather.
4. Other vehicles/drivers.
5. Impairment, inattention, intoxication.
6. Global warming potential.
7. Differences between men and women and between the young and the old.
8. Differing road types and associated conditions.

Each of these hazard categories has details that develop them as a liability.

The driver or traffic user is typically the focus of traffic risks. These users range from cyclists to vehicle operators to pedestrians; they encompass the vast majority of the populous. Animals, while not traffic users, are certainly traffic impactors as well as impacted by traffic. Animal–vehicle collisions happen every month of every year, and normally during commuting hours. Large game animals can destroy vehicle property and/or injure or kill vehicle occupants as well as themselves. Weather, while typically not a significant factor in occupant mortality, slows traffic and creates added hazards in the form of reduced visibility and unpredictable road conditions. Other drivers pose a unique risk. Modern economics is centered on the notion of rational thought and behavior by users, though actual experience is that each person acts independently and at varying levels of competence. This is an excellent analogy for the "other" drivers in traffic. These other drivers are such an impacting impediment that decades or traffic safety learning has focused efforts on defensive driving. Complicating this risk is that while a driver may have good behavior their conveyance may be substandard if impacted. Of particular concern and also the focus of many traffic safety trainings and roadway postings is the problem of impairment, inattention, and intoxication. More than 80% of traffic accidents are a result of these three problem areas. It is difficult to tell if another driver suffers from one of these, until it is too late. The tragedy in this is that these three focuses of the completely avoidable traffic hazards.

Mass transit, e-vehicles, and bio-friendly fuels aid in the reduction or mitigation of a currently growing global warming effect. This global warming potential measures the impact of emissions from (in this case) transportation. This is then translated into a per capita impact. Mass transit using biofuels, particularly a biodiesel 80% (B80) or greater, provides a significant benefit, approximated at 80% reduction in global warming emissions.

The "insurance institute" measures many factors required for insurance companies to provide coverage for both genders and across all available age groups. Their data analysis reveals women are safer drivers, men do most of the driving, and women are in more accidents than men. So how does this work? When a man is involved in an accident, he has a statistically higher probability, based on decades of measured behavior, to total a vehicle and to kill or seriously injure a vehicle occupant. Women related accidents, though more than men are typically deemed as property damage and/or minor injury. Like the difference between men and women, age groups also differ in accident type. Younger ages (<25 years old) are statistically involved in more accidents than older ages. Gender differences are also amplified by with the performance of younger drivers. The message here is that younger drivers and particularly male are statistically more likely to be involved in a total loss, fatality, or significant injury accident.

Roadways have many built-in hazards. These hazards are typically divided into four categories: (i) road condition (International Roughness Index [IRI] and the Present Serviceability Rating [PSR]); (ii) density; (iii) vegetated highway shoulders; and (iv) other impacts that transit or affect the road conditions.

TABLE 31.2
Road Condition Index.

International Roughness Index (IRI)					Present Serviceability Rating (PSR)				
<60	60–94	95–170	171–220	>220	>3.9	3.5–3.9	2.6–3.4	2.1–2.5	<=2.0
90 793	206 333	276 698	65 776	16 276	14 986	73 890	46 580	11 272	11 033

Source: Adapted from Department of Transportation; Bureau of Transportation Statistics (5).

The road condition criteria used in the Highway Performance Monitoring System includes two elements to determine ride quality: IRI rating less than 95 or a PSR rating greater than or equal to 3.5 indicates "good" ride quality; and IRI rating of less than or equal to 170 or a PSR rating for greater than or equal to 2.5 indicates "acceptable" ride quality. In the US, the average road condition index is expressed in miles as illustrated in Table 31.2.

With almost half of US roads with an unacceptable IRI and more than 22 k miles in non-serviceable condition, road conditions themselves become then a hazard amplifier.

Density is simply a measure of vehicle spacing per mile. As a user of the highway system, high density is seen in metropolitan areas, particularly during the primary commuting times. Lower density and better road conditions are ideal. Vegetated highway shoulders are approximated as the 30 ft either side of the farthest painted traffic markers. The dangers these shoulders introduce are as follows: if the shoulder has dense vegetation, they can be used as animal habitat; and if the shoulder is of weaker structural quality than the roadway, the shoulder can be lethal and contributes to the national crash statistics for vehicle rollover potential.

31.3 VEHICLES

Vehicle and types are segregated into several broad categories; privately owned, mass transit, and commercial vehicles. In addition to these three, there are myriad pedestrian types also interfacing on these transit corridors. Traffic types are also broadly segregated; urban neighborhood transit, urban non-highway transit, urban highway transit, rural highway transit, and service roads (i.e. forest service roads). As the vehicle types interface with the traffic types, various conditions exist as noted in Table 31.3.

In addition to the challenges, a driver faces in these varied traffic situations is the potential for commutes and super commutes. An average commute is approximated between 15 and 30 minutes, almost 10% of the working population commutes as a carpool. Super commutes are typically characterized as between 45 and 60 minutes and are associated with distance rather than traffic density, though density plays a role. Super-commutes are a growing trend with many drivers commuting 80–100 miles

TABLE 31.3
Transit Thoroughfare And Vehicle Interface Matrix.

	Neighborhood	Urban non-highway	Urban highway	Rural highway	Service roads
Private	Slow speeds and pedestrian traffic, inattention	Slow speeds and pedestrian traffic	Fast speeds, intoxication potential	Fast speeds, intoxication potential, agriculture vehicles, and animal–vehicle contact	Slow speeds, high variety of road condition, recreational vehicle, and animal–vehicle contact
Commercial	Limited use	High use, slow speeds, and pedestrian traffic, high police monitoring	Fast speeds, trained drivers	Fast speeds, trained drivers, agriculture vehicles, and animal–vehicle contact, limited police monitoring	Slow speeds, high variety of road condition, recreational vehicle, and animal–vehicle contact
Mass transit	Limited use	Limited use, slow speeds, and frequent stops	Limited use	Limited use	Limited use

each direction, daily. For example, three of the Department of Energy Laboratories have large shares of their working populous traveling 45–90 minutes each way daily. This commuting concept has a positive net effect when rotating carpools are optimized, giving the several vehicle passengers rest and allowing for some driver rotation. Super-commutes on the other hand pose an exhaustion and inattention risk. While carpooling can alleviate the risk proportion, ultimately physical fatigue and monotonous conditions set in. Commercial providers are combating these risk areas with enhanced driver training and with route management systems, so that a 90-minutes drive might be delivered to the driver as multiple shorter drives with destination spots between driving segments.

Average commuting in the United States was at approximately 80% personal vehicles, 10% carpools, and just over 5% public transportation in 2001. Workers with extreme commutes increased carpooling to almost 13% and public transportation by 23%. These trends have held for almost two decades, with almost 700k workers responding to the census bureau study in 2013.[1,2]

31.4 ECONOMICS OF TRAFFIC RISKS

Economics play several roles with regard to traffic risks as follows: (i) vehicle condition, (ii) roadway conditions, (iii) fueling costs and options, and (iv) accident costs. Vehicles are not designed to run forever without some or even significant maintenance and upkeep, though as vehicle prices and maintenance expenditures increase, vehicles fall into a deferred maintenance state, where the vehicle still operates, but less optimal than designed. This is a function of vehicle performance. Several examples illustrating this concern are as follows:

1. Imagine a person purchases a three-year-old vehicle with the latest safety features but does not service the vehicle as required or retains the vehicle beyond the service life of the vehicles safety feature components (i.e. airbags). When the vehicle contacts some accident-causing impediment, the safety features may not work as designed.

2. If a deferred maintenance strategy is employed for a particular vehicle, the probability of performing as designed reduces. For example, the manufacturer suggests a premium fuel grade, 3000 mile or three-month service, and tires for optimal performance. If the particular vehicle operator does not adhere to these standards, the vehicle will ultimately degrade to a less than desirable state of performance, and may contribute to problems outside the vehicle, like becoming the impediment another vehicle must deal with.

[1]2011 Data: US Workers Commuter Survey; Adapted from US Census Bureau, 1-year American Commuter Survey.
[2]2013 Data: Commuting to Work; Adapted from US Census Bureau, 2013 American Community Survey.

Roadway conditions are measured by the Department of Transportation and have an associated measure of road condition index. As the indices drop below 0.75, the roads become hazardous to well-maintained vehicles. US infrastructure is aged and is in the starting phases of upgrade and replacement. While some roads are improved, less transited roads fall into a deferred maintenance state and can trend to being an impediment for drivers.

Fueling costs and options are simply that some fuels have been shown to have a direct correlation to an increasing global warming potential. The vehicle itself is also a concern as fuel prices are growing and fuel economics are not keeping up.

Accident costs are illustrated as follows in Table 31.4.

The simple message here is that even minor accidents bear a cost, with fatality accidents having a significant expense as well as the ultimate expense of death. Larger vehicles and well-trained drivers help to mitigate accidents and fatality risks though do not eliminate them (Figure 31.3).

31.4.1 Risk Mitigation and Avoidance

Like most risk-avoidance adventures, behavior of the affected is the strongest indicator of risk reduction capability; for traffic users, following all postings and limits, driving for weather or traffic conditions, practicing defensive driving techniques (i.e. two-second following rule), and allowing space for others to act without the user being acted on. For commercial drivers, three added behavioral assists are provided in the form of Commercial Driver's License (CDL) training, added company specific safety training, and the potential to impact ones livelihood as a result of poor judgment and negligent behavior.

A secondary but important factor in safety performance for risk reduction/mitigation is the condition of the vehicle as compared to the condition of the

TABLE 31.4
US Estimated Economic Costs For Human Injuries And Fatalities.

Maximum human severity	Average cost of all injuries/fatalities ($)	Distribution % of all collisions	Distribution % of human incident accidents	Contribution cost to average human incident accident ($)
Possible injury	24 418	2.34	50.87	12 421
Evident injury	46 266	1.75	38.04	17 601
Incapacitating injury	231 332	0.47	10.22	23 636
Fatality	3 341 468	0.04	0.87	29 056
Totals		4.60	100.00	82 715

Source: Adapted from Refs. (6–8).

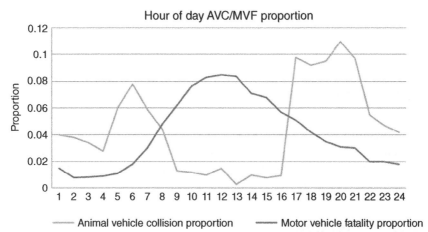

FIGURE 31.3   US Adapted Statistics from the National Crash Database, for HSIS States. *Source*: Adapted from (3).

transit corridor. Assuming the vehicle in question is in good and proper working order, if the task is transit forest service and other access roads, a high performance sports car is likely not the best vehicle for the task. Many manufactures provide vehicles that accommodate many road condition types well (i.e. sport utility vehicles).

Over the previous five decades, vehicles have become substantially safer for occupants, despite higher speeds and increased use. A few examples of these safety features are the three-point seat belt, air bags, antilock brakes, all-wheel drive, and crumple zones. Key to this is the public and government demanding (through measurement) safer, more resilient vehicles. Regarding this notion of a future where vehicles overcome behavior, artificial intelligence is beginning to be reliable enough for autonomous vehicles. These vehicles of the future are designed to avoid collisions with other vehicles, especially other autonomous vehicles and large objects (i.e. large game). Given the progress in safety-based improvement and behavioral improvement and acuity by users, the future should be even more exciting.

A final note on vehicle safety is the notion of mass transit. For example, weather flying or using a bus or train system, the probability of accidents dramatically decreases. For flying, the probability of death is one in 11 M flights and for bus users the probability is approximated at 1 in 36 M miles. The probability of death in a passenger vehicle is estimated at almost 1 death in 5000 uses. Mass transit provides the right conditions for the best traffic use results, weather air, boat, or land travel. The operators of these several vehicle modalities have professional training, certification, high-tech safety systems, and control points to aid in traffic management. Finally, given the income-impacting consequence of accident or death based on poor judgment and operator error, professional operators have many reinforcements for best driving behavior.

31.5 SUMMARY

Traffic is dangerous, death does occur, and hazards are real and unpredictable. How one can protect themselves? By practicing safe driving habits, use a vehicle appropriate for the task, be aware of surroundings, and assume all other drivers are out to get you, and finally, use mass transit as available.

Self-Check Questions

1. How do the statistics provided in this chapter compare with traffic accident statistics in your area?

2. Given that mass transit systems are overall safer, how might the concepts of mass transit be made more palatable to the driving population?

3. Using the data provided or data you can locate develop a comparative risk analysis of the driving a point of view (POV) versus a mass transit system.

4. POVs hitting deer is a major problem in the United States. How can these encounters be reduced? What have other countries done to reduce these encounters?

5. President Eisenhower had a big impact on traffic safety. What was it and how did it help cause the intersection of the two curves in Figure 31.1?

REFERENCES

1. United States Department of Energy, Energy Information Administration, Short-Term Energy Outlook (STEO) (2015). Independent Statistics & Analysis. US Government Report, Washington, DC.
2. United States Department of Transportation, Federal Highway Administration (2017). Highway Statistics (Washington, DC: Annual Issues), Table VM-202. http://www.fhwa.dot.gov/policyinformation/statistics.cfm (accessed January 2018).
3. United States Department of Transportation, National Highway Traffic Safety Administration (2017). National Crash Database. https://www.nhtsa.gov/research-data (accessed January 2018).
4. Heron, M. (2012). National Vital Statistics Report Deaths: Leading Causes for 2009; as cited by the United States Centers for Disease Control. Mortality in the United States. https://www.cdc.gov/nchs/products/databriefs/db293.htm (accessed January 2018).
5. Bureau of Transportation Statistics (2018). Road Condition, 2018. https://www.bts.gov/content/road-condition (accessed January 2018).
6. Transportation Research Board (2003). United States Department of Transportation and TRB Annual Meeting 2003. http://onlinepubs.trb.org/onlinepubs/general/2003_trb_annual_report.pdf (accessed January 2018).

7. United States Department of Transportation, Federal Highway Administration (2008). Wildlife-vehicle collision reduction study, report to congress, FHWA-HRT-08-034, Washington, DC (August 2008).
8. Khattak, AJ (2003). Human fatalities in animal-related highway crashes. Transportation Research Board 82nd Annual Meeting Compendium of Papers CD-ROM, Washington, DC (12–16 January 2003).

Acronyms

AFM	acute flaccid myelitis
AMTs	aviation maintenance technicians
ANSI	American National Standards Institute
APU	auxiliary power unit
ASEP	Accident Sequence Evaluation Program
ATHEANA	A Technique for Human Error Analysis
BA	British Airways
BES	Bulk Energy System
BLS	Bureau of Labor Statistics
BP	British Petroleum
BTU	British thermal unit
CAHR	Connectionism Assessment of Human Reliability
CAMEO	Computer-Aided Management of Emergency Operations
CFPB	Consumer Fraud Protection Bureau
CFR	Code of Federal Regulations
CHRIS	Chemical Hazards Response Information System
CIMA	Chemical Industry Mutual Aid Organization
CIT	Critical Incident Technique
COOP	Continuity of Operations Plan
COSO	Committee of Sponsoring Organizations of the Treadway Commission
CPSC	Consumer Product Safety Commission
CREAM	Cognitive Reliability and Error Analysis Method
CSB	Chemical Safety Board
CSRF	cross-site request forgery
D&D	decontamination and decommissioning
DDoS	distributed denial of service
DOE	Department of Energy

Risk Assessment: Tools, Techniques, and Their Applications, Second Edition. Lee T. Ostrom and Cheryl A. Wilhelmsen.
© 2019 John Wiley & Sons, Inc. Published 2019 by John Wiley & Sons, Inc.
Companion website: www.wiley.com/go/Ostrom/RiskAssessment_2e

DOS	denial of service
DVI	detailed visual inspection
ECCS	emergency core cooling system
ECL	energy conservation loop
EFIS	electronic flight instrument system
EM	emergency management
EMA	Emergency Management Agency
EPA	Environmental Protection Agency
EPZ	emergency planning zones
ERA	ecological risk assessment
ERG	Emergency Response Guidebook
ERM	enterprise risk management
FAA	Federal Aviation Administration
FEMA	Federal Emergency Management Agency
FMC	flight management computer
FMEA	failure mode and effects analysis
FMECA	failure mode, effects, and criticality analysis
FTA	fault tree analysis
FW	feed water
GDPR	General Data Protection Regulation
GEM	generic error modeling
GPS	Global Positioning System
GVI	general visual inspection
HAZMAT	hazardous materials
HAZOP	hazard and operability study
HEPs	human error probabilities
HERMIT	human error modeling/investigation tool
HEROS	human error rate assessment and optimizing system
HRA	human reliability analysis
IoT	Internet of things
IRGC	International Risk Governance Council
IRM	integrated risk management
IRMs	intermediate range monitors
ISO	International Organization for Standardization
LED	light-emitting diode
LOCA	loss-of-coolant accident
LPG	liquid petroleum gas
MITM	man in the middle
MMS	Minerals Management Service
MOC	management of change
MRO	maintenance, repair, and overhaul
MS	mode switch

MSDS	material safety data sheet
MTBF	mean time between failures
MTTF	mean time to failure
NARA	nuclear action reliability assessment
NASA	National Aeronautics and Space Administration
NDE	nondestructive examination
NDT	nondestructive testing
NRC	Nuclear Regulatory Commission
NSC	National Safety Council
NTSB	National Transportation Safety Board
OEM	original equipment manufacturer
OF	operational factor
OPA	Oil Pollution Act
OSHA	Occupational Safety and Health Administration
PHA	preliminary hazards analysis
PHZA	process hazard analysis
P&ID	piping and instrumentation
PM	periodic maintenance
POD	probability of detection
PPE	personal protective equipment
PPI	Production Plant, Inc.
PPID	positive perceived indication of damage
PRA	probabilistic risk assessments
PRM	project risk management
PSA	probabilistic safety assessments
PSF	performance shaping factor
PSM	process safety management
RAT	ram air turbine
RBS	risk breakdown structure
RCP	reactor coolant pump
RCRA	Resource Conservation and Recovery Act
RMP	risk management plan
SCADA	supervisory control and data acquisition
SCRAM	safety control rod axe man
SME	subject matter experts
SPAR-H	Simplified Plant Analysis Risk Human Reliability Assessment
SRM	source range monitor
STA	shift technical advisor
STEM	science, technology, engineering, and math
TAT	threat assessment team
TCDD	2,3,7,8-tetrachlorodibenzo-p-dioxin
THEA	Technique for Human Error Analysis

THERP	Technique for Human Error Rate Prediction
VPP	Voluntary Protection Program
WBS	work breakdown structure
WHO	World Health Organization
WVPP	workplace violence prevention programs
XSS	cross-site scripting

Glossary

Accident An unexpected and undesirable event, especially one resulting in damage or harm.

Acetaldehyde An organic chemical compound.

Acute flaccid myelitis (AFM) A rare but serious condition. It affects the nervous system, specifically the area of the spinal cord called gray matter, which causes the muscles and reflexes in the body to become weak.

Anomalies Deviation or departure from the normal or common order, form, or rule.

Autobiography Written by the individual himself or herself.

Basic event A fault or failure in an accident sequence that can occur, which has an impact on the overall outcome or the top event of a probabilistic risk assessment or fault tree analysis.

Bayesian analysis A statistical procedure that endeavors to estimate parameters of an underlying distribution based on the observed distribution.

Bioconcentration Uptake and accumulation of a substance from water alone.

Biographical study A study where the researcher writes and records the experiences of another person's life.

Biomagnification The increase in concentration of a substance such as the pesticide DDT.

BlackEnergy A malware program.

Boundary conditions The values or conditions that constrain a system.

Case study research A qualitative approach in which the investigator explores a bounded system (a case) or multiple bounded systems (cases) over time, through detailed, in-depth data collection involving multiple sources of information (e.g. observation interviews, audiovisual material, and documents and reports), and reports a case description and case-based themes.

Chloracne An acne-like eruption of blackheads, cysts, and pustules.

Closed loop Materials do not enter or leave a system.

Component System, job/person, part, tool, or other thing that performs the activities that make up the critical function.

Component failure An electronic or mechanical part of a system that ceases to work. In risk assessment terms, this unit has an impact on the success or failure of a system.

Risk Assessment: Tools, Techniques, and Their Applications, Second Edition. Lee T. Ostrom and Cheryl A. Wilhelmsen.
© 2019 John Wiley & Sons, Inc. Published 2019 by John Wiley & Sons, Inc.
Companion website: www.wiley.com/go/Ostrom/RiskAssessment_2e

Component fault   An electronic or mechanical part of a system that ceases to work or ceases to work correctly. In risk assessment terms, this unit has an impact on the success or failure of a system.

Conditional probability   A probability whose sample space has been limited to only those outcomes that fulfill a certain condition.

Consequences   The positive or negative outcomes of decisions, events, or processes.

Critical function   What has to be in place to achieve or maintain the mission.

Critical Incident Technique   A technique that has applicability to a wide range of risk assessments.

Cryptocurrency   Digital or virtual currency.

Cut set   A set of basic events that lead to the top event in a probabilistic risk assessment or fault tree.

Cybercrime   Vandalism, theft, fraud, extortion, data ransom, phishing.

Cyberterrorism   Nation-state conducted or sponsored threats.

Delphi process   A structured communication technique, originally developed as a systematic, interactive forecasting method, which relies on a panel of experts.

Dengue fever   A painful, debilitating mosquito-borne disease caused by any one of four closely related dengue viruses. These viruses are related to the viruses that cause West Nile infection and yellow fever.

Discrete distribution   A statistical distribution that has specific values.

DoS attack   The ability to interrupt the smart grid systems. This is called a denial of service.

Ebola virus disease   A rare and deadly disease most commonly affecting people and non-human primates (monkeys, gorillas, and chimpanzees). It is caused by an infection with a group of viruses within the genus *Ebolavirus*.

Emerging risks   A high level of uncertainty, in which frequency and potential impact of risks are difficult to assess. An emerging risk is typically characterized by a low frequency or, in other words, not likely to happen and has a high impact.

Enterprise risk management   The process of planning, organizing, leading, and controlling the activities of an organization in order to minimize the effects of risk on an organization's capital and earnings.

***Escherichia coli* (*E. coli*)**   Bacteria that normally live in the intestines of healthy people and animals. Most varieties of *E. coli* are harmless or cause relatively brief diarrhea. But a few particularly nasty strains, such as *E. coli* O157:H7, can cause severe abdominal cramps, bloody diarrhea, and vomiting.

Ethnographic research   This study describes learned and shared patterns of the group's behaviors, beliefs, and language.

Event tree   A graphical representation of the possible sequence of events that might occur following an event that initiates an accident.

Failure mode and effects analysis (FMEA)   A detailed document that identifies the ways in which a process or product can fail to meet critical requirements. It is a living document that lists all the possible causes of failure from which a list of items can be generated to determine types of controls or where changes in the procedures should be made to reduce or mitigate risk.

Failure mode, effects, and criticality analysis (FMECA)   The additional dimension of probability and criticality added to FMEA(s) by the prioritization of steps/sections of procedures that need to be changed or the process changed to reduce risk; pointing out where warnings, cautions, or notes need to be added in procedures; and pointing out where special precautions need to be taken or specialized teams/individuals need to perform tasks. The criticality is mainly a qualitative measure of how critical the failure to the process really is based on subject matter experts' opinion and based on probability of occurrence and/or on the consequence or effect.

Fault tree analysis   A form of safety analysis that assesses hardware safety to provide failure statistics and sensitivity analyses that indicate the possible effect of critical failures.

Gates   Logic structures in a fault tree that connect basic events.

Grounded theory research   Generates, or discovers, a theory.

Hazard   Any risk to which a worker is subject to as a direct result (in whole or in part) of his/her being employed.

Hazmat   Hazardous materials.

HAZOP   A hazard and operability study is a structured and systematic examination of a complex planned or existing process or operation in order to identify and evaluate problems that may represent risks to personnel or equipment.

Hierarchical task analysis   A broad approach used to represent the relationship between the tasks and the subtasks.

Human reliability analysis (HRA)   Used to analyze the human response to an equipment failure and any process or activity that involves humans is susceptible to human error. HRAs are used to quantify the probability of human errors and can be used to identify steps or activities in the process that can be targeted for changes that could reduce the probability of human error.

HVAC   Heating, ventilation, and air conditioning.

Hydrazine   A colorless, fuming, corrosive hygroscopic liquid, H_2NNH_2, used in jet and rocket fuels.

Intelligent automation   Machine learning, robotic process automation, which complements and augments human skills that can increase speed, precision, quality, and operational efficiency.

Internet of things (IoT)   Artificial intelligence and robotization.

Involuntary risks   Those associated with activities that happen to us without our prior consent or knowledge. Acts of nature such as being struck by lightning, fires, floods, tornados, and so on and exposure to environmental contaminants are examples of involuntary risks.

Link analysis   Identifies the relationships between the components of a system and represents the links between those components.

Methylmercury   A bioaccumulative environmental toxicant.

Minamata disease   A neurological syndrome caused by severe mercury poisoning.

Mission   Goal of process, organization, or task.

Monte Carlo analysis One specific multivariate modeling technique that allows researchers to run multiple trials and define all potential outcomes of an event or investment.

Nanoparticle A small object that behaves as a whole unit in terms of its transport and properties.

Narrative research A method in how and why we make meaning in our lives, a way to create and recreate our realities.

Newton's First Law Every action has an equal or greater reaction.

Nomenclature The terminology used in a particular science, art, activity, and so on.

Nominal value The value of a security that is set by the company issuing it, unrelated to market value.

Operations sequence diagrams Identifies the order in which the tasks are performed and identifies the relations between the person, equipment, and the time.

Oral history A compilation of events and causes, found in folklore, private situations, and single or multiple episodes.

Perception The process of interpreting sensory stimuli by filtering it through one's experiences and knowledge base.

Phenomenological description A phenomenological study describing the meaning for several individuals of their lived experiences of a concept or a phenomenon.

Preliminary hazard analysis A hazard analysis performed at the very beginning of a product or facility life cycle to determine the hazards.

Preliminary hazards list Hazards initially determined from an analysis.

Probabilistic risk assessment (PRA) Focuses on equipment failures and may include a section that discusses the probability of human failure being the initiating event.

Probability The likelihood that the event will occur.

Process mapping A technique that produces visual representations of the steps involved in industrial or other processes.

Process risk management It is the process used by project managers to minimize any potential problems that may negatively impact a project's timetable.

Process safety management A set of interrelated approaches to **managing** hazards associated with the **process** industries and is intended to reduce the frequency and severity of incidents resulting from releases of chemicals and other energy sources.

Qualitative analysis Nonquantitative analysis. An analysis that is descriptive in nature.

Qualitative research Methods that at least attempt to capture life as it is lived.

Quantitative analysis An analysis that seeks to determine the numerical value of something.

Reverse engineer The process of discovering the technological principles of a man-made device, object, or system through analysis of its structure, function, and operation. It often involves taking something (e.g. a mechanical device, electronic component, or software program) apart and analyzing its workings in detail to be used in maintenance or to try to make a new device or program that does the same thing without using or simply duplicating (without understanding) any part of the original.

Risk The potential for realization of unwanted, adverse consequences to human life, health, property, or the environment; estimation of risk is usually based on the expected value of the conditional probability of the event occurring times the consequence of the event given that it has occurred.

Risk analysis A detailed examination, including risk assessment, risk evaluation, and risk management alternatives, performed to understand the nature of unwanted, negative consequences to human life, health, property, or the environment; an analytical process to provide information regarding undesirable events; the process of quantification of the probabilities and expected consequences for identified risks.

Risk assessment The process of establishing information regarding acceptable levels of a risk and/or levels of risk for an individual, group, society, or the environment.

Risk control The application of the risk assessment evaluation.

Risk estimation The scientific determination of the characteristics of risks, usually in as quantitative a way as possible. These include the magnitude, spatial scale, duration, and intensity of adverse consequences and their associated probabilities as well as a description of the cause and effect links.

Risk evaluation A component of risk assessment in which judgments are made about the significance and acceptability of risk.

Risk homeostasis theory In any activity, people accept a certain level of subjectively estimated risk to their health, safety, and other things they value, in exchange for the benefits they hope to receive from that activity (transportation, work, eating, drinking, drug use, recreation, romance, sports, or whatever).

Risk identification Recognizing that a hazard exists and trying to define its characteristics. Often risks exist and are even measured for some time before their adverse consequences are recognized. In other cases, risk identification is a deliberate procedure to review and, it is hoped, anticipate possible hazards.

Risk perception An individual or group assessment of the potential for negative consequence.

Severity The degree of something undesirable.

Smart grid A new innovation of the current electric grid to provide power across our nation and the world.

Statistically nonverifiable Risks from involuntary activities that are based on limited data sets and mathematical equations.

Statistically verifiable Risks for voluntary or involuntary activities that have been determined from direct observation.

Support Utilities, materials, activities, or other items that support the components.

Target risk A specific level of risk an organization feels comfortable with and aims to achieve.

Task analysis Task analysis is any process of assessing what a user does and why, step by step, and using this information to design a new system or analyze an existing system.

Thematic analysis One of the most common forms of analysis in qualitative research. It emphasizes pinpointing, examining, and recording patterns (or "themes") within data.

Threat Source of danger.

Threat assessment A structured group process used to evaluate the risk posed by a student or another person, typically as a response to an actual or perceived threat or concerning behavior.

Timeline analysis Used to match up the process performance over time, which includes the task frequency, interactions with the other tasks, the worker(s), and the duration of the task.

TNT Chemical compound.

TOP events The event of interest in a probabilistic risk assessment or fault tree analysis to which all other basic events feed.

Undeveloped event Events with little information or no information or those that do not need to be developed because they concern things such as weather or other natural events.

Voluntary risks Those associated with activities that we decide to undertake (e.g. driving a car, riding a motorcycle, drinking, and driving).

Vulnerability A weakness in a system or human that is susceptible to harm.

Zika virus A mosquito-borne flavivirus that was first identified in Uganda in 1947 in monkeys. It was later identified in humans in 1952 in Uganda and the United Republic of Tanzania.

Index

Accident frequency, 122

Agent Orange, 29, 36, 51

Air Canada, 20, 206, 209

Airport, 112, 196, 199, 200, 201, 205, 287, 301, 306, 326

Aloha Flight, 4

Analysis phase, 41, 49, 54, 272

Analysis plan, 41, 42, 48

Aviation, 7, 55, 176, 177, 198, 255, 263, 273, 275, 317–328

Bathtub curve, 4, 87

Bayesian analysis, 80, 93, 108

Bhopal, 13, 29, 37, 466

Binning data, 117, 119, 142, 391

BLS Accident Rate, 122

BLS Disabling Injury Rate, 122

Boolean logic, 6, 320, 530

British Petroleum Oil Company (BP), 29, 446, 461, 468

Bunker Hill, 28

Butadiene, 465, 466

Case Study/Studies, 294, 268, 271–272, 275–280

   Chernobyl, 231

   Chipotle, 290

   Gracie Claire, 237

   Intruder, 295

   Keystone Pipeline, 245

   Mt Everest, 75

   Multipurpose Academic Building, 295, 311

   Panama Canal, 400, 407

   Pepsi in the Philippines, 415, 417

   Ramona, OK, 226

   Risk Framework for Aviation Maintenance, 487, 545

Center for Disease Control (CDC), 345

Challenger, 120, 121, 228

Chernobyl, 18, 23, 29, 37, 223, 231, 235

Chisso Corporation, 34, 36

Columbia Gas, 330, 331

Columbia Space Shuttle, 203, 204

Communications, 11, 15, 217, 270, 294, 386, 405, 435, 456

Community planning, 329, 333

Component fault, 187

Consequence, 2, 4–7, 11, 17–24, 37, 39, 54, 61, 64, 71, 147, 150, 154, 182, 217, 270, 294, 386, 405, 435, 456, 224, 228, 230, 234, 235, 251, 252, 253, 256, 257, 259, 262, 265, 273, 288, 289, 294, 300, 301, 307–309, 323, 326, 327, 359, 360, 385, 389, 390, 391, 392, 393, 396, 398, 414, 422, 425, 429, 430, 441, 442, 444, 467, 541, 555

Consumer Product Safety Commission (CPSC), 61

Continuity of operations plan (COOP), 214, 217

Control measures, 8, 147, 354, 357, 366

Printed and bound by CPI Group (UK) Ltd, Croydon, CR0 4YY

16/04/2025

14658538-0001